Western Channel
Hunter Bay
Norrie Bay
DOMINION EXPLORERS RADIO STATION
Boadway I.
Achook I.
Cornwall
Domex Bay
Rogers Lake
Hogarth Island
MacAlpine Channel
Rap.

From Cloddy Earth to Glittering Gold

by

George Norrie & Harry Norrie

TRINITY ENTERPRISE INC.
Saint John • New Brunswick • Canada

First Printing - January 2006

Canadian Cataloguing in Publication Data

George Norrie 1940 - ; Harry Norrie 1918 -

From Cloddy Earth to Glittering Gold

ISBN: 0-9739036-3-5

1. Norrie, James Paul, 1891-1945. 2. Gold mines and mining--Québec (Province)--Abitibi Region--History. 3. Mining engineers--Québec (Province)--Biography. 4. Prospectors--Québec (Province)--Biography. 5. Industrialists--Québec (Province)--Biography. I. Norrie, Harry. II. Title.

TN140.N62N67 2005a 338.2'092 C2005-905754-8

Editors: Yvonne Wilson, Chas Goguen, Catherine Thompson

Special Thanks, for help with final edit: Jonathan Franklin

Typesetter: Chas Goguen

Cover Design: John van der Woude

Printed and bound in Canada

TRINITY ENTERPRISE INC.
105 Prince William Street
Saint John, New Brunswick
Canada E2L 2B2
(506) 693-2252
info@trinityenterprise.ca
www.trinityenterprise.ca

Contents

Foreword

by Margaret Norrie McCain

My father died suddenly on October 9, 1945, following a massive heart attack.

It was a terrible shock to the whole family, but especially my mother. At thirty-nine, following twelve short years of marriage, she was alone with four young children. I was the oldest at eleven, followed by Judy, nine, George, five, and Roderick, three. There were also four older stepchildren.

The day after my father's death we drove to Montreal to catch the train for Truro, Nova Scotia for the funeral service and burial in the family plot. This wonderful man was buried beside his first wife, his parents, grandparents and great-grandparents. It was a traumatic experience for all of us. Yet in the midst of the enormity of her grief, my mother had to deal with uprooting her children from their home, school, friends and community; and then settling us into a whole new life and environment.

She also had to deal with the grief of her children accompanied by the insecurity that comes from losing the strong centre in one's life.

Fortunately we did have a place to go. Five years earlier, my parents had built a house on the Norrie family farm. From 1940 on, we spent our summers there. Although she was raised in a farming family in Sackville, New Brunswick, my mother knew nothing about the business of managing the large, modern dairy farm she inherited.

And she knew absolutely nothing about mining. Her married life had been spent raising babies and managing an active, busy household. My father's company, Inspiration Mining and Development Co. Ltd., was sold. Staff members were asked to pack up his files and ship them to Nova Scotia.

Several months later, six boxes – about four feet by two feet by one foot in size, wrapped in heavy brown paper and tied with thick ropes – arrived at our house in Nova Scotia. They held the contents of my father's offices. On my mother's instructions, they were taken up to the attic and placed in a far corner.

This attic, which was the equivalent of a third floor, was brightly lit, had high ceilings and was easily accessible up a wide staircase. It was a perfect rainy day play area for us children. Over the years, the attic was transformed into a school, a store, a hospital and, quite often, a theatre for our amateur theatricals – which were open to all the neighbourhood children. Certainly dozens of children came and went – my siblings and me; our friends; our children and later, our grandchildren.

All the while, the boxes sat in the corner unopened, untouched; not even the object of anyone's curiosity.

In 1983 my mother died – thirty-nine years after my father. The boxes still sat in the corner of the attic – unopened. But, by this time they had been joined by many other boxes. My mother was a "pack rat"; she threw nothing away! Every invoice, letter, card, report card, essay, piece of child's art – was saved in a drawer. When the drawer became so full it could no longer be closed, the contents went into a box. This was then taken up to the attic and stored beside my father's boxes. Eventually the boxes took over most of the attic space.

The chore of sorting through them all fell upon my siblings and me. It took weeks, but finally we had sorted, cleared, and divided all the contents of the boxes – all but my father's boxes. They still sat unopened in the far corner. I knew they contained mining material which was beyond my comprehension. So I instructed my brothers to open them and deal with whatever was inside as best they could, believing we would probably make a bonfire of the contents.

Time passed. In fact twelve more years went by. Then one stormy winter day George went to the attic to find a book. For some reason, the boxes captured his attention and he decided to open one – then two – then three.

He began to read! Very quickly he discovered that they contained fascinating and exciting material. He kept going, astounded by what he had found – for in those boxes was a complete, meticulous record of my father's whole career.

It quickly became apparent that my father, too, had saved every piece of information, every letter, every memo, every record of the development of the gold mining industry in northern Quebec. And everything was neatly filed in a well organized manner.

George had indeed discovered a "gold mine".

For the next few years, he devoted hours and hours to reading, studying and sorting all the material. It became his passion. After all he was only five when our father died so he knew nothing of his mining career. And Mother had only told us what she knew herself, which was only a few highlights.

As he read all the material, George not only learned about my father's important influence on the Canadian mining industry; he got to know the man he barely remembered.

In January, 2000, my father was inducted into the Canadian Mining Hall of Fame. The family asked George, as the person who had all the mining information, to accept the award and give a speech about our father.

As he put together notes for his message, he began to realize that a book needed to be written.

Initially our desire was to piece together the mining story as a record for our family. Then we asked Harry, with his amazing story-telling ability, to reach into the depths of his incredible memory to add the "meat and flavour".

For three years, these two terrific men have devoted countless hours to this project, collaborating on all aspects of the story. It has been a true labour of love; nonetheless, it has taken enormous time and energy for which the family is extremely grateful.

Finding Yvonne Wilson, our editor was a stroke of luck. When George gave her the material, she encouraged us to think beyond the scope of the family. It was her contention that the story was an important piece of Canadian history. We are very grateful for her inspiration, help and support. This book project has pulled the whole family back into the life of a beloved person taken from us much too early. Through it we feel connected to the father we hardly knew. This means a lot to us.

It is our hope that you will enjoy reading this story as much as we have enjoyed writing it.

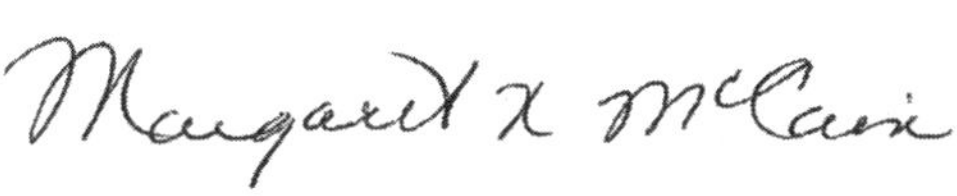

In memory of James Paul Norrie

1891-1945

J.P. Norrie

"The glorious sun stays in his course and plays the alchemist, turning with splendour of his precious eye the meagre cloddy earth to glittering gold."

- King John, Act 3, Scene 1

James P. Norrie is not an alchemist, but in turning "cloddy earth to glittering gold" he is the outstanding mine maker in the all time history of Quebec.

- Toronto *Globe & Mail*
November 14, 1944

Introduction

This is the story of a Canadian pioneer. A pioneer of Canadian mining, gold mining.

This is the story of a gold rush but a very different kind of gold rush from the usual, romantic version – a hard rock gold rush. The gold rushes to California, the Yukon, Victoria, the Rand were searches for placer gold (gold found in gravel beds of streams and rivers). That gold was available to anyone and everyone. This gold rush in northwestern Quebec required a whole new set of rules because the gold was deep underground in hard rock. Hard rock mining.

From his early days at Stadacona (1927) and Malartic Gold Mines to his final days at Croinor Gold Mines, J.P. Norrie developed more gold properties than any other man in Quebec mining. The list is long and each prospect or mine is a story in itself.

When he died in 1945 at the age of fifty-four, he was in charge of forty gold properties in northwestern Quebec.

"My husband," Margaret Norrie said in quiet pride, "was known as 'the father of the Malartics', that rich Depression-era nest of gold properties between Rouyn-Noranda and Val d'Or along the northern transcontinental line of the Canadian National Railways."

To put these developments into perspective: Herbert Hoover, the president of the United States, became famous for providing five thousand jobs on the Hoover Dam. Each one of J.P. Norrie's producing mines was the Canadian employment equivalent of this massive project!

Back in the early 1930s, when the soul-searing blast of Depression was wasting the land, the gold mines of northern Ontario and northwestern Quebec were credited with saving the Canadian economy from total disaster.

Each of the projects employed from fifty to four hundred men and supported many, many more. Communities were built by the unemployed and hungry overflow from the Canadian south and the U.S.

J.P.'s mining career was primarily in northwestern Quebec, but there was one romantic adventure to the Arctic, which would have an unforeseen and seemingly uncontrollable impact on the history of the human condition. To the present time, this has been just a family story.

Margaret Norrie wrote:

"...while not many people know it, and [my husband] was never given the recognition for it, he and two associates made the first uranium strike on the shores of Great Bear Lake, which since has become the federal properties of Eldorado Mining and Refining.

"They were flying in and out of the Great Bear country at the time, with Dominion Explorers, and James spotted pitchblende, knew there had to be uranium there, told Gilbert Labine about it and advised him to ship a sample for assay in Ottawa.

"He did, but if it hadn't been for my husband and his companions, Gilbert Labine would have passed up the whole thing."

In the early mines and in the Arctic, J.P. was just an employee, but a very valuable one. In the later mines, his mines, the Depression and the war imposed a perspective that few men ever experience. The operation and promotion of the mines and the public awareness of his successes drew him into board rooms and the political eye. He was led away from his first love, prospecting, and the love of that first discovery. He eventually returned, but it was too late. He died in 1945.

In this book each mine story is told primarily through correspondence between J.P. and the prospectors, financiers, politicians and employees involved. The contents of the letters show clearly the character of each writer. That's why the letters are included as written.

Many of the reports of the geology and mine development may seem tedious and repetitive, but they demonstrate just how much knowledge and imagination were necessary for success. The diplomacy and respect required to bring the prospectors, promoters and personalities together in the various projects is impressive.

Some of the personalities were prominent when the letters were written; others would gain prominence after J.P. had died. There are six Canadian Aviation Hall of Fame bush pilots and approximately fifteen Mining Hall of Fame associates referred to in these stories. (But many would never be heard of again.)

Isolation and the elements always had to be overcome before anything else.

Hard work, physical strength, tenacity and courage were the daily requirements of these men. J.P., however, was the driving force and their leader. It was his optimism, instinct and knowledge or his "nose for ore" that led them to success. More than anything else, trust, loyalty and bonds of friendship were fundamental. Each venture, big or small, demonstrates this fact without a doubt.

In the following accounts, J.P.'s sons and daughters, especially his son Harry, give each story a life that could only be provided by someone who was there, someone who knew that friendship, loyalty and a sense of humour were the reasons their father, his friends and associates could overcome the hardships they faced.

J.P. Norrie's saga begins back on the farm in Nova Scotia where he was born, where he would return periodically after he gained wealth and prominence. But it does not end with his death in a hotel room in Noranda at the age of fifty-four. J.P. Norrie's story lives on in the annals of the larger-than-life Canadians who pioneered the North.

George Norrie

Chapter 1

The Farm in Onslow

by Harry Norrie

There was hate in the house, a lot of hate. Annie hated Henry, Henry hated Annie, and they both ignored Grace.

One word alone describes Annie: "Self". But used as a prefix it does better: self-centred, self-indulgent, self-contained, self-seeking, selfish. Her only need for other people was to reflect herself. Annie's song was a long drawn-out praise of herself, and any song with that cadence is always wearying.

Annie was of the McDougal Clan. You don't hear much of them any more. Perhaps most of them have been murdered.

Annie was born and raised in Maitland, Nova Scotia, the daughter of a sea captain of sorts. As a brag, she told how she would run up and down the ship's rigging (hopefully in pantaloons), much to the delight of her father. There was a large family, many of whom were carried away by the disastrous flu epidemic of 1918. Along with Annie, four others survived: Harry, town clerk in Truro, boastful like Annie; Margaret, a farmer's wife in Great Village, Nova Scotia; Eva, married to Jim O'Brien, a sea captain, but later a hotel owner in San Francisco, California; and Frank, who lived in Winnipeg, but for some reason was never mentioned.

When Eva and Jim were both over eighty they bought twelve cylinders, drove across the continent to Halifax, and made it with verve. They arrived at the farm late one morning to take three of us for a day-long drive along the north shore of the Bay of Fundy. Jim was denied the pleasure of the outing for having offended Annie in some way. Eva quietly rebuked Annie but felt it not her place to intervene further. Thinking back on it now brings tears for Jim and not much of anything for Annie.

Annie had been a school teacher. On one posting, in Harbour Grace, Newfoundland, she found the love of her life but was denied perpetual happiness by some unrevealed happenstance. But when conversation got slack she would often remind Henry that she could have married a subaltern to God in Harbour Grace. Annie had one joke, her only joke, about a Newfoundland teaching experience: she asked her most brilliant student how to spell wagon. "Well," he said, "if it's a small wagon, you use one G. But if it's a large wagon, you use two G's."

History has long forgotten how Annie got her harpoon into Henry. He was not a drinking man, it couldn't have been that. With his superior intelligence, it couldn't have been by persuasion. But there is one way; yes, it could have been that. Some people with their devious ways can cause firmness and provide an opening that is neither invited or felt for, and Henry was a healthy, virile man. In the heat of the moment discretion, too late, comes with an afterthought.

But I am getting ahead of myself. Let me first tell you how Pauline, Virginia, Jim and I [Harry] came to live on the farm in North River, Nova Scotia (near Onslow). It was upon the sudden and tragic death of our mother, Hanna, in 1923.

Our father J.P. was working in Ontario and unable to care for four young children and sent us to be cared for by his mother and father. This was difficult for them as well, as Annie and Henry were in their sixties and not able to deal with children from five to eight years of age.

J.P.'s good friend Ned Rutledge had introduced him to Hanna Grace Hartling, who lived with the Rutledge family and worked in one of the city (Halifax) stores. At the time, J.P. was either assistant to the Deputy Minister of Mines (Nova Scotia) or Manager of the Dufferin Mine near the village of Port Dufferin, Nova Scotia.

In the fall of 1923 Hanna Grace went to the Victoria General Hospital in Halifax for treatment of a minor ailment but contracted either tetanus or meningitis in the hospital, from which she never recovered. I can vaguely remember standing with Grandma Jane and Virginia on the shore waving to her when she left.

I think the reason that so little is known about the Hartlings is that the death of Hanna was such a shock to J.P. It was very painful for him to talk about it. Pauline said that he explained it to her once in just that fashion.

Pauline and Jim had been sent to the farm in Onslow prior to her leaving for Halifax. I was five at the time but do not remember being told of her death. Shortly afterward Virginia and I were sent by boat "The Dufferin" to Halifax where we were met by our grandfather, Henry Norrie. We stayed that night with the Ned Rutledges and went by train to Truro on the following day. Virginia remembers when we arrived at the farm; no one paid any attention to us - we just stood there for a while until we gradually accepted the feeling of being a nonentity.

A few things, of little interest, I do remember. The first was when the bunch of us were in bed with our mother and I flooded the entire area, forcing a general review of the sleeping conditions.

There was an incident when I was four and Virginia five. We were heavily engaged in mixing mud pies where the only natural water flow was neither from lake nor river. At that time we became aware that there really is a difference.

From the moment I was born until I hate to remember, I was always supplied with voluminous amounts of body fluids, primarily the amber kind. Early on, this condition did not seem unusual. My attendants didn't complain, anyhow, until I arrived at the farm and was bedded down with brother Jim.

Perhaps it was the proximity to farm animals and their total disregard for location when discharging that imprinted on my mind and prompted unsolicited wetting. More likely it was Annie's shrill voice and her fixation on thumping me when told that I had done it again, last night, in bed. Jim didn't seem to care, even though he was afloat more often than on a dry straw mattress.

But now comes the puzzle: Why? Why did I not do it when travelling on trains? Never, never once did I abuse Mr. Pullman's marvellous invention. I have gone as far back as Hippocrates, Erasistratus, Galen. Galen particularly was puzzled how kidneys and surrounding stuff could hold such large amounts of water. Erasistratus, according to Galen, was simply out of it. Galen would have been absolutely amazed at my ability.

It wasn't until gun powder was discovered, with all those body parts lying around, that people started to figure out what was going on inside, and with water not having a lot of nourishment, what goes in must come out. We don't increase the volume much by absorption from the atmosphere either. But it was undoubtedly Mr. Pullman who taught me how to behave before I started mixing with fair company.

The pages following that I (Virginia) have written about our mother seem rather bare and stark but are all I can recall. One thing I am sure of is that she loved her children, and I remember her beautiful smile.

I can truthfully say I did not know Daddy all that well either, except through his letters which he wrote to me all the time I was in California until he died. He adored all his four young children and enjoyed them thoroughly. (Virginia married Raymond "Jake" Jacobsen, who came from San Francisco. They moved there in 1938.)

I can write volumes about Marg (J.P. Norrie's second wife after 1933) - but only little flashes come to mind about my mother. After she died and we went to live with Grandma Norrie, her name and her family were never mentioned to us children again. No wonder her life seems but a shadow to Harry and me. I don't have any memory of living like a family "day by day living" at all.

Until Marg came to us, there was no family - just four children growing up under strict supervision - care but no love ever, unless you call a bad three meals a day and chores love. I know that Grandma Norrie has been bitterly criticized for her lack of caring and affection but I don't blame her for that. She was the product of her own childhood, where her brothers were the shining lights. I know that I am a product of 84 years of living a life. There are many things I could feel bitter about, but I don't. I have a grand life and have been truly blessed and still am. One of my treasured blessings has been having you four children (the children of the second mother) in my life. Sometimes I wonder how I ever got so far away from "home" and then I know in my heart I am where God meant me to be: Psalm 39.

Our mother, Hanna Grace Hartling, was of Scottish and German heritage. Her mother was Scottish and her father was German. The original family name was Hirtle - later changed to Hartling to Anglicize it.

Mother was a beautiful woman. She was about five feet six inches tall and had a fine figure. Her hair was very dark and she wore it long, done up in a chignon on top of her head. She had lovely dreamy blue eyes and she was always smiling. Our brother Jim looked more like her than any of her other children. She was a Christian woman and in her marriage announcement it noted that she attended the Anglican Church and a bible study class there.

There are a few incidents that cling to my memory after these long, long years since her death. One day she was combing her hair, and Jim and Harry and I were standing close by watching her. She drew a long hair from her brush and said, "Oh, look, a grey hair." Jim said, "What does that mean?" Half in jest, she replied, "I guess I am getting old." Jim started to cry, and Harry and I looked at him with surprise. Mother took Jim in her arms and Harry and I clung to her as she comforted us. She was thirty-two years old.

One day she was ironing and she was wearing a long shirt that was scalloped along the hem. I asked her, "Mama, what is an operation?" And she said, "Oh, I just go there and they cut another scallop in the bottom of my shirt." That answer satisfied me and I went on with my play.

It was raining the day she left to go to Halifax to have surgery for some female problem after having four children. She was wearing a dark blue suit and a velvet hat with what looked like silver pussy willows on it. She kissed me and ran out to the vehicle that was picking her up. She turned and said, "Harry, come kiss Mama." He said, "I can't because I cut my toe." She hurried back and kissed him and hugged him. Then she was gone. We never saw her again.

It was still raining a week later. Harry and I were playing in a huge cabinet against a kitchen wall. The bottom of the cabinet had drawers. The top had doors behind which were shelves. Harry and I would climb up and lie on the shelves and close the doors. We were having lots of fun. In the middle of the morning a neighbour named Mrs. Hector Hartling came to the kitchen door and came in. She asked our grandma, "Where are the children? Send them outside." I knew something terrible had happened. Mrs. Hector had come with the word that our mother had died October 13, 1923. I crept into the pantry and heard her say, "Grace's last words were, 'Keep my children together'."

Harry and I stayed with Grandma Hartling through the winter until March when we were sent alone on the ship "Dufferin" to join Jim and Pauline - already living with Grandpa and Grandma Norrie. Our childhood had ended.

Love, Virginia

Little is known and little remembered of the Hartling side of the family. The Hartling home was on a short side road, about 500 feet off the main highway that runs through Port Dufferin. Hanna Grace Hartling was the daughter of Thomas and Jane Hartling. Thomas was an able bodied seaman, although at times, through narrative, he was elevated to sea captain. Jane's maiden name was Rutledge but she was not related to Ned Rutledge of Halifax, and that's about as far as I can go in that direction. She is buried in the cemetery on the off side of Port Dufferin. The grave of Thomas is unknown.

Others in the family were Garley, Harry, Bess and twins: Edward (Ned) and Susan Elizabeth.

Susan died at age five. Ned was mentally and physically impaired. I recall walking with him once from the house to the blacksmith shop on the corner of the road where it meets the main highway. When he became too much for Jane to look after, he was sent to an institution in Musquodoboit where he died at age sixty-three.

Garley became a harbour pilot in either New York or Boston. Virginia and I [Harry] are in disagreement about which harbour. I am of the opinion it was Boston but Virginia is adamant it was New York - whatever. But it is certain he piloted Woodrow Wilson into one or the other on his return after negotiating the Treaty of Versailles, ending the First World War.

Bess married and moved to British Columbia.

Harry lived in Lynn, Massachusetts and was employed at General Electric in Boston. With the '29 Depression he was laid off and came back to Canada with his American wife Ruth. He was hired by J.P. at Treadwell Yukon to drive the boat from the mine to Amos. They lived with us on the second floor of the manager's house.

I may have been named after Harry Hartling, Harry McDougal (Annie's brother), Harry Norrie (J.P.'s brother) or more likely Mephistopheles (the old Harry).

Now to get back to Annie and Henry and the hate in the house at Onslow.

Henry was a dour Scot. He didn't want to be dour, he was forced into it by his wife Annie, miscalculation, himself, life. Physically he was not an imposing figure, it was his mind that was superior. He was about five feet ten inches tall, carrying a couple of hundred pounds on a large frame, healthy, muscular and very strong. He had large callused hands; the back of either would demand obedience from a disobedient grandson. He smoked a pipe that was not pleasant to look at and difficult to forget and when stoked with cut-plug tobacco, not nice to be around. When this pleasure was not called for, he chewed tobacco, equally unpleasant but for a different reason. Henry talked very little, not wanting to or unable to overspeak Annie, who would inevitably be berating or giving direction to someone.

Henry's farm, located a few miles northwest of Truro, consisted of about one hundred acres, bounded by a river and a road and a couple of rickety fences. The terrain was flat and the soil productive, if properly treated. The farm usually supported nine or ten milking cows, a couple of spavined horses, and in prosperous times a pig or two. A few geese and chickens claimed by Annie just about rounded out the farm's animal life.

If even a small cow stumbled into any part of one of the fences, a gaping hole would appear, permitting the cow to escape and lead the rest of the herd to greener pastures. Then the cry would be

The Norrie family farm, Truro, NS, 1832-1940

sounded: "Go get the cows!" Morning, noon or night, most likely in the middle of the night, they would have to be brought back to home turf.

Although Henry was called a farmer, he never was and the land showed it. All his life he wanted to be a lawyer and with that longing he studied politics and the law at every unoccupied moment, mostly after supper in the evening until bedtime at eleven o'clock. He read everything that would help him in this field (rather than the other kind of field), including the Halifax Chronicle, the World's Work and other books and magazines. His son (J.P.) said that he knew more about world events and politics than any other man alive; perhaps a little oversaid, but close.

Henry did his legal work and that of his neighbours on an old, scuffed-up roll-top desk. I was endlessly fascinated by that desk with its little pigeonholes full of odds and ends of Henry's papers. Sometimes, when he was deep in thought, working on whatever, I would, unknown to him, stop at his shoulder, peer into those little holes, and wish that I could be alone there to fish through all of those secrets. I would give a bag of money now to own that old desk and all that it contained.

Neighbouring poor farmers would come to Henry to have their wills and deeds made up in proper form, which they would then take to a notary public, who would sanctify them for a buck or less, depending on the look on the guy's face and the amount of dirt on his boots. Today's lawyers feel in your pockets to determine their charges. And don't smile or the price goes up.

Along with Henry's knowledge of the law, animal midwifery and gentle executions ("farm wise", a misnomer if there ever was one), he also carried out autopsies on all farm animals that died from unknown causes. One morning Henry went to his work and found a dead cow at its place in the barn. It was no great loss since it was a mangy beast anyway that dribbled only a small measure of milk to Henry's benefit. Still he autopsied it before shoving it into a pit for burial, but found no disorder in any of the usual organs.

Next morning he was on the road in search of another cow to even out his small herd. He returned in the evening with a fine animal, the best he ever owned. It was a very large, black-and-white Holstein with an udder about the size of a forty-five gallon drum, and daily gave just about that amount of milk. It came with no papers, but obviously there was superior blood in its background, and it answered up to Henry's brag when he spoke of it. Henry always made sure it was well fed and he took it on himself to take the milk from it while Jim and I drained the other cows, working out to three or four each.

Henry's usually stoic demeanour was elevated for the next six months – until tragedy struck! At a Sunday dinner, while we sat around the dining-room table enjoying a bowl of Grace's home-made tomato soup (as with breakfast's oatmeal porridge, this was regular nourishment at all Sunday dinners; one of Annie's regimens), Henry turned to me and asked if I could see the big cow in a pasture quite some distance from the farmhouse. I told him I could make out all the other cows, but not that one. This, for some reason, agitated Henry. (What if you can't see a cow a half mile away?) He went to the telephone and called a neighbour who lived close by the pasture, and asked him to send one of his kids to run out and check on the cow. In a few minutes, the report came back: "The cow's dead in the pasture." Henry was greatly disturbed; this would be a particularly careful autopsy.

The following day, he and the farmhand, Perley McNutt, dug the grave, then went to work on the animal. Everything was checked out. Nothing. No obvious diseases or injury. Then, just before pushing the cow into the pit, Perley, on a whim, picked up the heart and slashed at it with his knife. The knife passed halfway through the heart and stopped. Further dissection found a straight, four-inch piece of heavy gauge wire exactly in the centre of the heart with neither end extending to the outside. It is not uncommon to find all manner of small foreign objects in a cow's stomach: bones, rocks, nails, metal... But this animal had eaten its own death.

The barn was Henry's refuge from Annie. She didn't venture there, only to the chicken coop for her eggs. I came to understand the reality of an old barn from Henry. There is something about an old barn on a dusty country road that tends to draw photographers to stop and take a picture. Old houses have their pictures taken too, but, it seems, less frequently. Perhaps, with an old house, the photographer has a feeling of unwelcome intrusion, but we often see pictures of old barns in books, magazines and other print forms.

To have lived with an old barn... How many farmers are at the birth of a child? This progression is carried on in a bright, off-farm room, and not with the farmer present in the room. He can sympathize with his wife's pain, but he does not feel it.

But in the barn he is there: he had better be there when one of his animals is giving birth, and here his anxiety is part of the animal's labour.

The first few months of his child's needs are almost totally attended to by his wife. He can look on and build a love unequalled by any other, but in the barn he is totally involved. His newfound animals are part of his wealth; a wealth that protects his wife and child.

At first he doesn't have a lot to do with the feeding of his new calf or colt; they come with a built-in road map. But when it comes time to persuade them away to a different feed source, it is his business to show them how. It is a far more odious job to keep them clean; just consider the volume.

If he's lucky, the life span of an animal is about a third of his own, and all that time he is their teacher, doctor and provider, which means he must always maintain his barns, fields and crops: unending labour and a different kind of love than that shared with his wife and family.

Unlike an urban boy, a farm lad will tell you very early on what it means when an animal of one breed takes a fancy to another of different breed and sex, and will, after a few months, help his father administer to the result; while urban youths lie to each other about that which they know not and keep on lying and not knowing until another of *their* breed and opposite sex shows them how.

Oh! Yes, an old barn has seen life from every angle, even to the death of the farmer. The photographer doesn't know, but unseen, this is what he is filming.

The North River separated Henry's farm from that of his neighbours to the northwest. In the spring when torrents of water and blocks of ice slammed into a central section of Henry's land they would gouge out five or six feet of his most valuable acreage, and what Henry lost, his neighbour gained. This went on year after year. Henry would build breakwaters, locally called brusheaps, with massive logs and tons of rocks, and each year the relentless river would carry them away along with more of Henry's property.

Henry tried to reason with the neighbour to find a solution, but the neighbour was no fool. Nature was adding to his land year in and year out, and given enough time he would pretty much put Henry out of business as a farmer.

Finally in extreme desperation Henry acted on his own; he threw a well-designed breakwater diagonally across the river, shooting the flow directly onto the land of his neighbour. The result was effective.

It caught the attention of the neighbour, who immediately hurried off to seek legal counsel. The counsel made a visual appraisal of Henry's worth and quickly figured that with little effort he could end up with ninety-five percent of it, the balance going to his client. Henry was haled into court just in time to hear the judge declare, "You lose, you pay." Pay?? What with??

Sure enough, the neighbour got everything. Henry was forced into bankruptcy but Annie strode to the rescue. She borrowed enough money to save the place, but forever after she would have Henry know it was her farm.

Henry Norrie

In those days bankruptcy was just one level above suicide. Henry had been a regular church goer, but never again until they carried him there in a box and planted him years later. Bankruptcy, then, was everlasting, whereas today you can do it over and over, piling your cash up in a numbered account in Europe or in a bag under the bed. Still, Henry's other neighbours liked and respected him and elected him to County Council for a period of eight years.

While Henry had never read John Locke, David Hume or Immanuel Kant, he had his own philosophy boiled down to his essence. I think Henry would say that every man is a world to himself, in his own orbit. He can talk to you all day about who he is; the first six hours are little white lies, the last six, after he thinks he has your confidence, are bold-faced lies. You will know less about him when the sun goes down than when it came up. If we could know people we would have perfect government. But then again, if we knew people there would be fewer of us; most would have to be shot.

For Henry, give a man your word, smile and be kind to him, and most of all admire his skin colour like when you see a nice picture.

On a bleak fall day, weather closing in, Henry had to harvest a field of oats. Two wagons were needed, his own and that of a neighbour. Toward the end of the day the helper on Henry's wagon noticed him acting strangely, making odd, slurred comments unrelated to the work at hand, but they continued on for another three hours until the field was cleared and the grain stored for the winter. After sheltering and feeding his animals, Henry went to supper but didn't make it. He went directly to bed. The following day a doctor was called and after a short examination declared that Henry had sustained a major stroke. He lived for many more years, but he would never again be a farmer or a lawyer.

Model T Ford purchased by J. P. for his mother, Annie. Harry is sitting on the running board

On a spring day, early in the old age of Annie and Henry, their best remaining son came home from his work in northern Canada with a Model T Ford and presented it as a gift to Annie. We should pause here and consider the reasoning behind this gift, considering the age and ability of the recipient. The only answer can be simple lack of forethought in an effort by a grateful son to ease the life of a parent.

The Model T Ford was not, at any time, easy to handle, particularly when the front wheels took on a life of their own and, defying persuasion from the driver, would wander off at any angle other than that suggested by the road. Another peculiarity of the front wheels was what was called "shimmying", a condition that can't be described, only experienced.

On a trip to Truro with Pauline and Virginia, travelling through the centre of town, Annie was confronted with violent shimmying. The car left the street, went up on the sidewalk and fatally injured a lady pedestrian. Extensive investigation found Annie not responsible. (The irresponsible part was that she was driving the vehicle in the first place.)

On a cold winter's day with few farm chores to worry about, Henry sat by the kitchen stove in his rocking chair. Annie, having her breakfast, was as usual flicking verbal darts at him about some nonsense or other. Annie picked at Henry endlessly like a crow at road kill, but Henry paid not the slightest attention. Deep in thought, smoking his pipe in his rocking chair, he was mentally forming new laws that not any government, unfortunately (municipal, provincial, or federal) would ever hear.

That day he seemed to be fooling around with his tongue inside his mouth. After a while he put a thumb and forefinger into his mouth and pulled out his last tooth. From then on he would not bother with teeth. It didn't affect his expression very much, except it made him look like he'd been hit in the face with a broad axe.

I was in school in '35 or '36 when I received a telegram from Mother Marg that Grandpa, Henry James Norrie, had died. The message didn't mention the cause of death but I was later told it was from gas pressure on the heart. I don't think an autopsy was carried out so I think it was more likely from another stroke. Either way, I don't suppose it makes a lot of difference. They planted him anyway in the Onslow Island cemetery. At the time, he was living with Annie and Grace in the little house that Dad had put up to replace the hundred and fifty plus year-old house where we four children were, what you might call, "raised", "brought up", whatever. There is a picture of Henry sitting in a rocking chair, outdoors with the rubble of the old house all around him. He has a full, imbecilic smile on his face, so different from his early years.

Gramma Annie Norrie

Persistently, during his last years, before Dad's success, he worried about who would take over when he was gone. Now he knew. Years before, one night when he and I were milking the cows and Virginia and I were the only two children still on the farm, he said to me, "You're going to take over the farm, aren't you, Harry?" I told him, "No, I'm not." I think that broke him up a bit. He never mentioned it again, but he still worried.

There's a picture around somewhere of Annie sitting on the porch of the little house at a time when she was the sole survivor of her family left on the farm. Annie died in the Truro hospital. I don't think she died "from" anything, she just ran out of people to pester.

In the ten years I was on the farm I was in Annie's bedroom once - the morning after Henry had his first stroke.

I was in the guest room once when one of Annie's sons came home for a visit. None of the "children" were allowed in the "living room" except for Virginia, who had to go there to dust often. She hated it!

Annie and Henry produced four children. The eldest, our father James Paul, the subject of this story, was known in Nova Scotia as Paul but in northern Canada as Jim, a very much admired mining engineer. Harry, a handsome young man, would die at age 21 from an accident sustained while working at the Round House near Truro. Then Joe and Grace.

The first two of Annie's four children, born over a range of twelve years, were very handsome; the third not too bad. The last child, a girl, was named Grace. The Oxford Reference Dictionary gives a long definition of the word grace, but our Grace didn't seem to fit any of it, except in spirit. When the messy things were cleared away after birth, it became pretty certain that this child was not going to go anywhere. Two small eyes, a large nose and a couple of ears placed at random. Where beauty counts most, it was absent. This girl child, on one to ten, was pushing ten if not beyond.

Physically her body was cylindrical, with large dimension, standing on two large legs with a head placed somewhere on top, rotating more than 180 degrees but less than 270 degrees, recording everything of interest in a cranial notebook. With built in kindness, happy, never morose, eager to help, her only put-off was her unfortunate body, and God knows she didn't deserve it. God must have been overworked at the moment of her conception and forgot to arrange things in their usual order.

Grace's early education was at a rural school a short distance from the farm. Following that, she attended and completed her education at the Academy in Truro. After high school graduation she remained on the farm. When the registry of birth was signed, an invisible registry of servitude was also signed. Grace would work for Annie, no one else. Never in her lifetime would Grace travel beyond fifty miles of the centre stake. But in that short radius she knew everything about everyone, mostly in the form of gossip, and that at a time when there was no telephone, radio, or TV, and a slow postal service. Still, with a two-cent stamp and a horse drawn wagon, the mail was faster than it is today.

The implication that Grace was not bright was allowed to whisper, was even encouraged, and of all the cruel things that were permitted to happen to her, that, by far, was the worst. She was bright, well above average, if there is an average. No magazine or books came into the house except those ordered by Henry, which were of no interest to Grace. But to keep up with political events, Henry read the Daily Chronicle, and this newspaper Grace would read a few days after Henry had finished with it, if time permitted.

Her only excursion from the farm was to church on Sunday, a momentary release, but I don't think she believed a word of it. She had Annie's word endlessly.

Grace's job was to keep the house clean (spotless), do all the cooking, assist nieces and nephews with their studies, milk cows when regular milkers were sick or absent, clean the stables when nephews pretended sickness, weed a small garden, work in the hay fields when threatening weather demanded urgency.

One of Grace's duties was to maintain a kitchen garden, an area in the range of fifty feet by fifty feet adjacent to Annie's pretty little lawn. With the weight of the winter's snow having compacted the soil, Henry and his horses, with a disk harrow, would loosen things up, then forget about it. Grace would form the rows, plant peas, beans, carrots, lettuce; keep it weeded; harvest in the fall and preserve the excess.

I don't know where it came from or what it was doing in that environment but occasionally Grace brought out a harp, a small wooden contraption with a few strings. Placing it on a flat surface she would strum on it, not with concert capability but pleasing more or less. At other times she would read to us. I remember one of the stories was about a guy called Enoch Arden; sad, but not out of place.

Grace never had a boyfriend, but a few advances were not repulsed, one of which was revealed to Annie and Henry by a family member. That caused such a violent storm of multiple voices that even far-away neighbours must have heard and enjoyed it, but it caused terror in lesser members of the family.

It may have been a period of monetary imbalance on the farm, when it was decided that Grace should get a job. She was employed by Eastern Hat and Cap. The job lasted for just hours when she rushed home weeping bitterly. The inevitable happened, her fellow workers teased and abused her mercilessly, drove her from the place, not a friend in a factory full of workers. Even with Annie's persistence she would not be budged, she would stay at home; if employment was needed off the farm, let Annie do it. She would not go back to that inhuman torment. If it had happened in this day and age, Grace would have been a perfect customer for an AK47 and not need instruction in its use.

After Annie and Henry had died, it got to a point where I suppose Grace got underfoot and out of place. With kindness, Mother Marg rented rooms for her in Onslow, part of a mile west of the bridge that crosses the North River, below the farm.

Through the years since early childhood I have been continuously overflown by a Guardian Angel. Anyway, at a point in time, one among many, a windfall fell on me with a passel of money. Before the passel ran out I bought a "rag top" Oldsmobile and drove to Truro to show off. I stopped to see Grace and found her at peace, residing in her own apartment with several old friends close by. I think she was happier than she had ever been - she had no restriction on money and was under obligation to no one. Later Marg provided a small house for her, beside the dairy.

In the end Grace was affected by some form of cancer and died in the Truro hospital.

In the early twenties, bumper wheat crops on the Prairies forced growers to import labour. A number of Nova Scotians with empty pockets, including Henry, were eager to oblige. "Riding the rods" and "blind baggage" they made their way west. It was here that Henry found, along with a little money, the western horse. Henry fell in love with the western horse and thereafter would always try to have a team of them on his Nova Scotia farm.

After years of hard work a horse gets old and useless. You love that old horse but he has to be put down. With regret and sadness you lead him to the glue factory.

On one occasion, Henry, back from the glue factory, was in need of a horse. A vendor assured him that he was buying a western horse, but when the deal was done, Henry found that he had been hoodwinked. He had bought an Ontario horse. (There are some mean men in Nova Scotia too.) Now Ontario horses, like some of the people, are ornery, mean, overfed and arrogant. But with the passage of time Henry had the last laugh. For the price of one horse, Henry had bought two horses. The Ontario horse, unknown to the vendor, was with foal. A couple of years more and Henry had for himself a magnificent stallion. But, as I said, the Ontario horse was mean, downright hazardous to life and limb. Once when Henry went up close to make some harness adjustment, the horse was lying in wait. The brute let go with an iron-clad hoof, fortunately just grazing Henry's leg but leaving him with a limp for several days. The horse's intent was to kill, at least to take away a leg.

Henry heated his little house with a furnace in the basement and a cook stove in the kitchen, fueled with dry wood. In the slack winter months he would buy a few spruce logs from a neighbour who owned a wood lot. But Henry would have to cut and snig the logs himself. In this way he would keep the next winter's wolf from the door.

One winter was upon him and Henry had still to snig his firewood. With the last load, in cresting a hill, disaster fell. Just before the weight of the load would have prevented the ornery Ontario horse from kicking, it kicked. The leg came down on the off-side of the canter pole forcing the horse to stumble and fall, pulling the other horse down on top, breaking the offending leg. Any broken bone on a farm animal is serious: a broken leg is irreparable - fatal. Henry was forced into the role of executioner; the animal would not be allowed to suffer. A blow with a sturdy wooden club applied with force just below and in front of the ears and Henry was again in need of a horse. (Some years later Henry's farm, combined with several adjacent farms, would be home for a number of noble Pecherons, but not now, not yet.)

Next morning he set off aboard his remaining horse to buy another. He was gone all day. Henry was a fixture on the place and when he was not there little nerves in the body suddenly started sparking. But then, late in the evening he came home with a new horse and put it in the empty stall. We were eager to pass judgement on Henry's acquisition but it was bedtime and Annie's rigid regimen would not be overruled, up early would be soon enough. First off, we asked Henry, "You got a new horse?" "Yes." "Where?" "Folly Lake." "Who from?" "Mr. Fajarly." Well, that settled the name for the new horse. It lived and was buried named Mr. Fajarly (pronounced M i s t a Fa Jarly).

That horse's imperfections, too, would last for the rest of his life and a lot of ours. Regardless of how much food he ate, which was a lot, he never gained or lost weight, not an ounce. If you threw your hat at him, it would hang on. And he was always changing - subtly, when you weren't watching, never when you were. For instance, he seemed to be able to change at will any bone or group of bones in his body. You could look at him, look away, and he had changed, but you never knew how or why. Strange! He was always different.

He stood on four large posts with huge hairy feet; a Clydesdale's feet are dainty in comparison. His main body was a misshapen mass of bones with a light, chocolate-brown-coloured, wrinkled hairy rug thrown over it. His head and neck could fit on either end. That might have been one of his secrets, you never knew. He stood stolidly with his neck never higher than parallel to the ground, usually lower. But he didn't drool! His eyes were never more than half open, but when you looked deeply into those eyes they talked to you. In there was a little round ball of devil humour, perhaps with wings; just don't mess with me!

Regardless of what Mr. Fajarly was attached to, he would move it, be it the Eiffel Tower. When he was drawing a load teamed, it didn't matter what the other horse was doing, he would draw it too, and usually did.

Annie had her beautiful, little, daily-curry-combed pony and a polished carriage. Henry had his carriage too: splayed wheels on iron rims supporting a bench, drawn by Mr. Fajarly. No two wheels on Henry's old wagon ever ran in the same direction.

One day Henry needed something only available in Truro and I was invited along. We hitched Mr. Fajarly to Henry's carriage and got aboard. Henry pulled on the reins and said, "Ha-a-a, git-up." There was just a moment of hesitation, then suddenly without a flicker of warning, Mr. Fajarly, his eyes wide open, lifted his tail straight in the air, opened that vast channel of endless decomposition and let off a thunderous, blasting crepitation: a lighted match would have destroyed the farm. The lifting tail, I think, gave the warning to Henry, who jumped and instantly fled; but I, being less alert, stayed for a second, then just in time catapulted out the other side. With that, and I think Mr. Fajarly knew we were away from danger, he up with both feet and nailed that carriage into total destruction: kicked it all to Hell. The loss of the wagon was incidental, but even now that moment of furious destruction is frightening. It changed my impression of Mr. Fajarly. That little devil of humour I had seen in his eyes was accompanied by his pet, the many-headed Cerberus, the dog that Dante and Virgil found in the third circle, the dog that guards the gates of Hell.

Henry stood on one side of the yard and I on other side with Mr. Fajarly in the middle, returned now to his normal state of repose, head down, tail down, eyes half closed. Gradually Henry eased over to Mr. Fajarly to determine what had caused him to react so volcanically. The problem was soon solved; a long bloody gash in Mr. Fajarly's rug. Further inspection found the tongue of an ancient buckle in that decayed old harness, rusted open, straight out, and when Henry pulled on the reins the buckle cut Mr. Fajarly, releasing the dog from Hell. (Just for once, we were allowed Annie's carriage to go to town – but not her horse.)

With that sudden kick perhaps Mr. Fajarly had just been waiting for an opportune time to repay a gross injustice that had been done to him a few days earlier, that came to purposeful event in the following manner. From some unknown source, a large truck tire found its way onto the farm, and with the dearth of other toys such as little wagons, bicycles etc. that old worn out truck tire was sometimes a source of amusement for a small boy. Set in motion in an upright position and hit with a short stick it could be wheeled about the farmyard for no other purpose than just doing it, and since the water well with its wooden pump was situated on a small rise above the general level of the yard, the tire could be sent speeding along in almost any direction just for the simple pleasure of seeing it crash into something, preferably something that Annie cherished, like a flock of her prized chickens. But on the day in question, when Mr. Fajarly was alone in the yard leisurely cropping the odd blade of grass, his large rear end was a beckoning target. Myself being of a bland and gentle nature, I would never have thought to harm Mista Fajarly in any way. But when the seeds of jeopardy are planted in a juvenile mind by an older sister (Virginia) the weeds of malice soon take over.

Starting on the modest little hill by the pump, the tire purposefully sat erect and sighted in. With a smart push and the force of gravity it rolled quickly away and soon slammed into Mista Fajarly's most tender area, between his hind legs. Even though early in his life his potential for noble horsehood had been cruelly reduced by the power of two, it was still a place that could not tolerate a lot of abuse. With the impact, his eyes flew open, his head jerked erect, his back arched to the limit and his hind legs elevated to the point of tip-toe. He was forced with much difficulty to extricate himself from the mean torture machine, with the offence made even more obnoxious by the sight of two brats now rolling around with glee on the little hill. Yes, he could have been remembering that heartless incident when he let go with that well-placed kick on Annie's manicured lawn.

Along with all the other animals, there was a dog on the farm, a large, black, misshapen Newfoundland dog, hated by everyone but Annie. Any extra food, cake, cookies, pie, whatever would be thrown to that dog. Anything that would fill a little space in a cavernous kid would go to that dog. The dog's name was Bobby. Wherever, whenever a dog is to be named, it should not be called Bobby. I was not born with hate: Bobby did it.

In front of the house, Annie had sectioned off, with a sturdy wire fence, a little patch of green that under Annie's supervision had always to be kept properly trimmed with a push lawn mower. A large gate in the fence allowed passage of a wagonload of wood for the furnace in the basement. Otherwise the gate was always closed. The farm animals, horses and cattle, would stand in the yard and gaze at that lovely patch of green; until one day the gate somehow was left open. Mr. Fajarly was alone in the yard. A little fresh grass was just the thing for a momentary separation of his many bones. He was at it! A short time later I came out of the barn, and with permanent instructions imprinted in my head to forestall any damage to Annie's lawn, I went to shoo away Mr. Fajarly. But to give him the maximum pleasure, I slowly wandered in behind him. Just as I instructed him to leave, that goddamned dog rushed up with a ferocious bark right in Mr. Fajarly's ear. Mr. Fajarly, momentarily confused, corked up one huge foot and fired, catching me mid-centre. Just a few inches closer and I would have been airborne over Truro, trying to recall the text of our minister's last Sunday sermon.

Normally, though, we children could play around, under, and over that wonderful animal. Even now Virginia or I may mention the words Mr. Fajarly and just giggle.

In the evening after we had milked the cows, the milk, in two large metal cans was brought into the kitchen where the cream would slowly rise to the top, and when combined with the morning's milk, would be placed by the side of the road to be picked up and delivered to Borden's Milk Co. in Truro.

The bedroom for Jim and me was over the "old kitchen". One morning I awoke to find myself alone, but after a minute or two Jim came back into the room. I asked him where he had been but got no reply. We put on our clothes and went down the stairs to the kitchen to wait for Henry to go with us to milk the cows prior to having our breakfast, always a bowl of oatmeal porridge.

On this particular morning, when we entered the kitchen I noticed drops of milk in a line on the floor. I said to Jim, "I wonder who did that." He said, "I did." He had come down and scoffed a glass of milk.

As far as us four kids were concerned it was an act of parliament: don't touch food of any nature except at mealtime, and then in limited amounts and always under circumnavigating eyes of Annie, along with unending instruction about proper manners and the grave hazard of certain foods to children's guts. The long proscribed list included tea, coffee, toast, pepper, pickles, vinegar, salt (used only in small amounts), eggs (except for two each on Easter Sunday morning). This list was particularly maddening with all those chickens Annie had running around the yard dropping gobs of grey stuff that got between the toes of bare feet. But beyond restrictions taking the pleasure out of eating, we were amply fed, and clothed.

To get on with Jim's milk caper. We were allowed one glass of milk each at each meal and obviously Jim had broken the law. With the drops still on the floor and Henry and Grace approaching from different directions, with panic in every fibre, I waited: the theft would have to be reported to Annie.

Certain it was that Jim would be beaten some time after the sun went down; twelve, fourteen hours of mental torture. Almost as much for me as for Jim. When I look back at it now, the question remains, why the hell did he do it? Any dumb little thief will cover his tracks, and Jim was very far from dumb. Unanswered - unanswerable.

The darkening of the setting sun only increased the darkness of the previous hours. Jim and I, along with Pauline and Virginia, finished our studies at the kitchen table and went to bed at the mandated hour, 8.00 PM sharp. There was little if any conversation. Jim and I got into our night shirts and climbed into bed against each other, quaking with nerves. The hour arrived: the summons came. "Jim!" Down he went to Annie. I was already crying. I could hear the strokes against him: slam - slam - slam in drawn out succession. He came back and got into bed, crying hard but seemingly with grit in it. After a while our crying stopped. This simple act of torture was over, but there would be more to come, for other acts of defiance, real, imagined or otherwise. The torture machine was a thick section of horse harness with two brass rivets in the hammering end, that was struck with force.

For two weeks to come, in the morning and in the evening, I would examine the welts on Jim's back, two inches wide, black and blue from the top of his legs to the tip of his shoulders. The damage would gradually lessen, until the next time.

After three quarters of a century, still, why?

Perhaps there are some among us who can't be broken; damaged but not broken. And perhaps there are some who will defy to test their defiance.

No one knew about Jim; least of all Jim.

In wild strawberry season, we would be sent to pick and bring home as many strawberries as four kids could find. Any change in the redness of our lips would bring reproof. But Jim, in some way, would generally return with less than half a cup and still show no evidence of theft. In this way it was supposed that we showed Annie that she was much appreciated around the place. The only reward for our effort, and picking strawberries can be tough on the back at any age, was an undersized strawberry shortcake, soon eliminated. The balance of the season and into the summer we were supposed to be delighted that Annie enjoyed strawberry jam with her breakfast at about eleven o'clock each morning.

Potatoes were another thing. They were planted in a small part of a field not far from the house. Potatoes, high energy, were a large part of the main meal of the day, and at least a little of another meal. It was for that reason that an area larger than the kitchen garden was required to sustain a family of seven, four of whom were bottomless.

Seed potatoes were a portion of those left over from the past year's crop. They were sliced in a way that would leave two eyes in each seed, always two eyes. A small plow would form a depression where the seeds were dropped, to be covered by hand or with the use of a cultivator.

In too short a time after the plants had formed, the potato bugs zeroed in to exercise their gluttony and it was time to exterminate the pests. Pesticides came in two forms: a chemical called Paris Green, a compound of copper and arsenic that was dusted onto the plants from a container attached to a fireside-sized bellows. The other, children's hands, that would pick the bugs from the plants and drop them into an old tomato can. When the can was full, it was taken to a flat area away from the potatoes and dumped onto the flat place, where the bugs were mulched, mangled, and mutilated by unclad feet, or on lucky days feet covered by worn out sneakers. At season's end the potatoes were ploughed up, collected and dumped through a small window into the basement of the house for easy access to the kitchen.

Perhaps it was the annual whiff of Paris Green that changed the Norries' DNA and made them so large, powerful and knowledgeable.

One Sunday when I was twelve, I met Roy McKay in a small room off the main lecture area in the Onslow United Church. He told me he was organizing a group of young guys called Trail Rangers and asked me if I would like to join them. Sounded great. I put the question to Annie and amazingly she approved. To threaten to quit milking cows if I wasn't allowed to join would have been certain death. But I had to be home by nine o'clock.

Some weeks earlier I had found an old bicycle of Dad's in the attic under a pile of sawdust. With previously unrecognized mechanical skills I got it working and spent the unoccupied moment riding around the farmyard.

Meeting with Roy McKay and the other boys was a great pleasure; meeting with anyone away from the farm was a pleasure and a rare event - that is, except for the forced daily exercise of going to school.

Annie's demanition (demanition: a new word; with the same meaning as demand, but with sparks!) was to be home at nine o'clock - not later. I arrived back at the farm on my newly acquired mode of transport at 9:10 PM. All doors, windows and mouse holes were locked, barred and camouflaged. That scared the hell out of me. The intent was that I should find shelter from the night's witches and werewolves with the animals in the barn and I was about to do just that when the first call to rebel fired up in a corner of a previously passive mind. The proposition now was to figure out a way into that dark old house. First I went around to Grace's second storey window and bounced a few rocks off it. No response. The anger in my guts was boiling pretty good by now and I was determined to find a way in.

Then the advocate of all known evil showed his face and brightened my mind. I picked up the bicycle and jammed it into the space between the shed and the house, climbed up on the seat and shinnied onto the roof of the house. I salamandered over the edge of the roof, and hand over hand, worked my way up to the peak, then around the chimney to a short length of ladder that Henry had left in place after he had repaired a section of the roof. "Sh-h-h-h." I eased the ladder over to the skylight that provided weak light to my room over the kitchen. Mind you, all this anger-suppressing activity had to be carried out in complete silence. Elsewise Annie would have been out throwing rocks up at me. I grabbed the skylight, and with a firm yank broke the string that held it shut. I was in! Before going to sleep I went over in my mind what I would do when Henry called to wake me in the morning. I had the feeling that he probably did not know that Annie had blocked all the normal points of entry after he had gone to bed.

Soon enough I was awakened by the sounds of Henry and Grace when they came to the kitchen in the morning. Henry came to the foot of the stairs leading to my room and called, "Harry" and again, "Harry." This time for sure, no answer. Then I heard him say to Grace, "That boy didn't come home last night." This, to me, confirmed that he hadn't known that Annie had fortified the place. There were no more words between Grace and Henry.

Quieter than the proverbial mouse I got up, put on my clothes and went down the stairs. When I opened the door to enter the kitchen, I scared the hell out of two people. Henry turned with a shocked look on his face and said, "Why didn't you answer when I called?" I said, "Why didn't you let me in last night?" From that day to this not another word was ever spoken about it. But I sometimes wondered what must have rattled around in Grace and Annie's minds asking the question "How did he get in?" Henry knew, but I don't think he ever told anyone: he saw the bicycle between the buildings when he left the house that morning just before I moved it, when I followed after him and before anyone else considered the workings of a secret, devious mind. I told Virginia, of course, but a well placed mental fulcrum could not have wedged it from her.

Chapter 2

The First Summer in the North, 1927

by Harry Norrie

Grammy, Annie McDougall Norrie, drove us to the train station with her horse and buggy to begin a journey from Truro to Northern Quebec that could have been planned by an avid scout master or an imaginative tour guide. Jim and I stood aside as Annie pleaded with the ticket agent to accept our tickets and allow us to board the train. That done, we followed her to the Pullman car that would be our home and playpen for the next sixteen hours.

At the foot of the Pullman steps we were met by the train conductor and the Pullman porter where Annie exhorted each of them to keep a wary eye on us, but if they lost us it would be considered an oversight only and there would be no reward. Whereupon Annie departed sans kisses, hugs, handshakes or pats on the head. We were on parole for the next two months, she was gone, and we were in the care of two strangers, one white and one black, a good sign.

Four trains passed through Truro each day, two going north and two south. One train was called the "Ocean Limited", the other the "Maritime Express". The Limited stopped at every town, village and large gathering. The Express stopped only at larger centres of population. Its real purpose was to establish that Canada was a land trying to become a nation. In those days people lived on farms. Politicians, bureaucrats and lawyers lived in cities and travelled on the Express.

The Pullman car, invented by an American, George Pullman, was a marvel for its time. First the smell, a mixture of steam and coal smoke that drifted back from the engine. A smell that remains fixed in memory. At one end of the car was the men's room, sometimes called the "smoking room" and at the other end was the ladies' room, out of bounds for all small boys and always a source of mystery. Next to the ladies' room were two self-contained compartments where any number of unseen pleasures could be carried out. The main central area of the car was divided into seating sections running along each side with a rather narrow aisle between. Each section had two seats facing each other, ample room for four people, and smoking was not allowed. Pullman cars (the seating area) and churches were probably the only places in the world, at that time, where smoking was forbidden. But the genius of this work was the way the Pullman car would be changed into a sleeper by the Pullman porter.

His first move was to collapse the back and seat to form a platform for a bed. Then with a key device he allowed the ceiling to drop down, which would become the upper berth. During the day it was a place of storage for mattresses, pillows and other paraphernalia required for peaceful sleeping. When he had finished his work each seating area had become two completely separate, comfortable beds, an upper and a lower, draped on the aisle side with heavy curtains. The comfort factor might also be determined by the age, size, marital harmony and number of occupants. For the needs of two small boys the upper berth was ample.

I remember nothing of the train's conductor, but the Pullman porter "Big Tom" is still in my memory. He was a very large, not fat, just large black man and his colour had not been diluted. "Big Tom" was the first black person I had ever met. He left me with a continuing respect for black people in general. I have often wondered how self-proclaimed super-white people can disparage the intelligence, spirituality and vast humour of black people.

Only once have I had a confrontation with a black man. On a street in Port au Prince, Haiti, I was bumped and angrily told in French that I was a "Goddamned white man". I did not press the point, but if I could have I would have gladly and quietly changed colour. That night back in my hotel room I thought about that "bump" and gradually came to feel: "By God, it's about time a black man should start pushing whites around, after the barbarous way whites have always treated blacks." I described that event to black friends in Toronto and to a man, without exception, they were shocked that a black man would do such a thing. Think about it for a minute. Has ever a black man dared or even bothered to bump a white on a street in Toronto, Montreal or Halifax and make it known that that is exactly what he intended to do?

"Big Tom" had a large round face dominated by happy eyes and a permanent smile filled with large very white teeth. He was our shepherd on this part of our trip, and would be again, several times, in later years.

While I had not known any black people, I had heard a little about them, and a pestering question came to my mind that I wanted very much to put to Tom: Had he ever been a slave? My opportunity came the next morning when he was sitting alone in an empty section taking a rest. The question didn't phase him at all. He just gave me a big smile and an evasive answer and left me quite chagrined at being so bold. But I am getting ahead of myself. Let's go back to Truro.

At some time in the spring of 1927 arrangements were made between northern Quebec and Upper Onslow for Jim and me to be with our father for the summer holidays. Our father, yes, but still a man we did not know at all, having seen him only for short irregular periods in the forepart of our lives.

Then one night in early July we were scoured down in the family bathtub, a movable galvanized metal laundry tub. The scourer was Aunt Grace, chief body washer and food processor.

We might as well have gone without a bath since next morning there were chores still to be done. Starting at 6:00 AM while Grandpa fed the animals, Jim and I started milking the cows, sitting on a low three-legged stool and resting our heads against the cows' flanks, more for comfort than leverage, while regiments of cow lice paraded out onto our heads as we stripped the unprotesting cows.

That done we repaired to the kitchen to augment our health and energy with large bowls of oatmeal porridge. Then back to the barn to continue with the many chores remaining to be done.

Referring back to the business of cow lice, it was not an uncommon condition for farmers of the day to have them in their hair. I can still clearly see a particular farm boy that I sat behind at school and whose neck I would watch for numerous lice parading here and there. That boy later became Nova Scotia's Minister of Agriculture.

So here we are on the train, Jim and I, two little angels, pensive and apprehensive, but each with devil emergent. The train took us around the wide bend out of Truro, past the creosote plant, Bible Hill, Marshall's Corner, where we could see the Norrie farm buildings in the middle distance, then up the long grade toward Debert. We sat quietly, perhaps unsure that we were well away from Annie's domination. Maybe she was riding "blind baggage" and would reappear at any moment.

A short distance north of Debert we received the first great thrill of our journey. We were "up in the air", we were "airborne".

Happening to look out the window, Jim saw that the train was a frightful distance off the ground and with fear-shock on his face brought it to my attention. We could still hear the click clack of the wheels on the rails but craning our necks we could not see the road bed. The train was travelling very slowly and it was quite some time before we had crossed the trestle and were again on solid ground, where the train increased its speed. For two kids who had not, since coming to the farm, been further afield than Truro nor higher off the ground than a horse's back, it was a hair-raising experience. In future years, as we passed that way we always waited for the "trestle".

"Blind baggage" or not it was time to start exploring our surroundings. Pensiveness had departed and apprehension had given over to curiosity. I was eight years old and would probably have sat unmoving in the same spot all the way to Quebec City, but Jim, enjoying the advanced age of ten, was much more adventurous.

We found at one end of the Pullman an opening obstructed by a heavy curtain. The smell alone would verify this as the men's smoking room. With two heads around the curtain, we discovered quite a large room with a couple of leather-upholstered easy chairs and a bench running along one wall. Continued inspection found self-supporting ashtrays, a brass spittoon, wash basin with wall mirror and another door. Behind this door we found a small compartment with a sign affixed to the wall. "Do not flush toilet while the train is standing in the station" made sense, considering the splash effect, and satisfied our concern about where to go when necessity demanded.

In the passageway around the men's room we found a small crevice in a convenient spot available to all passengers. In this crevice was a faucet and a container full of small collapsible envelopes. Little time was required for us to discover the purpose of this arrangement. "Ice cold drinking water." With all the excitement of the day and a sudden onset of unquenchable thirst, we found it necessary to make trip after trip, back and forth to the water cooler, resulting in a substantial pool of water sloshing between our seat and this marvellous new type of water well.

In relatively short time we had completely examined all of our car.

The ladies' room was out of bounds and the two self-contained compartments, in this early stage of our lives, were of little importance, but there was still that long expanse, the rest of the train.

A passenger train might be thought of as a movable village. Second and third class: slums; Pullman: Rosedale, Forest Hill, the Bridal Path. The engine: heat, light, transport. The engineer, the mayor. The brakeman, conductor, porter: town councillors. Yes, and in the third class car next to the baggage car was a small store where one could buy cigarettes, sandwiches, magazines, post cards, and candy.

Jim had been supplied with a then large amount of cash, perhaps $10. I don't remember, but enough to compel us to run back and forth to the "store" while I'm sure we became an irritant to all other passengers. Before nightfall we knew our train from the rear-end observation deck to the far end of the last third-class car where a metal grill denied further exploration. (Only once in all the years have I been in a train locomotive and that was in New Mexico where I arranged to buy the cap from the head of an engineer on the Santa Fe, in the event that I might someday need to run a model train set up.)

Now with our explorations for the day completed we were ready to go to bed. Big Tom was making up our car. The upper berth, our bed, was lowered onto its enlarged bicycle chain supports, the seats were rearranged to provide the lower berth, sheets and pillows installed and the whole section secluded by heavy drapes. "Big Tom" secured a ladder in place and up we went.

Our bed on the farm was a cloth envelope filled with dusty straw, replaced once a year at the end of harvest. This was a luxury: white pressed sheets, fluffy blankets and a comfortable mattress wide enough for two skinny kids. A fish-net type thing hanging from a low wall was obviously a place to store day wear, and a rather long narrow window could provide an uncomfortable view of the passing countryside but was not much used due to the swaying of the train that so much encouraged sleep. But there was still one other little thing. That button on the wall that said "Press to call for assistance". We each fingered it caressingly for quite a while until... I'm sure it was Jim who pressed too hard and Big Tom was demanded away from making up the rest of the berths. Our "sorry" forced a frown but later a chuckle as he worked farther down the car. In the morning after we had muscled into our clothes, we didn't call for the ladder. We jumped down.

The cow lice, not able to survive without the comfort of a cow, were now all stone cold dead in the upper berth. So, after limited exposure to the men's room, we were ready for the dining car.

Located between the Pullman cars and the second class cars, it too was a wonderful invention. We had passed this way several times the previous night but paid little attention, having at the time more pressing explorations to be carried out. The head man (what's he called in a dining car?) met us at the door with more than the usual disdain, having to seat two non-bribing, smelly farm kids. But seat us he did, across from each other, at a table for two, beside a very large window where we could appreciate the fast-moving landscape as well as the activity around platforms while the train was standing in the station.

On the farm it could have been written in blood on the wall: Breakfast is a bowl of oatmeal porridge, with all other meals an afterthought. Now we were seated in a first class dining car waited on by a young man all in white with a napkin over his left arm and looking as though he would die from remorse if he displeased us in any way. This was certainly not from our noble bearing or dress, only from the firm demands of his stern CNR employer and a courtesy offered to all and sundry. Jim and I did not know that we had arrived. Still our manners were proper after years under the unblinking gaze from the glacial brooding eyes of Annie.

The dining car was of equal length to all the other passenger-train cars. It had tables for four people along one wall and for two on the other. This arrangement provided for the waiters to transport their large trays of food on a vertically extended arm and on feet well choreographed to overcome the incessant swaying of the car. I never witnessed an accident. They must have been trained by Phineus T. Barnam on a slack high wire.

The table was covered by a crisp white tablecloth and had a sugar bowl, salt and pepper shakers, water pitcher, all made of silver and with the ever-present CNR stamped, printed or embroidered everywhere.

We could see the chef with his tall white hat in his galley at the far end of the car: a whirling dervish, gaudy, dancing through the smoke and flame of that narrow space, turning out dishes that could have been advertisements for a present-day Julia Child. Years later there would be Antoine's, Top of the Mark, 21 but none better than our chef. But who am I? My knowledge of cuisine culture, then and now; it could just as well have been run-of-the-road-kill.

A thin leather folder was presented to us. From this Jim read the menu, for what to us seemed a royal feast, while I rippled in glee with low rumbling laughter: corn flakes, grape nuts, bacon, eggs (boiled, fried, poached, scrambled - whatever) toast, jam, and no oatmeal porridge. Over a short space of time we satisfied ourselves with much of the above, then cartwheeled back to our seats in the Pullman. But before I go further I must impress on you that this is where I was introduced to Grape Nuts. To this day they have something like an aphrodisiac effect on me; even more, something spiritual! I never needed to become a hippie - I had Grape Nuts.

When we finished our breakfast, the waiter removed the dirty dishes but then returned with bowls of water that he set in front of each of us. "Now what the hell is that for?" (Oh yes, we were well versed in vulgar language having listened attentively to Henry, who did very well with small-time, lower-case curse words.) Too young and untravelled to figure out the natural use of these lovely little bowls and having little use for the water, it would be some years before we would use them to cleanse sticky fingers.

During the night, while we slept more soundly than usual with the swaying of the train and a soft mattress, we had passed through New Brunswick and were then well into Quebec, and having satisfied our curiosity about the mysteries of that part of the train available to us, we decided to expand our field to the station platform at the next stop.

As we wandered around, Jim happened to notice the name of the place on a sign hanging from a station wall: Trois Pistoles - Three Pistols. The bloody place to me is armed cowboys, Indians, brigands. We fingered our non-existing side arms and, just in time, jumped back onto the train, to evermore remember the name of that nice little town on the south shore of the St. Lawrence River. Trois Pistoles.

We arrived in Levis some time before noon. I don't think we had any luggage. We had nothing to lug, and why carry an empty suitcase when one pocket was sufficiently large to carry our possessions.

We were met in Levis by Aurel Bourgette. How he found us in that milling mob of people as the train emptied, I do not know, probably by sniffing. Aurel did secretarial work for Dad and had been sent to intercept us, repair any damage and take us to northern Quebec. His home was in Levis so it only made sense that he would be the one.

Aurel was tall, thin, French, very handsome and a wonderful man. He was also frail and sensitive to a point where, due to incessant teasing by Dad's other employees, he had, eventually, to be discharged. I think Dad was able to get him a job at Sherritt Gordon in Manitoba where, out in a canoe, he capsized and drowned. I saw him once, alone in a canoe on a lake near Rouyn, and wondered then what kept him upright even for a short time. His family in Levis must have been shattered, especially his mother. He was not a bushman.

Aurel took us to his home in Levis where we were introduced to his mother, a classic example of a French mère. She was of medium height, rotund, rosy-cheeked and eternally cheerful: a perfect model for Monet or Rubens. Her immediate reaction was to cuddle and feed. There were numerous other members of the family but since they could not communicate with us in English they soon found other entertainment. To put it in the vernacular, they could not crepitate in English, nor us in French.

Late in the afternoon we crossed the St. Lawrence by ferry where, at Central Station in the lower town, at about 8:00 PM we boarded the train for northern Quebec. There was nothing of note between Quebec City and Senneterre, only small, uninteresting towns separated by endless dark forests; and travelling by that train was restrictive, cramped and boring even under a kind but watchful eye.

We were met in Senneterre by Dad, his brother Joe, and his girlfriend, Jean Hitchman. Once again, no hugs, kisses, handshakes, or pats on the head; we were just there. I was beginning to think this shunning was for the same reason that Aurel was able to find us so easily, at the station in Levis: smell!

But then, after we reboarded the train to continue on to Tachereau in a car of the second class section, and though I was still very young at the time, a strange feeling came over me that has been in the fog of my mind ever since. The reason for it was probably that I was now in the shelter of a parent; the air, even in mid-summer, was cool, very clean and smelled of the forest. A fitting adjective would be salubrious. Perhaps a wayward angel just happened to pass by and smile on us. That remains an amorphous moment in my life.

At Tachereau, a short distance to the west of Amos, we changed trains for a spur line to Rouyn-Noranda. Well, Noranda the town was not there yet, just the mine, with the smelter under construction and the first smokestack only halfway up. Noranda the town would come later.

The Americans didn't have a patent on frontiers. Rouyn was frontier with all the accoutrements to make it so (that's a good word, I looked it up), guns, knives, whiskey, fists, gambling, bad boys and good girls. The American frontier was about cows; the Canadian frontier about minerals and gold, and there would be other mining towns besides Rouyn: Malartic, Val d'Or, Cadillac. The town of Norrie was more a post office than a town. There was also Pascalis but it burned down in a bush fire and was forgotten.

On our frontier we didn't ride horses; we walked or went by canoe, or walked and carried our canoe. Footwear was a misnomer; you wore knee-high boots or went around with dirty feet.

One day I saw Dr. Rudd in his car, a Rio, stuck fast in the mud in front of Albert's Hotel, waiting for a horse or an ox to pull him out. The sidewalks were made of boards and you walked there or were in danger of appearing on the missing persons list.

And there were murders. Dr. Rudd told me about a man who had been stabbed. The doctor went in and fixed him up but then a year later he stabbed another guy, killing him, and they swung him for it. Another fellow, slashed from shoulder to hip, too drunk to be anesthetized, was sewed up without a murmur with the old crowfoot stitch, and sent on his way.

In 1935 the police brought a guy to Youville Hospital and put him in the next bed to mine. He'd been shot in Chibougamau by the son of a well-known Montreal lawyer. He died a couple of days later. Lawyers and bureaucrats tried to keep the case going but eventually the police moved in and spoiled it all. Jim and I were not a part of it then but we would be later.

In the summer of 1927, from Rouyn Jim and I walked seven miles west to Stadacona Gold Mines Ltd. where Dad was employed as manager. The road, suitable for horse-drawn vehicles only, was a passageway cut through the bush. It was not graded in any way except in very low wet spots where logs were placed side by side to create what was known as a corduroy road to prevent wagons from sinking completely out of sight. On those roads you walked or had your brains bumped out.

Stadacona consisted of three, small, tar paper-covered shacks and a diamond drill. There was an office, a bunk house for the crew and a cookhouse/dining room. Jim and I slept at the rear of the cook house. I wet the bed, frequently, and would for some time to come, perhaps as an act of revenge for never being hugged.

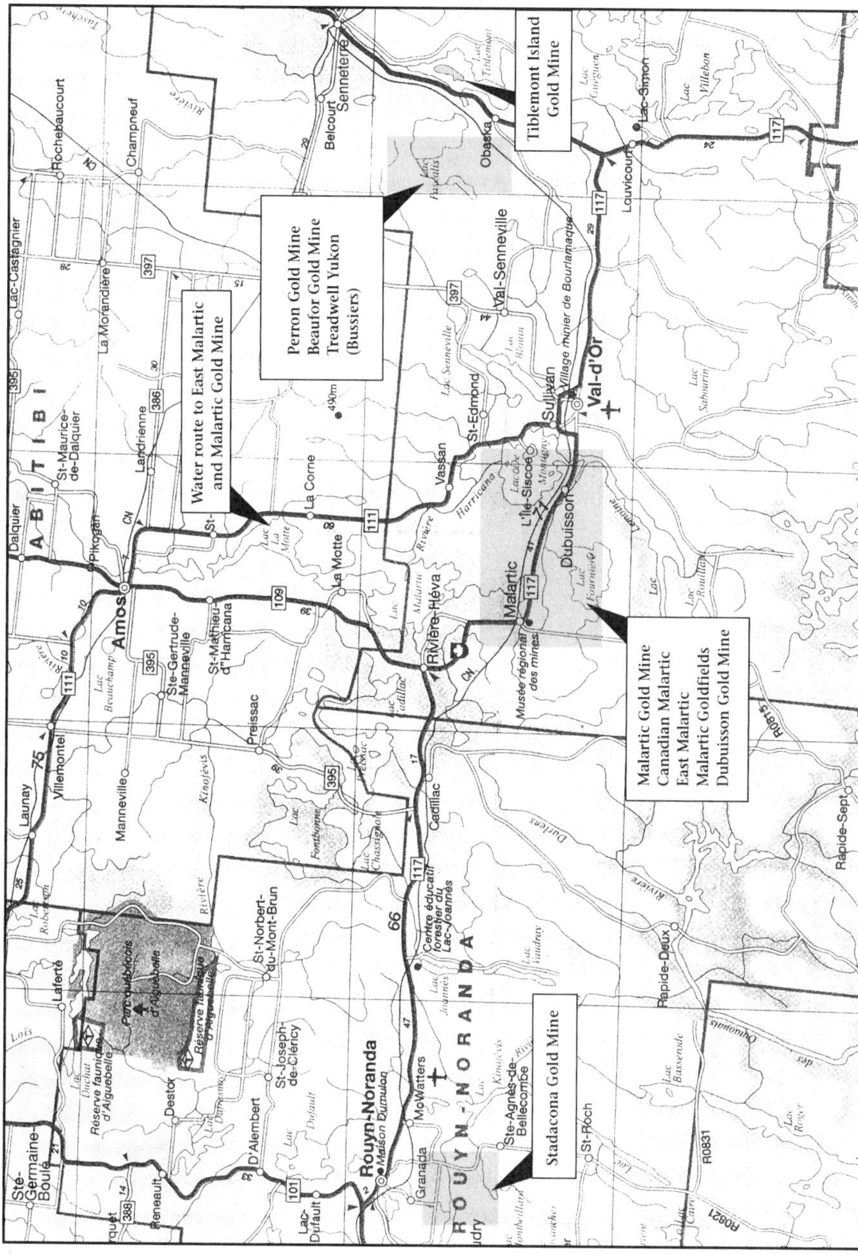

Map of the Abitibi region of Quebec today showing the location of many of the old mines

Stadacona 1927

by George Norrie

The Stadacona Gold Mine was near the Ed Horne ore body which soon became the giant Noranda Mines. Stadacona was an enterprise belonging to R.C. Gamble, who had also started the Granada and Senator-Rouyn Gold Mines.

J.P. was the first manager of the Stadacona, which had been incorporated in 1925 as Stadacona Rouyn Mines Ltd. It was reorganized in 1931, placed in liquidation in 1939, and sold by the liquidator in 1945 to Stadacona Mines 1944 Ltd.

Although J.P. sank the shaft and developed the first ore body, he was never very enthusiastic about the future of the mine, both financially and geologically, and moved to Malartic Gold Mines in August of 1927. The Stadacona remained in production until the mid 1950s but, although paying dividends for several years, was never much more than a break-even enterprise.

In Feb. 1939 J.P. advised his friend Harry Carmichael, a director of Inspiration, on recent developments at Stadacona:

...I believe recent developments on the bottom level of Stadacona are the best so far in the mine. However, they were apparently not good enough for Ventures to exercise an option which they held on 200,000 shares of stock.

This mine was recently examined by Ed Futterer who told me personally the mine could not make money especially under present management.

In the Stadacona mine high grade pockets are frequently found. These are always given much publicity to help boost the stock. I believe on the average these pockets do not mean much. Under present circumstances I do not believe Stadacona a good buy, anyway I would not buy it myself.

I got to thinking that Scots must believe that hugging and kissing caused urination, and forbore.

Jean Hitchman was said to be the cook, I think because someone supposed that Jim or I would pass this information to Annie when we returned to the farm, and that this fiction would relieve Dad of any recrimination. I think we failed. As a matter of fact, I know we failed, and Dad was side-salvoed from a distance for quite a long time, but with little effect.

Joe Richard (pronounced Reech - ore) was the real cook and a very good one. He was a French Canadian guy from New Brunswick, of medium height, balding head and with a great sense of humour. It was reported that he could not return to New Brunswick because he was an escaped convict. The story was that he had been caught selling illegal booze - moonshine - but had escaped on his way to prison by jumping from the train when his guard allowed him to use the men's room unattended. He simply opened a window and was gone. Jean may have assisted him in the cookhouse if the impulse arose, which I feel quite certain rarely did.

Joe followed Dad from job to job except for a period of time at Malartic Gold Mines where Len Newton was cook. Mr. Newton had two daughters... Well, I won't go into that just now.

At Stadacona Jim and I were totally unrestricted and when that happens to small boys - well, even large boys - they sometimes become a bit unruly, and that certainly happened to Jim and me. We feared the bush, so that was automatically out of bounds, but around camp, daytime was playtime. The cookhouse was in bounds and we ravaged it relentlessly. All of the foods denied us on the farm because they would cripple our physical and mental development were here in abundance: cookies, cakes, toast, eggs, pepper, vinegar, sugar, pickles (ah! pickles), H.P. sauce.

Stadacona was not really a mine, just a mine in waiting: claims being geologized and diamond drilled. It would become a mine later under different management; and later two of its promoters would go to jail to repent their naughty ways and improve their promotional skills.

One evening after supper, the pump supplying water to the drill failed, and Dad was called on to medicate. Jim and I followed along, a short distance through the bush. A number of men stood around while Dad tried to resuscitate the dead machine. After a while the pump coughed and water flowed from an attached hose. One of the men, a devil incarnate, whispered to me, "Grab the hose and point it at your father." I turned to stone, but the devil pushed, prodded and persisted until I took the hose, pointed it at Dad and wet him down thoroughly. The world stood still! Dad didn't move. I didn't move, while the water flowed. Then Dad exploded. I took off for about six feet. He caught me and tossed me in the air with a forgiving laugh, and declared me to be a devil. When I landed I looked around to point to the real devil, but like other devils I have known he was gone.

It was complimentary for Dad that many of his men followed him from job to job. Men like Joe Richard, the cook, Tony Mazuk, who was almost a member of the family, Tom Chequin and many others, but memory fails.

Dad's lifelong introduction to a change of venue was an instantaneous, "Get packed; we're leaving," and so it was one day in July. J.P., Jean, Jim and I left Stadacona for Toronto. First the seven-mile walk to Rouyn, then southward by car to what was called "the landing" where we boarded a rather large boat for the trip to Haileybury. Following a chain of small lakes and rivers we entered Lake Timiskaming on the eastern shore, then crossed to Haileybury. The cabin of the boat, a fully enclosed room where we were the only paying passengers, was also occupied by a two-man crew who manoeuvred the craft by a steering wheel at the front end of the cabin.

In the early afternoon a crew member broke out a case of beer and offered one to Dad and Jean but was refused with disgust, both being abstainers. But as the beer was consumed by the boatmen, Dad became more and more agitated, probably because of the increasing boisterousness of the men and their casual handling of the boat. We landed in Haileybury, however, in the early evening without mishap and went directly to the train station and boarded the train for Toronto, but without Jean, who decided to remain in Haileybury with friends.

I don't recall where we stayed in Toronto but I remember having lunch one day with General Hogarth and a Colonel MacAlpine, who I supposed were friends and financial men associated with J.P. I think it was at this time that Dad decided to leave Stadacona and go to work as manager of Malartic Gold Mines Ltd.

Here I would like to give an illustration of how my life keeps going around in circles. We went one day to Suite 301, 69 Yonge Street to meet a geologist with whom Dad had business interests. At that time 69 Yonge Street was equipped with very old-fashioned hydraulic elevators, the doors of which were simple, open, iron grill works, similar to those you would find on bank buildings. The grill work was noisily shoved aside to permit entry to the car.

We boarded an elevator that was already partly full so that we stood just behind the grill work. The operator pulled the lever and we jerked into motion, but with that sudden start my legs crumpled and I flattened to the floor with one of my hands projecting through the grill work. As we neared the second floor, just in time Dad reached down, grabbed a handful of collar and pants and set me on my feet. One of my early embarrassments!

The entrance to suite 301 was immediately adjacent to the left of the elevators. On entering we could see that the suite consisted of two small rooms. In the outer room two chairs occupied one wall. Dad pointed to the chairs and said, "Sit. Stay." He was ushered into the other room where he stayed for a great long time while Jim and I had nothing to do but sit and watch our legs grow longer. Eventually Dad reemerged and we returned to our hotel room.

Now here is where the circle closes. More than thirty years later, when I had moved to Toronto and was operating from a house rented from C.A. Pitts on Crescent Road, I happened one day to be walking along King Street when I came upon an old friend, Harold McNally. He invited me to his office. After we had talked for a while he said that he was giving up this space and moving to the office of a friend that he had business dealings with. Arrangements were made and I took over the office he was leaving. Rent $70 per month.

Then the day after I had acquired a mining property located in the Coxheath Hills just to the west of Sydney, Nova Scotia and was sitting in my new office reading a geological report on the property, I happened to notice the date, address and signature affixed to the report: 1927, Suite 301, 69 Yonge Street. My new office was the same one that Dad, Jim and I had visited so long before.

There is one other little thing about those old elevators that I would like to tell. After I had been in the building for a while I became quite friendly with one of the elevator operators, George by name. I had been noticing that George clumped around awkwardly on heavy, ill-shaped shoes. Every time I got into his elevator I couldn't help looking at George's shoes. Finally one day I could resist no longer, so I said, "Excuse me, George, for my impertinence, but what is wrong with your feet?" "Oh," he replied, "I got the front half shot off by a shell from a German battleship in World War Two. I was serving on the H.M.S. Ajax during the battle of the River Plata. The Germans broke off the fight and ran for cover. There they scuttled their ship, the Graf Spee, rather than surrender."

The old hydraulic elevators were dismantled and sent to haywire heaven and replaced by nice new electric ones while I rented in the building.

After Dad completed his work in Toronto, we went on to Brantford where he and a partner, I think it was a fellow called Billy Cram, had started an automobile battery factory but had failed. The reason for this side trip was to pick up a Model A Ford that Dad owned and had stored in the old factory building before he took the job managing Stadacona. The Model A was a one-seat convertible with a rumble seat, equipped with a new battery that Dad had brought along. The car started immediately and we were on our return trip to Rouyn, with the three of us sitting in the front seat.

I think we stayed overnight in North Bay and continued the next day on what was then called the Ferguson Highway. North of Latchford the road had recently been re-covered with a thick layer of gravel so that we drove through deep ruts. We had not gone far before we suddenly came up behind a number of cars parked in the middle of the road. Dad got out and questioned a person standing close by. "What's going on?" The fellow answered, "There's a car stuck in the mud on the other side of the bridge." Jim and I had gotten out of our car and could see the bridge over a narrow stream. The floor of the bridge had been torn up leaving a sub floor of large beams. I could see a large mud hole in the road on the far side of the bridge, but where was the car that was said to be stuck there? There was no car that I could see. OK! Now follow me closely, this becomes more interesting. On the evening before, a motorist was driving along the highway when suddenly his car became stuck in a mud hole. By rocking the car back and forth trying to dislodge it, he made it sink to the running boards. The driver got out to see what he could do about his problem, and while he watched, his car sank deeper and deeper to the middle of the radiator. He hurried away to get help, but when he returned, his car was gone out of sight and it's still there today, fifteen feet below the surface of the road - quicksand, muskeg, whatever!

The floor of the bridge had been torn up by a road crew to combine with logs and brush to build a temporary road around the mud hole. This work had been almost completed before we arrived so that we were not long delayed before we could continue on into Kirkland Lake. The highway between Kirkland Lake and Rouyn was still under construction, so Dad stashed the Model A and we took the freight train from there. That is, we took a freight. Dad loved freight trains. It was not unknown for him to flag down a freight train and climb into the caboose to get to his destination. Before arriving in Kirkland Lake I had become car sick, so it was with much relief that I could lie down in the caboose on the way to Rouyn.

Before leaving Kirkland Lake, we were walking along the main street when Dad stopped to talk with a prosperous looking older man accompanied by two other men. I remember that that older man was Sir Harry Oakes, staker and principal owner of Lake Shore Mines Ltd., the largest gold mine in the area.

When we arrived back at Stadacona, Jean was already there, having found her own way back. It was not long, however, before Dad pulled his little trick that was becoming very familiar: "Get packed; we're leaving." This time we were leaving for good. Many years later I returned for a short visit when Stadacona was a producing gold mine.

There is another small issue that may be worth mentioning here: I think Dad had started to feel that all was not quite right with the financing of Stadacona and had gone to Toronto in search of other employment and got it. As it turned out, one of the promoters of the company eventually went to jail for selling more stock than the capital structure of the company permitted. Again years later, I met this same man, unaware of his former activities and with a result not to my benefit, but that is outside this story.

Stadacona was a little-known gold mine. It lasted to the mid 50s, a marginal operation at best. Nevertheless every mine required a tremendous effort just to produce one ounce of gold and many never even reached that stage of development.

The stories associated with each mine are many and most have drifted into a distant history never to be repeated or heard by anyone again. At the very least, even Stadacona kept men working during the Depression.

This story is my own and I heard it more than sixty years after it took place. I was working for Day and Ross, a freight trucking company, in Truro Nova Scotia in 1989. An old man used to come to our terminal every three months or so to ship a pallet with a jack hammer strapped on it. I never paid much attention until one day I noticed he was wearing an iron ring on his right little finger. This is the ring worn by all Canadian engineers, so I asked him, "What kind of an engineer are you?" He replied, "A mining engineer." This immediately drew my attention. "Where were you a mining engineer?" He replied, "Northwestern Quebec." I jumped at this. "My father, J.P. Norrie, worked in northwestern Quebec." Well, he immediately began to laugh and said, "I have to tell you this story."

He said, "My name is Ivan Nelson, from here in Truro, and I took over as manager of the Stadacona mine from your father. One day I was looking through the files and came across a letter to your father from the mine directors. They were very, very upset that he had bought a shotgun and that he had paid $20 for it, a great deal of money in those days.

"Your father's answer was short and to the point. 'There was a bear marauding through the camp and endangering the men. I bought a shotgun for $20, shot the bear, and fed the men on it for a week. J.P. Norrie."

Chapter 3

Malartic Gold Mines

by George Norrie

J.P. Norrie wrote the following historical sketch of Malartic Gold Mines:

For several years prior to 1923, spectacular gold specimens were frequently exhibited by First Nations people coming from the area south of Amos. At this time it was known that the Porcupine-Kirkland Lake gold belt extended at least as far as Siscoe, which had been discovered some years previously, and thus it was logical to expect that gold occurred in the intervening Malartic area.

As a matter of fact one gold find had already been made in the Malartic district. However, it was definite that the samples found by the Natives did not come from this discovery.

Native Canadians have not proven to be good prospectors. However, some important metal discoveries have been located because of samples brought out by them. Among them may be mentioned Sherritt-Gordon and Rose Lake.

For some reason not generally known, it is difficult to get a Native Canadian to disclose the source of any ore samples he may find. Malartic proved no exception to this rule, and the first real discovery in the Malartic area was left to a most ambitious, non-Native prospector by the name of Wilbur D. Gouldie.

Gold samples from a new district generally enliven the interest of any prospector, and in Gouldie's case it took little of this, in addition to the great drive of the man, to settle the Malartic gold question. In the summer of 1923 he set out in earnest to find the source of the hidden "Indian" gold, which he did in quick time.

The story goes that in his hurried travel through the dense Malartic bush, Gouldie came to a squared tree from which led a blazed trail. Looking down the trail Gouldie could see natural gold shining in the summer sun. Some 500 feet from the tree the gold mystery was solved. The tree had been blazed and the trail cut by the Native who had brought in the sample. Gouldie staked the ground and this started one of the greatest staking rushes in northwestern Quebec.

At this time Carrol, Kennedy and Sladen, three budding prospectors, arrived on the scene. They staked the ground adjoining Gouldie's own strike, and in the following year made an important gold discovery which is now the Canadian Malartic. Another part of the ore body is on Sladen Malartic, who are now installing a large mining plant.

In 1924 the Carrol-Kennedy-Sladen claims were optioned to Porcupine Goldfields Mining & Finance Co. Ltd., who carried out a diamond drilling program which uncovered large ore bodies hitherto unknown.

Due to some hitch in agreements, the Porcupine Co. discontinued work on the property in the fall of 1925 and it lay idle until the fall of 1927 when the Malartic Gold Mining Co. was formed to sink a shaft and further develop the property with drifts, crosscuts and diamond drilling. This work proved the existence of large bodies of 0.25 oz ore which, at $20 an ounce, was thought to be marginal. Upon the report of a prominent American engineer the property was again closed in the fall of 1929. This engineer, not acquainted with gold mining in the pre-Cambrian shield, failed to recognize the fact that the Malartic ore bodies could be cheaply mined and milled, and that at $20 the ore was not really marginal.

Gouldie's discovery in the Malartic area did not bring to him the fame and fortune of some others, as work on his original discovery has not proved an economic ore body to date, although it is expected that further work may turn up something of importance.

1928-1929

When J.P. Norrie came to the Malartic area in early 1928, it was nothing but dense bush and wilderness, where in a few short years gold mines would be appearing everywhere. There were no roads, no power, no telephone except where wires were strung from tree to tree, mine camp to rail centre, no doctors, no supplies of any kind, just muskeg and swamps. The only means of transportation was by water on rivers and lakes or by dog teams and horses.

J.P. had spent the previous year opening the Stadacona mine seven miles south of Rouyn-Noranda for R.C. Gamble, which was also in dense bush country. He had also spent time as underground superintendent at the Sylvanite Gold Mine. His experience at these mines would be invaluable in opening the Malartic Gold Mine. J.P. at this time did not own the mine although he acquired as many shares in the venture as he could afford. He was working for the Ventures Group which had been formed by Thayer Lindsley in 1928 to consolidate his mining interests. Sherritt Gordon Mines, Sudbury Basin, and Malartic Gold Mines were the first companies included in the Ventures group. In later years Thayer Lindsley, through Ventures, would become partners with J.P. in most of J.P's enterprises.

In order to explore the gold claims themselves, transportation routes had to be established to supply the camps for the men required to carry out the exploration. These supplies had to come from Amos, the closest rail centre, which was 45 miles away. The Harricanaw River, Lac Malartic and Lac la Motte was the primary water route, and food, machinery and other supplies were transported as far as possible by water. Roads then had to be built from the water to the camps several miles away. Once the supply route was established, only then did the search for gold reach full operation.

C.D. Keading, located in Toronto, was the mine consultant engineer in charge of the exploration at Malartic Gold Mines, and J.P. was required to keep him informed by weekly reports on all operations. J.P.'s report of July 7, 1928 showed the number of employees to be 55 and gave a good example of the support required to keep 16 miners working underground:

Stripping and trenching was continued on the John Rogers claims. Interesting veins have been located, but so far nothing carrying values has been found. The shack to house the geologists was completed. Cross section work on the Rogers claim was done. Dr. James and Dr. Murdock have been doing geological work on the Rogers claims.

Placing of corduroy on the road to the landing [where the boats were unloaded] was continued.

55 men were employed as follows:

Mechanic	1	Underground	16
Construction	2	Trenching & Blasting	3
Bush Cutting	1	Hoistmen	2
Cookery & Camp	4	Deckmen	2
Teamsters	1	Cage Tenders	2
Diamond Drills	5	Blacksmiths	2
Line Cutting	4	Sampling & Assaying	3
Transport	2	Office and Supervision	3
Firemen	2		

Estimated expenditure for July:

Payroll	$8,000
John Bailes & Sons	2,000
P. Burns & Co.	400
Can. Ingersoll Rand	300
Diamonds	1,000
A.A. Drouin	150
Gardner Denver Co.	200
T.A. Lalonde Ltd.	150
St. Onge & Co.	300
Canadian National Railways	200
Frank Blais	300
Miscellaneous	1,000
Total:	14,000

These reports reviewed the steady supply of timber that was essential for mine and camp development. Timber was required for the mine and camp buildings, for corduroy roads, for cooking and steam powered boilers, and for the shafts, drifts and crosscuts. Mines in the Malartic area were fortunate to have a good supply of timber close by:

> We will have to cut approximately 2000 cords of wood before January 1. With this in view three men were started June 1.

Strong healthy horses and good teamsters were absolutely essential. All operations stopped when the horses stopped, as water for the drills and mine operations often had to be hauled by horse. All supplies from the boat landing were hauled by horse, even the bulky supply of hay for the horses themselves.

These weekly reports were acceptable communication for normal information and planning decisions. However, there were often results from drilling and exploration that were confidential and required quick action. Telegrams were used, but the question of secrecy was always uppermost in mines. Good news or bad, both affected public confidence in the mine, and investment was an important source of capital for exploration. Keading's letter to J.P. in May 1928 explained the use of codes to be used in these telegrams to prevent information leakage by excited telegraph operators:

> Will you wire us drill settings whenever you start a new hole, giving the position by the drift or cross cut and the distance from one of the survey points on your last assay plan - in this connection I am going to ask you to use a simple code that we are using with some of the other mines, a copy of which I enclose. You can use this as it is, or if you want to amplify a message you can use the Bedford MacNeil, a copy of which we are mailing you. In the Bedford, I have indicated a few extra code words by which you can further disguise a message. For instance, if you use the word "yachtsman", which means read the fifth word below, at the head of your message, we will count down five words from the code word you send, and therefore,

in sending, you select the word which you want to send and count five words up. The use of the word "yachtsman" at the head of the code means every word sent will be handled in this way. As a rule, a few code words will serve to disguise a message sufficiently and this will greatly simplify your work and also decoding here...

Let me know as soon as the roads are passable. I am looking forward to a visit with you in the near future.

Basic concerns such as the health of mine employees were obviously not simple to deal with. In the early stages of development they had no doctors. If they were lucky, they might have a dentist. Also injury would usually cause costly delays:

The diamond setter met with an accident to his eyes which delayed both drills for two shifts...

If an injury was serious, it would be necessary to transport the person to Amos or Rouyn or another mining camp, often under very difficult conditions. The mining camps shared any medical staff they might have. One mine manager asked for J.P.'s help:

I sent a message to your dentist about a week ago asking him to come over and fix a tooth for Mrs. MacPherson which he said he would. She would go there if it was possible but could not walk that far over these trails. Will you please round him up and send him over, otherwise I will have to send her to Rouyn as it is becoming more difficult each day for her to eat...

Natural elements were just a fact of life in this isolation and had to be dealt with. But there were others. Fire was one, often caused by the men themselves, where almost nothing could be done. Bush fires were just avoided, if possible, and ran their course without a contest. Fires in the camp and the mine were almost the same, with little to be done once the fires had a head start. This report describes one mine fire in April of 1929:

Fire at the west shaft: on the evening of April 10 the gasoline driven plant was destroyed by fire, the origin of which is unknown. The fire apparently started in the power house, while everyone was at supper, and was well away by the time it was discovered. A barrel of lubricating oil and a barrel of cable dressing which were right beside the power house, and the gasoline from the tank compressor, made it burn very rapidly, so that the power house was a complete loss. The fire was got under control before it could spread to the blacksmith shop or to any of the new machinery, which was parked nearby...

Shaft sinking and plant installation: east shaft; the work of installing the machinery has gone on steadily during the period. The boilers, hoist and steam compressor are all set up on the foundations. The diesel foundations are still held up waiting for the tile pipes for the air intakes.

West shaft; the shaft is down to a depth of 135 feet, and the first level station is in at about 15 feet. Underground work was stopped by the fire which put the plant out of operation. So the shaft crew are now being used to install the machinery for the permanent plant.

Payroll:

Machine & blacksmith shop	9
Carpenter shop	13
Sawmill and logging	12
Foundations & mach. inst.	27
Clearing, hauling wood & cutting ice	17
Gravel pit & hauling gravel	3
Tractor drivers	2
Radio operators	1
Assay & survey departments	4
Cookery and camps	10
Office and supervision	4

There are also 22 shaft men and 31 diamond drillers boarding at our camp. Since shaft work has been temporarily shut down, the 22 shaft men are being used on installation work on the west plant.

102 men were employed in the month of April. This whole operation was obviously very expensive.

The April 1928 estimate of expenditure was:

Payroll	$12,000
Shaft sinking	2,500
Diamond drilling	10,000
Brown & Moore hardware	2,200
Pas Lumber Co.	2,200
Monarch Lumber Co. Cement	510
Coal	1,000
Misc	1,000
K&E Engineering supplies	61
Miscellaneous supplies	500
Total:	32,571

This was a very large outlay of cash for one month's exploration at no return, as no gold was being recovered from the exploration shaft or the drifts and crosscuts. The ore body that had been indicated by surface trenching and diamond drilling had not been reached as yet underground, and all expenses could only be recovered if, in fact, the gold ore body was there. Otherwise all these cash outlays would be lost for all investors.

Malartic Gold Mines, as we have said, was controlled by Ventures who had interests in several other large properties in Ontario and western Quebec, such as Sherritt Gordon Mines. As with many exploration programs, particularly in Quebec, cash flowed from the producing mines in Ontario, in the form of dividends, into exploration projects. In this case Sherritt Gordon Mines were partners with Malartic Gold Mines. This helped defray expenses but there was no less pressure to find the gold ore body.

The other partners were the public shareholders. In the 1920s, and before the stock market crash and subsequent depression, many of the local businessmen and citizens had considerable sums at stake in gambles such as Malartic Gold Mines. These were times of great optimism and everyone in Quebec was looking for the same success that had been found in Ontario. Many of the small investors were French Canadian, and the following is a list of investors in 1929 just in Amos, Quebec:

Hector Authier	15,000
Frank Blais Ltd.	1,000
Leonidas Boisvert	800
Louis Brochie	1,000
Gontrand Brunet	10,000
Mrs. Lucille Brunet	2,000
P.X. Cossette	200
Phillias Cossette	1,700
Marthe Dessert	200
Yvon Dessert	200
François Dorval	3,000
J. Ulric Dumnot	1,500
E.H. Gamble	500
William F. James	75
Jamieson & Parker	100
Jean B. Jourdain	2,000
Harold S. Kennedy (vendor)	17,714
La Reine Mine Ltd.	15,000
Bertha Lessard	200
T.L. Lynch	100
Louis Massicotte	3,000
Albert Morasse (priest)	500
Rev. A.J.B. Rioux	1,500
Rosario Roux (mine recorder)	800
Mrs. Eugenie St. Cyr	100
William G. Ward	500

Many investors came from Rouyn, Siscoe, La Sarre and La Tuque as well.

Rumours, gossip and confidential leaks of information were always difficult to control, as we have seen. However, results had to be given to the public investors on a regular basis. Negative information was always discouraging, such as this news item in the *Mail & Empire* of Toronto, Oct. 16, 1928:

Official information regarding results being obtained by diamond drilling on the Malartic Gold Mines confirms the early reports coming down from Amos. A diamond drill operating from the 375 foot level station and pointed at an angle intended to intersect the downward continuation of the main deposit at around 500 feet in depth from surface, revealed only about seven feet in width of vein structure. Those in charge are at a loss to understand as yet this change, a width of 78 feet at 375 feet in depth giving place to 7 feet in width at 100 feet deeper. As yet the diamond drill hole has not been surveyed, but in going the short distance in question, it might be deflected sufficiently to miss its objective. While this result is discouraging yet the past records of diamond drilling would suggest caution in arriving at a decision. It may be that two diamond drill cores drawn in the early days of the Lake Shore Mines were from holes directed right through what is now the richest number two deposit. Nevertheless, a poor diamond drill core under the circumstances existing at Malartic is disquieting until such time as further work discloses further information.

In spite of bad news reports such as this, exploration operations continued through the winter and spring of 1929 but by July the decision was made to discontinue drifting and crosscutting. The ore body had been located and a stock pile of low grade ore had been separated from the rock hoisted from the drifts and crosscuts. But before a decision was made to mill the ore, Pope Yeatman, an independent consultant, was brought in to give his assessment of the future of the mine. As the price of gold was still only $20 per oz. and the milling methods used on gold ore in the Malartic area were still not efficient, he therefore decided that the profitability of the mine did not seem likely. This opinion was not shared by J.P. or Yeatman's assistant, who both maintained that the mine could make a profit. Yeatman, however, gave his report and recommended that work cease. On August 20, 1929, the mine shut down except for two pumpmen to keep it from flooding.

When the mine closed, all employees were out of work and the value of the Malartic stock became almost nothing. With the stock market crash of Oct. 1929 just ahead, and then ten years of the Great Depression, this would be devastating, particularly to the small local investors. Nobody could foresee the turnaround in gold values which would allow the Malartic mine to open again in three years under the new name of Canadian Malartic Gold Mine. The shares would again be worth something.

J.P. had invested all his personal wealth in the mine and was now broke, along with everyone else. At the end of every story, however, there is always the beginning of a new story. Deep in the Malartic Gold Mine a dream was beginning. J.P. had always been employed by someone as a mining engineer: Porcupine, Sylvanite and now Malartic Gold Mines. It must have been the damp, dark drudgery of the underground, but for whatever reason J.P. wanted to own his own mine.

The Ventures group he worked for had the perfect model around which he could bring his dream to action. While he was managing the mine he was staking properties for himself. He knew all the prospectors in the area and with Lee Coffin he was staking the Senore and East Malartic properties. He also knew other prospectors, such as St. Barb Sladen, as well as the developments of all the early mines in the Malartic region and eastward, such as Siscoe and Stabell. When he lost all his money with the Malartic shutdown, he would maintain his claims with Lee Coffin. It would take several years, but the story of these claims would be his dream come true.

Every cloud has a silver lining; the claims would be one, but first J.P. was off on one of the greatest adventures of his life.

Chapter 4

Malartic Gold Mines 1927-1929

by Harry Norrie

Leaving Stadacona in August of 1927 we went by train from Rouyn to Amos, where we stayed in the Transcontinental Hotel, a three-storey wooden structure managed and owned, I believe, by Mr. Frank Blais.

We called it northern Quebec but even at four hundred miles north of Montreal it was still southern Quebec with the northern tip of Quebec still a thousand miles or more further north at about 45 deg. 30 min. N.

When J.P. arrived in Quebec in 1927, the area between Rouyn and Senneterre and beyond in all directions was a solid blanket of bush populated by scattered bands of Cree who picked up a few furs and looked for a place they would later call their own. A few venturesome prospectors, not sure what they were looking for, had entered the area too. If they went north from the CNR they certainly were out of place. It was J.P.'s contention that no gold mines would be found north of the tracks and he was right; that is, until you get to Chibougomau about three hundred miles further north.

But along the tracks and to the south, there was gold.

The area is flat with few undulations rising above three or four hundred feet. Three northward flowing streams drain the area. A few miles east of Rouyn is the Kenojexis. The Malartic region is drained by the Harricanaw, and Senneterre by the Bell. These rivers and tributary streams provided prospectors and developers with highways better than the land-based ones to come.

In 1927 a spur CNR rail line connected Tachereau, on the main CNR line, to Noranda but there was no highway connection to the "outside" in any direction. With the discovery of gold at Cadillac, Malartic and Val d'Or, however, rapid development took place. By 1935 highway, power and rail lines were nearing completion and towns were being built.

In 1927, Malartic Gold Mines Ltd., where Dad was now the manager, was, like Stadacona, not a mine, just a prospect located some forty-five miles south of Amos. It was Dad's job to find the mine, if there was one there at all; but it was on a geological structure on which many mines would be found, some of which would carry Dad's fingerprint.

To get to the area we travelled from Amos by freighter canoe up the Harricanaw River, across Lac la Motte, and into the Malartic River.

A short distance up the Malartic River, passage was obstructed by a series of rapids, and it was at this point that Dad decided to blast the river bottom and construct a dam to deepen the river which would allow boats to come within four miles of the mine site even though this made it necessary to transship supplies from one boat to another around the dam.

A short distance from the dam site, on a high bank above the river, was a group of three old log shacks that had been built earlier by a logging operation, and it was here that Dad made camp for the dam construction crew.

A large canvas tent was erected for the men while Jean, my brother Jim and I were housed in an old stove-in cabin. Some of the debris was pushed aside, a tarpaulin laid down, and with a couple of blankets our bed was ready. In all my life I have had only one worse bed and that was on the south shore of Grassett Lake where in an old shack I was forced to sleep one night on a bunk made of rough poles cut from the nearby forest, without a pad. The builder of that contraption should be sentenced to accompany Virgil and Dante Alighieri on their next trip to Hell.

The "cook house" was a huge open fire with cooking pots hanging from an iron rod. The dining table was made of rough boards with attached benches. Joe Richard had come away with Dad and acted as cook. It was one of his duties to keep two pails of water available for any of the men who were in need of a drink. They were placed just inside the door, now only a large open space, in one of the less weathered log shacks, and it was these water pails that introduced me to the night of mayhem. And "copy cat" mayhem that was soon to come.

But first I must introduce Mike, one of Dad's men. Mike was a Canadian Native. He said he was of the Dogrib tribe. If he was, what was he doing in Cree country a thousand miles from home? But it could have been so, perhaps he was Dogrib; it adds little to the story in any event. He was of just above average height, built like a brick and very strong. It was said that he could lift anything loose on one end. To me, because he was Native he was the most interesting of all Dad's crew, and I followed him everywhere.

Some of the men, along with Dad, worked in waist-deep, ice-cold water deepening the channel by drilling holes and blasting them with dynamite. It was the first time I had seen dynamite used and it was great fun to stand at a safe distance, hear the roar, and watch the geysers of water shoot up.

Mike's job was with another group of men cutting logs and carrying them to where the dam was under construction. Jim and I would follow at his heels back and forth while he was at his work.

I had heard many stories about the white man's war with Indians and was eager to put the question to him. Finally one day, as he sat resting on a windfall, I asked him: "You ever been in an Indian war?"

"Oh ya, lots of times," he answered.

"Ever kill anyone?"

"Oh sure, thousands and thousands," and he rolled around laughing.

Be that as it may, and I'm sorry, but he was the cause of all the trouble that was soon to follow. One evening just at twilight I went to get a drink of water, but just as I was about to put the dipper in the water I noticed a sheen of oil on the water. I was puzzled by this but I put the dipper down and was about to turn away when I noticed the figure of a man silhouetted in the light of a window behind him. I wouldn't know until a little later that it was Mike standing there and it was he who put the oil in the water pails.

Jim and I were sound asleep when suddenly a hell of a racket jolted us awake and started us staring out one of the windowless openings. Two guys were running around the campfire cursing at each other while the rest of the crew stood watching. One of the contestants was "Joe the cook," the other "Indian Mike," and it was obvious that Mike was the aggressor. Dad and Jean had been quite a distance away but when they heard the racket Dad came raging in with his legs rotating like the wheels of an old CNR locomotive on a steep grade. Just as Indian Mike was about to let go at Joe with a fistful of knuckle bombs, Dad arrived – and you're right – the fist exploded in Dad's eye. The fight was over.

You'll want to know what caused all this. Well, I'll tell you. It was not previously known to anyone, least of all Dad, that Mike was given to fits of madness. For some reason he had taken a hatred for Joe, who more than most men was quiet, fun-loving and pleasant to be with. As it happened, Joe's bed was next to Mike's, and it was while they were lying there that night that Mike started pushing Joe, demanding more room, starting the fight. It had also been Mike's demented idea that putting oil in the water pails would cause trouble for Joe. As with many fights, it's not the contestants that get hurt, more probably it's the peacemakers and so it was with Dad.

But now we have the case of the "copy cat" mayhem to deal with. It seems a quirk of nature that some people, although only witnesses to a fight, will find rage boiling in their gut and a fierce desire to participate even though they are in no way involved. Enter Jim. On returning to bed, he found it necessary to enlarge his sleeping quarters and started pushing me around. Even though I offered little resistance, he hauled off and delivered a miniature of the fist bomb that Mike had handed to Dad, directly in my eye!

I don't think Jim took any great pleasure in pounding me. He rolled over and went to sleep, and so the battlefield was cleared and there was peace in our corner of the bush for the balance of the night.

Early in the morning it was pack-up time again as Dad hustled Jean and Jim and me into the canoe, bound for Amos out of harm's way, to summon the RCMP to come and collect a wayward Native. Like many crazies, Mike was a nice guy when his demons were asleep, but a distant bush camp was no place for him, better that he be in a confined area.

Back walking on the streets of Amos Dad took a little time to parade around with me at his side, each with a multicoloured "shiner". With a skewed smile on his face and merriment in his eyes, I could tell there was a little devil running around inside him and that he was quite satisfied with his circumstances. The jokes and jibes levelled at him by his friends, "Hey, Jim. Pick on someone your own size;" or "He pretty nearly got the best of you, Jim," added to his merriment and stretched the buttons on my shirt with pride.

From August 1927 to August 1929, J.P. set up and ran the exploration program at Malartic Gold Mines Ltd. During what was left of the summer of 1927 he would also have to open a land route, forty miles to the railhead in Amos, since the water route would soon be frozen over. He would have to create a small village in a wilderness area to support a crew of up to fifty men. His chief worry was the health of his men where even a toothache was serious.

The winter road was not a great difficulty; simply knock down a few trees and smooth the base with snow and ice. Telephone communication was established by stringing a wire through the trees and trying to disregard the static.

Mining camp with J. P. and dog. Visible are bunkhouse, dining hall, and warehouses

Teams of horses hauled in the mining machinery: boiler, saw mill, compressor hoist and everything else that went into making a village and a mine.

The camp buildings: headframe, hoist house, cookery, ice house, and bunkhouses were all made of logs or lumber shaped by the saw mill. But there were two buildings that stood out. One, a bunkhouse for the office staff, two storeys high, was formed from squared logs. (A large, two-storey log building is quite unusual. Generally they're not more than one storey and, at that, not too sound or of long duration.) The mine staff, except J.P., lived on the ground floor. The top floor was used for storage.

The other: J.P. if nothing else was very practical. As the underground work progressed, empty dynamite boxes piled up. The little wheels in his head could almost be heard turning as the pile grew. What to do with them? All so similar, rectangular and strong. Suddenly the light flashed: fill them with sawdust and put up a building. Done! A two-storey building with more than ample office space on the ground floor and two bedrooms on the top floor, which J.P. occupied.

Building constructed of empty dynamite boxes

The story of that building followed J.P. all his life. Certainly any dignitary visiting the camp commented on it, and always with a feeling of tickle on the ribs. It was a fine looking building, unique in all the world. A picture of it should have been sent to Frank Lloyd Wright.

One of the pointer boats used for transportation to several mines.

The water route to the mine was by way of the Harricanaw River, Lac la Motte and the Malartic River to the landing dock four miles from the mine. When the highway was completed to the village of Lac la Motte, J.P. arranged to leave his car with a farmer while he was in the bush. Have you ever heard of a car called a Durant? Well, that's what it was, a Durant.

Lac la Motte is quite shallow and requires a boat of particular shape. The front end is pointed and extends forward out over the water.

It seems reasonable that that is why it is called a pointer. In a stiff breeze, with its sharp front end the pointer bites a chunk out of each wave and spits it out in the troughs, permitting a rather smooth ride. Our pointer was equipped with an old beatup car motor, chugging but resolute.

A brave soul tried one day, as we watched, to cross the lake in a storm in a motor-driven canoe, but found it difficult to keep the front end down and to keep the canoe from up-ending. He slowed the motor and came to shore where he found a large round rock to place in the front end to level the canoe. He started out again but the front end still came up. Then the rock rolled to the back end and sank the canoe.

Across Lac la Motte and a short distance up the Malartic River, materials and people would transship around the dam to a large freighter canoe. From the dam to the boat landing was nine miles. The trail from the landing was an intermittently gravelled road. Somehow J.P. snagged a Model T Ford onto the property and used it to travel back and forth from the mine to the landing.

Sometimes one person on a train will cause extreme aggravation to all the other passengers; similar to what Jim and I did on our first trip up north. On a trip from Montreal J.P. found himself confronted with just such a situation. A short, French-speaking little guy was raising hell, obviously in a state of joyful intoxication, at the point that exists just prior to when you fall on your face fast asleep.

J.P. hated booze, especially when the inebriated otherwise showed signs of above average intelligence. Sleep cured, the little hell raiser fell into conversation with J.P. and gradually J.P.'s story of an exotic lifestyle intoxicated the little guy in another way. He revealed to J.P. that he was Eugene Audette, a dentist from Montreal, and that he had undergone terrible things, the memory of which could only be tolerated when mixed with copious amounts of booze. But it seemed to J.P. that the booze rather than the memories was the offender. Anyway he persuaded the dentist to join him for a while in his exotic life. The dentist agreed and followed J.P. to the mine where he proved to be a marvellous little person wrapped up in a small package. We all quickly got to love him. He was full of humour, ability, intellect and what the French call *joie de vivre*.

Once J.P. had him at the mine, he was unable to leave. J.P. wouldn't let him go. He couldn't walk out and he wasn't allowed on

Dr. Eugene Audette, mine dentist and family friend

a boat. Alcohol was not permitted on any of J.P.'s properties so he was always sober. Except for once saving Pauline's life, he didn't seem to do very much. But with the nearest doctor more than forty miles away through the bush, it must have been a great comfort to J.P. to have a person with at least some medical knowledge on stand-by. It probably wasn't the exotic life of the north bush that enticed the dentist, but whatever it was he stayed with J.P. for many years at various mines.

On a sunny August day, early in the morning, Pauline and Jean decided to go for a stroll, carrying a bit of lunch. They arrived at the landing and spent the entire day there. At about four o'clock, J.P. became worried and sent the Model T to find them. When they arrived back at the mine it was obvious that Pauline was very sick. She sat on the steps of a building and had a terrible nosebleed. Dr. Eugene Audette was called and with Dad's help carried her to bed where she fell into a deep coma. For an hour or more she lay on her back with the doctor visually monitoring her. Suddenly the doctor noticed a condition that no one but a doctor would recognize; her tongue was falling back cutting off her breathing. With some difficulty he forced her mouth open and drew her tongue forward.

With that she regained consciousness. She remained in bed the remainder of the night without much discomfort. In the morning we all left for Rouyn where Dr. M.S.K. Rudd diagnosed that she had been exposed for too long a time to direct sunlight. Without Dr. Audette she certainly would have died.

Life in the bush was never gentle. Still it was not unappealing. It forced the industrious to contend with hardship and feel better about it. The slack were excess baggage and would not be carried. If you were not self-reliant you were not welcome. Weakness would soon be made apparent and would be shunned. But those who tried, it would reward with riches and an inner sense of glory. It would prepare them for any obstruction that lesser individuals always place as stumbling blocks against the determined.

With the winter road open, J.P was still not content with the pace of an ordinary horse. He stopped off at the Blue Bonnets race track in Montreal and bought for himself a team of first

Two old racehorses purchased at Blue Bonnets Raceway, Montreal. They hauled freight to the mine.

class, top of the line race horses. Strangers on the road only thought they saw a team of horses fly by. Before the invention of the afterburner that was about as fast as he could go.

Bill Youmans, teamster and camp maintenance

Bill Youmans was his driver and horse caretaker. Bill was born on a farm near Tweed, Ontario but he didn't hang around there very long. He joined the army in the First World War but was sent to the Queen Charlotte Islands to cut Sitka spruce for the Royal Navy. From there he drifted into mining but never became a miner. His work was on the surface as a handyman, and he loved horses.

He was as quick with his tongue as he was with horses. When he saw a very thin horse one day he said, "Look at that poor horse. He's so thin you could read the *Toronto Star* through him." Talking about a friend he said, "That guy is so awkward he'd fall off a floating township." And there were dozens more, much to the delight of those in his company. Unfortunately for someone, he never married.

Bill Youmans was not so much a prospector as he was a camp keeper. With a small crew he was the cook. He cut the wood and carried the water, but he never appeared to be working. At the end of the day it was always a surprise to see the amount of firewood he had neatly piled ready for use. The canvas tents securely tied down and equipped with a Quebec heater would keep you warm for the entire night, at least until Bill got up at six o'clock to re-stoke the stove, always in his long woollen underwear regardless of the season.

He was immaculately clean. Every Sunday he washed his outer clothing and boiled his underwear in a bucket especially for that purpose, an example few others in the crew followed.

Left alone for a short spell at one camp, he tamed a couple of Canada jays to come and feed from his open hand. Their perch while waiting their turn could be his head or his shoulders. He took great delight in showing this ability to returning camp mates.

Bill Youmans quit the bush when he was over seventy and retired to a beer parlour in Kirkland Lake.

J.P.'s other means of transportation was a nine-dog dog team. I guess after he got the race horses he didn't need the dogs but they stayed around the mine and always came to greet him when he was about. He belonged to them. In the summer when we went to Amos by boat, the dogs always followed him as far as the dam. We would see them swimming across the river back and forth on the bends in the river after the boat had passed. The smallest dog in the pack was a female and she let it be known that she was the lead dog.

A hole in the ground with its black unseen depth will draw almost anyone, especially children. After everything within the radius of the property has been looked at, overturned, stoned, that black hole is still there and drawing like a magnet. A simple metal bar keeps the foolish from falling in but it doesn't warn you to look up for other possible dangers.

Leaning over the bar, looking into the darkness I suddenly heard Tom Chequin holler "Get away from there." Tom came over, put his hand on my shoulder, and said, "Always look up. The cage may be up there and when it comes down it could cut your head off." A lesson I never forgot, but Tom did.

Tom was the "bull gang" foreman. A bull gang is a bunch of misfits who have to do, day or night, without complaint, what other men refuse to do any time. A barrel of oil may have to be moved from here to there. A thousand cases of dynamite may have to be placed out of reach of an oncoming forest fire, or a building moved to a place where it will be useful instead of standing empty. Call Tom.

At another mine, the day that Tom forgot to look up he found that a mine-car rail leading into the shaft had been broken and needed to be replaced. Rather than call a couple of guys from the bull gang he decided to do it himself. The simple solution was to measure the distance with a tape measure, cut off a chunk of rail and nail it into place. Instead he found a short piece of rail which was a little too long for what was required. He butted it onto the end of the sound rail with the intention of marking the end now extending out over the shaft. But as he was pulling it into position the cage came down and hit the part extending out over the shaft. Tom's end came up and slammed him under the chin, killing him instantly with a broken neck. J.P. was in his office in Amos when he heard of Tom's death. Along with a lot of others he wept, but the memory of a good man remained.

Tom was native to Gaspé. He was part French and part Irish, and with that was a mixture of poetry, song and industry. He could fix anything and laugh while he was doing it; give him a piece of baling wire and he'd build you an aircraft. He was lean and bony, quite tall, unshaven and untidy. There was always a long stubble on his face. He had an impediment in his speech, and when he had his usual pinch of Copenhagen snuff tucked under his lower lip - watch out! If you asked him a question, stand back, the spray would qualify you for a measles ward. When he was confronted with the results of careless workmen he could say quite distinctly, "B-b-b-blue Jesus. Now look what you gone and done." He was loved and admired by every man in his crew and many others.

One day when I was eleven, with his usual conservation of words Dad said, "Come with me." I followed him to the shaft house where he put a miner's helmet on my head and handed me a light. Another guy joined us and the three of us got on the cage. This was my first trip underground. Dad and the other man didn't seem greatly concerned. That helped to put a cap on my fear, just below the panic button.

The object of the trip, I guess, was to examine the geology and mineral content of the rock as the work progressed, but that was of little interest to me. We walked along the drift for a short distance until we came to a couple of men fooling around with a machine attached to a vertical metal bar. J.P. and his other companion spent a little time examining the face of the drift, then backed away. Dad gave a nod of his head and one of the machine men turned a lever, and all hell broke loose. With the machine running and the bit hammering into the rock, the noise in that small space was horrible. Everything about me was quivering, including my miner's helmet and my boots. In short order I started to cry.

Now the object of crying is to create a great noise with the hope of attracting the attention of some compassionate person in the immediate vicinity, but there was no way I could sound above that damn machine. I looked at Dad for help but he was just standing there watching the machine working a hole into the rock. After what seemed the best part of a week, really only a minute or two, I gathered a little part of my inner brain that wasn't quaking, and figured, "This isn't doing any good at all," and stopped crying.

Back on the surface I was a man among men.

Len Newton was a small man with fingers highly educated in food culture. With one assistant he prepared the food for everyone at the mine, including all dogs and passers-by. The cook house was divided into two rooms: Len's room and a dining room, where long tables were weighed down with Len's inventions. One or two tables were segregated for office staff but open to inspection by the common working man. Len made sure that all were common in his kitchen.

Len had two lovely daughters, one too old and removed to be of much interest to two small boys; but the young one, Mazie, nearer our age, created an image worth pursuing. In a moment of intended secrecy, the two small boys tried, without success, to find out just how boys and girls differed in physical make up. The following astonishment inflicted by reasonable elders was not serious but lasting in memory, leaving the boys to keep guessing for many more years.

But then eons later, in Toronto, a wife came home inquiring of her husband, "Just who is Mazie Newton?" Can you imagine that she had happened to enter a restaurant alone, in search of a casual, refreshing cup of coffee, where the friendly waitress, unpressed by an excess of other customers, fell into conversation with the wife and eventually inquired her new friend's name. When informed, the startled waitress said, "Do you know a guy called Harry Norrie?" Loud noises. No bruises. A short explanation and the incident was forgotten.

The kitchen garbage disposal unit was a large box set on two wooden runners and parked a short distance from the kitchen's back door. When the box was full, it was hauled to the dump by a team of horses. One evening coming out from the cookery I noticed the dogs in a group barking furiously at something I couldn't see from my vantage point. The dogs would bark and then retreat for a few yards. To find out what their problem was I went back into the kitchen and looked out the back window. A large black bear was feeding in the garbage. It paid not the slightest attention to the dogs, and at its own time ambled back into the bush. From then on it came back every night at almost the exact same hour and became tame enough for Len Newton to feed it honey on the end of a long, long stick. It was still making its nightly visit when the mine closed but shortly after had to be destroyed by a caretaker when it broke into the kitchen and stole the caretaker's provisions.

One of J.P.'s dogs was a large black Newfoundlander called Oscar. On a very dark night Charlie Brazel of the office staff returned to the bunk house and found the dog sitting on the front steps. But when he went to pat it on the head it made a "Whoof" not common to a dog and ran back into the bush. Before going to his room Charlie paid a visit to the outhouse.

On a trip to Rouyn with Pauline, Virginia, Jean, and Mazie Newton, Dad took rooms at the Radio Hotel. Virginia and Mazie were placed in one room but for a short stay did not need the use of a chest of drawers common to all hotel bedrooms. There was no call to look inside or examine this particular piece of furniture. Late in the evening when they were asleep, a knock came at their door. No answer. Another knock. No answer. But now, with persistence, another knock. Virginia and Mazie woke up and by this time were shivering and whispering. What to do! Mazie insisted that they should not answer, but Virginia went and stood inside the locked door and inquired of the intruder, "What do you want?"

"Open the door."

"No."

"OK... Look, there's a revolver in the top drawer of the chest of drawers. Get it and pass it out to me." And she did!

One day John W. "Turk" Mackenzie, a young engineer from Baddeck, N.S., working in the mine office, coaxed a voice out of the old Model T and invited us kids on a trip to the landing. On the return trip the little fiend that held permanent residence behind Jim's heart moved up and sparkled from his eyes.

But first: to start a Model T you first have to choke it by pulling on a little wire that runs from the carburettor to a fixed position just to the west of the radiator. Pull the choke a few times and then crank the motor furiously until the car starts. On a good day, the motor should show a little life; otherwise forget about it until God is in your corner.

Without that little fiend giggling behind Jim's eyes, "Turk" would never have known that once the Model T started you could choke it to death by pulling that little wire. Perhaps he still doesn't know.

Transportation from landing to Malartic gold mine - Model T Ford

On the trip, Jim and I had to sit outside on the front mud guards; there certainly wasn't room in the car for more than three people. We were chugging along merrily when suddenly the motor quit. I could hear the terrible blasphemy coming from Turk's lips as he got out to whip the crank again, but I already knew most of the curses and it didn't greatly alter my spiritual position.

A half mile down the road we were again stalled. Again with fresh curses gradually getting louder, now some in French, Turk had to crank. Haltingly we arrived back at camp. It wasn't until in bed that night Jim told me what he had done. It was devilishly simple. While sitting on the mud guard he reached down with his toes to that little choke wire, secured it between his toes and pulled. Only after his little fiend was home and asleep did Jim feel secure in telling me of his devilry.

In the summer of 1929 the underground evaluation of Malartic Gold Mines was still in progress with no decision having been made to go into production. It seems very likely that with gold selling at about $20 per ounce, it was just not practical, particularly since there was still no all-weather overland road. Work was discontinued and the mine closed in September, 1929.

Early in his presidency of the USA, however, Franklin Delano Roosevelt made an open offer to the world to buy gold for the US Treasury at almost twice the 1929 price. Malartic Gold Mines was reorganized to Canadian Malartic Gold Mine and was reopened. It went into production and survived for many years although the grade was very low. With excellent management it made a modest profit, and it taught J.P. Norrie that there were other mines yet to be found in that area.

Along about fifty years ago I was driving from Louisbourg, N.S., to Sydney. Rounding a bend in the highway I came upon an old man trying to hitch a ride, but not in the authorized stance of a true professional hitchhiker. That is, he should have been standing erect, head held back, with moist beseeching eyes. His left arm should have been down at his side, his right arm extended horizontally with the thumb held upward, prominently: all in all looking like half a crucifixion.

No. This old guy was standing a little bent over, his head lowered, with frosty eyes looking from under frosted eyebrows. His left hand was in his pocket while his right arm was held vertically over his head with the hand rotating rapidly in circles, like an ex-Conservative Member of Parliament seeking reelection all the while professing to be the single author of the Rights and Freedoms Act.

His clothing looked overworked, drab but clean. The hat on the back of his head seemed a permanent fixture. His hair had a greyish colour like the underside of a storm cloud. Tufts of whiskers here and there suggested he had little converse with barbers. On a fair day, he looked as if he'd been rained on.

With a momentary spark of sympathy I stopped. He opened the door and sidled into the car. After a period of casual conversation he asked what I worked at. I told him mining and prospecting. With that he went into a spit-spattering diatribe about mining and how back in '29 he had lost all his money, gone broke investing in a mining company called Malartic Gold Mines Ltd. I very nearly fell out of the car. I don't know what his violence would have been had I told him that I was the son of the manager of that mine at that time. I doubt that he would have been appeased had I told him that J.P. had gone broke at the same time for exactly the same reason.

On a couple of occasions J.P. related how he had gone broke three times, once before and once again after '29.

I drove the old man to his home in Sydney - grey like himself. He eased out of the car, all bent over as though looking for earthworms in tall grass on a sand dune. (If that seems not to coalesce it's not intended to.)

I went to say a prayer for the old man - the Lord's prayer. But then I thought of the first two words - Our Father - and what might have been.

Chapter 5

The Arctic Adventure, 1930

by George Norrie

Dominion Explorers was owned by Colonel C.D.H. MacAlpine, and J.P.'s old boss Thayer Lindsley was a director of the company. When Dominion Explorers decided to send an expedition to Great Bear Lake in the spring of 1930, J.P. was out of work and was hired as chief engineer of the 20-man group. The Arctic adventure was on.

Ventures Ltd. was a mining company set up in 1928 to promote mining properties. It became one of the largest development companies in the world when Thayer Lindsley was president. Three other prominent mining men were associated as directors of the company: Joe Errington, Gen. Donald Hogarth and Col. MacAlpine.

At the same time as Ventures was set up, Col. MacAlpine had incorporated Dominion Explorers with himself as president and with Errington, Hogarth and Lindsley as directors.

The 1930 Great Bear Lake expedition had a harrowing preview of how unforgiving conditions can be in the north when another Dominion Explorers' expedition in 1929 ran into some serious trouble.

In August, 1929 MacAlpine had left on an expedition to explore around Great Bear Lake and the region to the northeast. His expedition was lost in the Arctic and an air search on a grand scale was organized to search for him and his party.

In the summer of 1929, J.P. was still closing down Malartic Gold Mines. He therefore missed being one of the MacAlpine party but was included in the ground effort that began in late winter, 1930.

MacAlpine flew north from Winnipeg. The purpose of the expedition was to prospect the barren lands for minerals and to explore, chart and name many of the lakes, rivers and other features which they encountered.

Dominion Explorers and Ventures funded most of the expedition but the federal government was involved as they provided supply ships and provisions.

One aircraft, a Fairchild (CFAO), was owned by Dominion Explorers and piloted by Stan McMillan with Alex Milne as mechanic (engineer). The other aircraft (CASP) was leased from Western Canada Airways and was piloted by G.A. Thompson with D.A. Goodwin as mechanic. Ed Boadway, a mining engineer and pilot, and Richard Pearce, editor of the *Northern Miner* newspaper were also members of the group. Richard Pearce's personal diary records every detail of this harrowing adventure. [Pearce's account was published in 1930 as a book under the title *Marooned in the Arctic*.]

A series of accidents left the MacAlpine party marooned on the edge of Arctic waters (Queen Maud Gulf) out of contact with civilization and with the Arctic winter about to close in. With no shelter and only limited supplies of food and clothing they had to depend on the Inuit for survival.

Pearce's diary picks up the story:

Sept. 15: House building is going ahead at a fast rate, which is a good thing, because it now seems sure that we will have to stay here until freeze-up. The walls of the structure are up four feet and we hope to have the roof on tomorrow.

This has been a tiring day for all of us, carrying frozen sod for the mud house and gathering fuel. Eight or nine hours' hard going, and only two scanty meals, could hardly be considered a satisfactory day. …

We are now satisfied that the Eskimos will take us out when travelling is possible, and not before. Our new Eskimo showed McMillan a four-gallon can of gas in his cache, but says there is no more in the neighbourhood. He also told us there was no gas nor any supplies at Ellice River, the abandoned post we flew over twenty or thirty minutes before we spotted Joe's camp and landed.

The party fortunately had plenty of ammunition and for the first two weeks, until they migrated south, were able to shoot many ptarmigan. The ptarmigan flew at high speed and combined with their white camouflage were very difficult targets. Again the party were fortunate as Boadway turned out to be a superb hunter and ground squirrels were plentiful until the early part of October when the snow became deeper. The main diet, however, was dried fish given to the party by the Inuit. The party had a small and dwindling ration of white man's food for variety.

Not knowing how much fish the Eskimos would be able to supply us with we voluntarily went on very strict rations from September 10 to October 12. Checking our consumption afterwards it worked out at 6 ½ oz of food per man each day.

Alex Milne recalled later that as for the Inuit their generosity saved the day. They provided fish and plenty of seal blubber. In doling out fish, they would smell the portions, then give the best ones to the visitors while keeping the bad-smelling ones for themselves. Alex related how eating Inuit fare took some getting used to and that seal blubber was at first violently rejected by the stomach.

Late in September more Inuit came to the camp. One of them spoke English and he informed the party they were located on Queen Maud Gulf, well east of Bathurst Inlet, and the community across the open water was Cambridge Bay. On October 21 the lost expedition finally set out for Cambridge Bay.

October 21: Bob suggested there would be no diary today because the writer would be too tired to bring it up, but here it is. We are about 25 miles from the old mud shack, camped opposite the end of Melbourne Island. We found out that dog-team travelling is not like it appears in the movies, with speeding dogs carrying mushers along. The real stuff is largely foot work, real hard going when the snow is a little soft.

The worst of today's experience was that the dogs started off at an awful clip, and kept us going for a while. We were afraid they would keep it up all day, but they soon slowed up. We left at 9.30, had a cup of tea at noon and went on until 4.30. Not a bad day for greenhorns to put in; but boy, how tired we are!

Bad weather and the fact that the water was still not frozen delayed the party and it wasn't until November 2 that the journey continued.

November 2: It is early in the morning, and there is a conference of Eskimos in our igloo. They appear to be talking over the trip today. We had a wonderful breakfast of three cups of tea, a half bannock, and a big slice of boiled land trout. Zero hour approaches, and I hope we spend the night at the Cambridge Bay post.

What a tough day! We spent hour after hour trying to pick a course through a large ice flow, wandering this way and that to find the best going. Sleds were tugged and pushed over ice hummocks that we uninitiated thought were impossible to cross. More than once we tumbled into holes between ice cakes. It was all nerve-wracking and tiring.

Narrow leads of unsafe ice had to be crossed, but before we ventured on them the Eskimos tested them with spears. Alice, the wife of the Eskimo with whom the colonel, Baker and I were travelling, took her boy of three after he had been tossed from the sled several times, stripped him naked and tied him on her back under her clothing. This did not stop her from continuing a man's share of the work of urging the dogs and helping tug the sleds over the most difficult spots.

Once while crossing weak ice, apparently frozen only the night before, she broke through to her knees. Instead of trying to scramble out—which would probably have meant that she would sink deeper—she lay flat until one of the Eskimos rescued her. On we went for a quarter of a mile, when Alice changed into dry clothing out in the open without any apparent great discomfort, although the temperature was well below zero.

At 4 PM we struck thin ice. The Eskimos could find no way to cross it. It was then a matter of parking for the night on an ice floe, hoping that in the morning the going would be safe. Some of the party feared that the strong wind then blowing would carry the floe down the straits and perhaps break it up. The Eskimos assured us, however, that the wind would die down during the night.

November 3: This is the night of nights. We are at Cambridge Bay. Messages have gone out to our families and we can go to bed in peace. But what a day! It started at 5 AM with a very scanty breakfast and then a race across thin ice. The colonel, who was in great discomfort, got a ride partway over the first five-mile stretch, but the rest of us had to run, dodging the dark green spots in the ice that we could see wavering under the men ahead.

By the time we got on safe ice it was a weary lot of travellers who followed the sleds. Our legs were tired, sore and chapped, and bitter wind with a temperature of 27 below zero bit into our lungs. Toes, faces, and fingers were frozen.

While the expedition was fighting its way out, at least 12 aircraft and some of the most famous bush pilots in Canadian aviation took part in the search. Many of the pilots and crews did not have extensive experience flying these uncharted barren lands and, under these conditions and with limited fuel supplies, there were several dangerous and costly accidents. Punch Dickens, Roy Brown, Bertie Hollick-Kenyon, Bill Spence and Charlie Sutton were pilots who did have experience in these conditions and began searching September 26. Other pilots were in charge of backup materials and supplies for the searching pilots: Vance [full name not known], Pat Reid, Andy Cruickshank, Ken Dewar and T.W. Siers. In large part, Thayer Lindsley personally financed this rescue mission.

When the weather permitted, three planes, under the direction of H.G. Blanchet, searched the barrens for the marooned party, and one of them flew over Dease Point where the two Dominion Explorers' aircraft were buried in snow. On November 5 an Inuit arrived at the Dominion

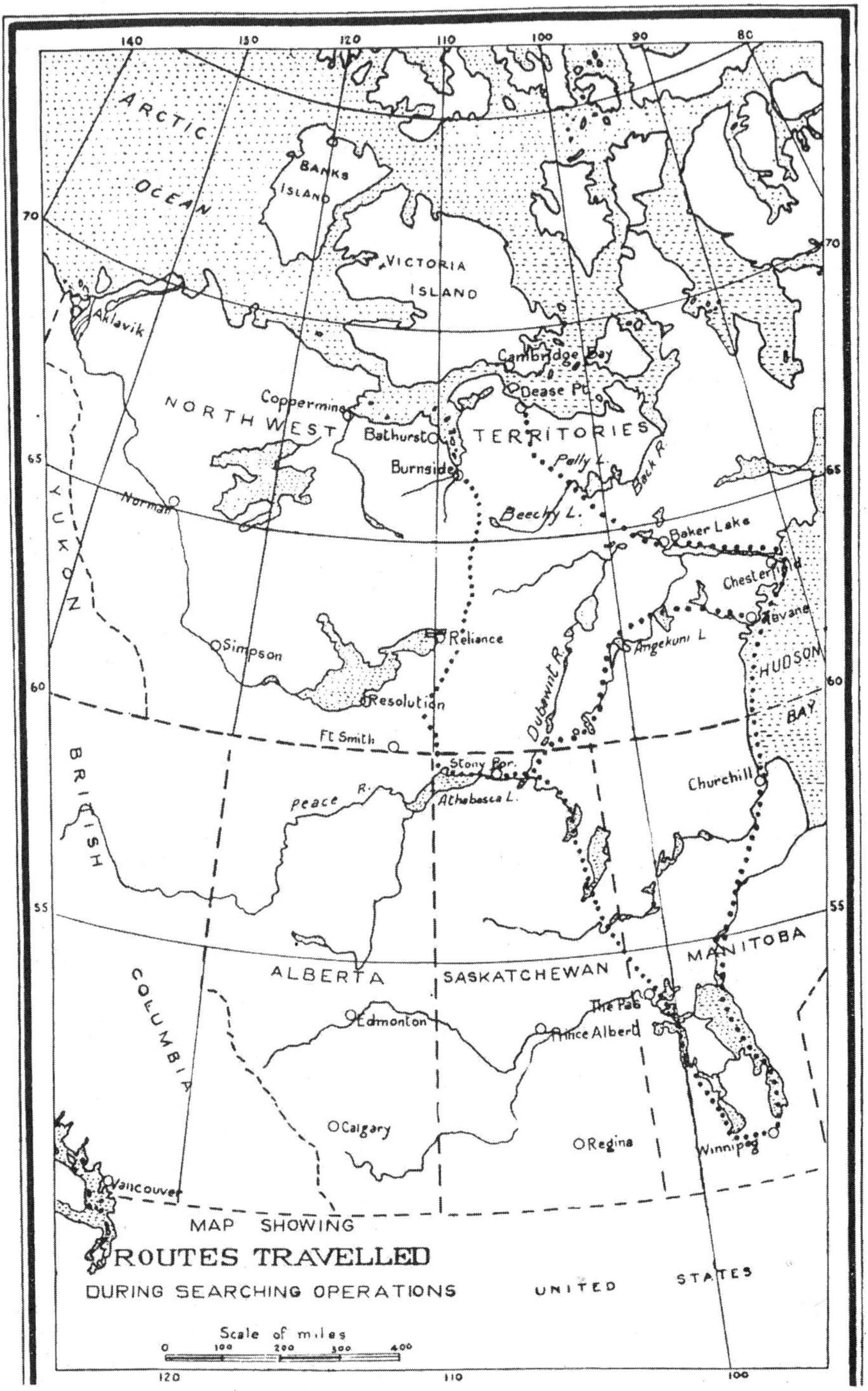

The dotted lines show the routes taken by the searchers in their efforts to locate the lost MacAlpine party [from Marooned in the Arctic*].*

Explorers' post at Burnside River and informed them that the MacAlpine party had reached Cambridge Bay. The search was over but the rescue was far from complete.

Four planes were used to bring the expedition back to Winnipeg. Bill Spence's Fairchild was wrecked at Musk Ox Lake while Roy Brown's Fokker was wrecked at Alymer Lake. The expedition finally arrived at Cranberry Portage December 4.

Pilot G.A. Thompson later wrote a tribute to the Inuit:

> ...if we had been their own children they could not have mothered us with greater care. When luxuries such as tobacco, tea, and sugar ran low they shared their meagre store with us. Even our lack of suitable clothing that caused us the greatest discomfort they did their best to supply.

This expedition accounts for J.P.'s tremendous respect for the Inuit and his fear of bush flying, that lasted the rest of his life. Anyone reading Pearce's diary can easily relate to J.P.'s feelings. Nevertheless by January 1, 1930 he was working for Dominion Explorers and helping to organize another expedition to Great Bear Lake. Three of the members of the marooned expedition would be a part of this Great Bear Lake mining venture: Ed Boadway, Stan McMillan and Alex Milne. This expedition would be headed by Thayer Lindsley with John Rogers as chief geologist and J.P. as on-site mining engineer and manager of the 20-man party at Great Bear Lake.

Inuit skin tent

Eskimo mother and babe. Dressed in beaded caribou skin. Her Sunday best. The Eskimo always smile. - J.P. Norrie

Baker Lake in winter

Winter in Tavane, Rankin Inlet

Tavane, Rankin Inlet, in the summer

Ed Boadway, J. P. Norrie, and Inuit at Great Bear Lake

From J.P.'s Great Bear Lake Diary

The day to day details of J.P.'s activities during this expedition come from his personal diary and give another perspective on the expedition. Combined with his previous official report a clearer picture of the Great Bear Lake expedition is achieved.

The diary was required primarily to record his geological observations. That is what each member was up there for and all interesting possibilities were to be noted. Copper was the mineral of primary interest since many copper occurrences had been discovered on earlier expeditions.

The geography was particularly important as many of the lakes, rivers, bays, channels and islands were largely unknown and not named. Several of these would be named after members of this expedition, including J.P.

The importance of water routes to access the country when ice had gone is obvious from the diary notes. Even so, to walk 20 miles a day over this wilderness was not unusual. As part of the management of the expedition, J.P. kept a record of the location of all parties, and their locations and their discoveries are noted.

Dependence on the bush plane to move prospecting parties from one location to another is also clear. The bush planes kept all groups in touch with J.P. and with each other; this was their only communication. The planes were kept busy; they moved samples and supplies when they were not moving the prospectors.

J.P.'s diary gives clear pictures of many other points of interest: the weather, the animals, the mosquitoes, and ice and water conditions. Many other details of much less importance to the expedition are also included. When Bellachey was treed by a mother bear with two cubs, it didn't mean much to the purpose of the expedition, but it sure did to Bellachey and the others. There were many areas up there with only spindly trees, or no trees at all!

Although J.P.'s diary gives many specific details, it is not a complete account of the events of the expedition. The most interesting evidence for today's reader is the discovery of pitchblende (the mineral source of uranium). Its discovery was a surprise to the expedition.

The only known use of uranium and its radioactive properties at that time was for the treatment of cancer. The Manhattan Project and the atomic bomb were some years in the future.

The accepted account of Gilbert Labine's discovery of pitchblende at Great Bear Lake has been blended from several sources and from it questions arise.

J.P. was never acknowledged as a participant in this discovery but his diary identifies his location every day and shows that he was at Labine Point on May 19, 1930, and met Labine and St. Paul on many occasions during the last two weeks of May, 1930. It does not record J.P.'s first "sketchy" observation, but it was probably so speculative that it didn't seem important. The diary does record Lindsley, Rogers and J.P. examining uraninite (uranium) on July 9, 1930, before the assay results came from Ottawa.

Many of the specifics that we would like to know today do not appear in his diary. The diary does, however, support and give credibility to his family's story that it was he who pointed out the deposit and advised Gilbert Labine to send samples to Ottawa for assay.

J.P. left Rouyn, Quebec by train at noon on Feb. 11, 1930, for Edmonton, Alberta. Transportation to the mines in northern Quebec and Ontario had always been a headache. Each trip into Stadacona or

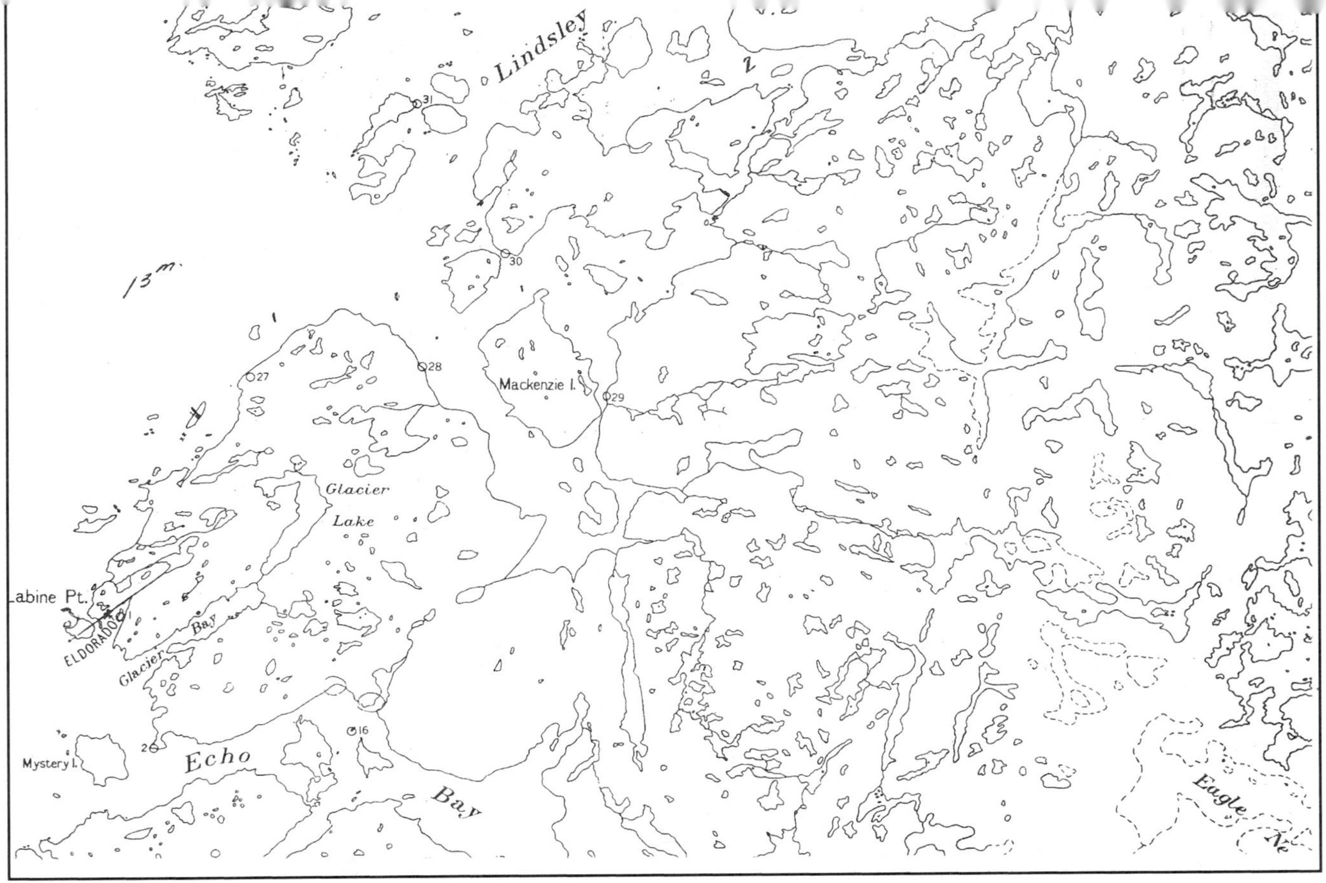

Map shows the area covered by the second Great Bear Lake expedition. Note Norrie Bay (at top right), MacAlpine Channel and Lindsley Bay (middle) and Labine Point (at bottom left).

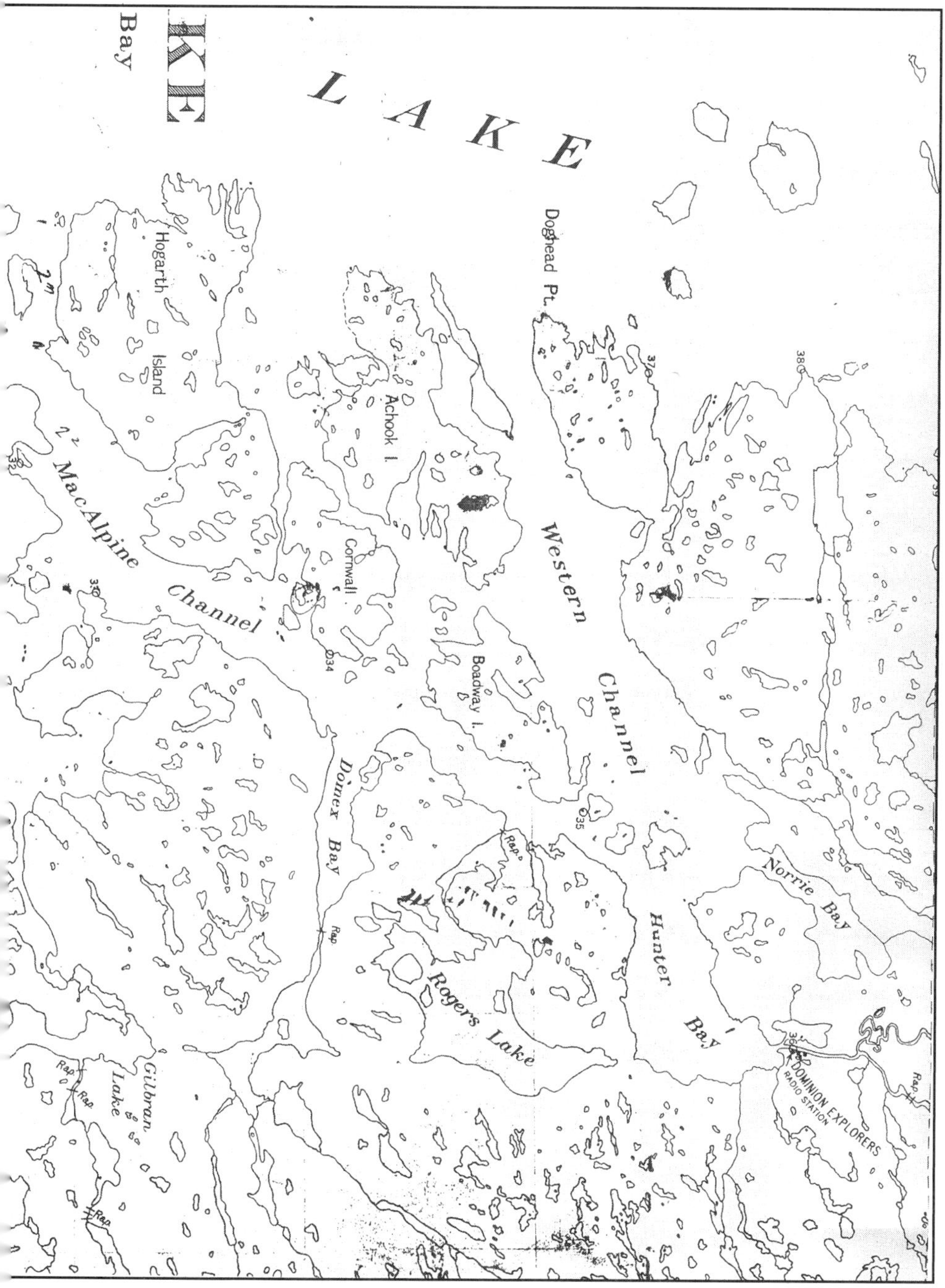
LAKE
Bay
Doghead Pt.
Hogarth Island
Achook I.
Cornwall
MacAlpine Channel
Western Channel
Boadway I.
Doimex Bay
Norrie Bay
Hunter Bay
Rogers Lake
Gilbran Lake
DOMINION EXPLORERS
RADIO STATION
Rap.

Malartic Gold Mines was an adventure. But travelling across the country by rail amazingly took only two and one-half days. The train isn't much more efficient today.

Arriving in Edmonton Feb. 13, 1930, J.P.'s first stop was the Royal George Hotel. This was certainly one of the finest hotels in Edmonton and two of the prospectors, Charlton and Clark, had already arrived. Bellachey came two days later. All prospectors stay at the best hotels!

J.P. was the manager of the expedition and had to buy supplies for the northern winter immediately. Hudson's Bay Co. and Northern Traders were the two main suppliers in Edmonton. Then, as soon as supplies had been purchased, the members of the expedition took the train to Waterways. Waterways was the end of rail transportation and the jumping-off point for all northwestern expeditions. Bush planes took over at Waterways.

Arriving in Waterways Feb. 16 J.P. and the others stayed ten days at Colonel Cornwall's house waiting for the three Dominion Explorers' planes. Colonel Cornwall was a friend of Col. MacAlpine and had been involved in previous Dominion Explorers' expeditions. There were too many supplies for the three company planes to transport easily, so Commercial Airways was contracted to take the overload of supplies and possibly an advance party of prospectors to Great Bear Lake.

Any expedition arriving in a small community starts rumours flying. An expedition heading north at that time of year with three planes of their own would certainly raise questions. J.P. heard rumours that their competition, Boland and Murdoff, were going to Great Bear Lake to stake claims and that Boland had already been there earlier that winter staking claims. This information gave a sense of urgency to get at least some of the men to Great Bear Lake. J.P. arranged with Kelly of Commercial Airways to take Charlton and Clark to Great Bear at once to begin staking.

Feb. 23: Had visit in evening from Murdoff, Becker, May and Kelly. Only divulged Kelly's destination to him alone. Becker and May wanted to know destination of plane, but did not tell them. Arranged for plane to leave Monday at $45 per flying hour, and $90 per day for each flying day. Supplies were bought from Burton's at Waterways, and Perry's at McMurray, for two men for a month. Total plane load, including men and kits: 775 lbs.

Wop May would later fly in fuel and supplies and he would be told the destination then. Secrecy was important. J.P. didn't want Becker and Murdoff to know their destination. One of the reasons they were prospecting in the middle of the winter under such conditions was that they wanted to get ahead of the competition for staking claims.

Wop May was one of Canada's most famous bush pilots by this time. He was a young pilot during the First World War and flew one of his first missions for the RAF in Roy Brown's squadron. Brown was also Canadian. During the early part of the mission May's plane developed an engine problem and he left the squadron to return to base. Baron Von Richtofen "The Red Baron" came in behind him to shoot him down. But then Roy Brown came in behind the Red Baron, and history has given Brown the credit for shooting down Von Richtofen. Three Australian artillery units were firing at the Red Baron from the ground but he was killed by one bullet. Brown has the credit for shooting him down, and Wop May was the bait!

J.P. and Bellachey waited the next ten days in Waterways for the three company planes to arrive from the east. On March 2, J.P. went to Fort Franklin with Wop May and arrived at McMurray that afternoon. Fort Franklin was not directly on the route but it was an opportunity for Kelly and May to observe flying conditions at Hunter Bay. Kelly landed on the Sloan River at the end of Hunter Bay, where the Dominion Explorers' camp would be located. Landing conditions were good and the route and stopovers from Yellowknife to Great Bear were also in good condition.

The cache of gas at each stopover was the major concern of the pilots of the north. Supplies were kept at several locations, including Yellowknife, Rae, Hottah Lake and Great Bear Lake. The gas might belong to N.A.M.E. (Northern Aerial Minerals Exploration), Dominion Explorers, Consolidated Mining & Smelting, whoever had planes in the north at the time and whenever gas was taken, the pilot recorded the amount and replaced it at the next visit. Otherwise planes could be stranded in the cold. Pilots would have to wait to be rescued by another plane if they could survive at all.

Took 308 gallons N.A.M.E. gas from Rae. Left 72 at northeast corner of Hottah Lake. Still owe N.A.M.E. 108 gallons. Cache at Yellowknife 5 miles up river from village. Gas at Rae 6 cases at 4 gallons - 24 gallons. Gas at Yellowknife 40 - 45 gallon drums, 100 - 10 gallon cases, oil 30 - 4 gallon cases Mobil Oil. B.B. 1 Johnson motor. 2 canoes and paddles. 1 cooking outfit.

Welcome to the North

On Mar. 9, the three company planes arrived from Lake Laronge and landed between McMurray and Waterways; the Waterways strip was covered with slush, preventing landing. The planes were loaded with supplies, and the 800 lbs. overload was to be taken in by Wop May and Kelly. The three Dominion Explorers' pilots were Spence, McMillan and Boadway. Spence was the senior pilot on the expedition. The mechanics were Milne, Wheeler and Lorrie. McMillan, Boadway and Milne had been part of the marooned expedition, Sept. to Dec. 1929, and were now going back to the north in March! Each plane had one pilot and one mechanic. This was normal procedure in northern flying. It took an experienced mechanic to keep the planes in the air and repair them when they crashed in the wilderness.

The trip to Great Bear Lake would seem at first to be without planning or rather haphazard. Nevertheless, it was all based on gas and the weight of supplies at each stop.

On Mar. 11 the three planes took off in the morning. Spence had trouble getting off as his load was too heavy. They all landed safely at their first stop, Fitzgerald, at 12:30 PM. They took off from Fitzgerald at 3:00 PM. The weather was cold (30 deg. below zero) and the landing at Rae was rough due to hard snow. The next day they took three loads of gas from Yellowknife to a cache at Moose Lake, then went to Resolution for gas and supplies to replace what they had taken from Yellowknife.

Mar. 14: Spence and Boadway arrived from Resolution with supplies. McMillan went to Fitzgerald and arrived with Kelly's load and 200 lbs. supplies. Spence and Boadway freighting gas from Yellowknife to Rae in afternoon. 1,500 lbs. rations now assembled at Rae, but due to shortage of gas, only 1,200 lbs. can be taken in to Great Bear Lake.

Mar. 15 Spence and McMillan took off for Bear Lake with 1,600 lbs. and 80 gal. of gas. Boadway was freighting gas from Yellowknife. On Sunday Mar. 16 it was very windy and Boadway had to lay up because of a frozen gas filter. Spence arrived back from Bear at noon but McMillan did not return due to heavy fog at Bear. Spence did take 1,200 lbs. of supplies back to Bear. The next four days at Rae were extremely stormy and nothing moved but McMillan did arrive from Hunter Bay.

On Mar. 21 Boadway continued to ferry gas from Yellowknife. The next day McMillan made one trip to Yellowknife and Boadway fixed his carburettor. The next several days the three planes ferried supplies and gas to Yellowknife from Fitzgerald and on to Hunter Bay.

Mar. 25: Spence, McMillan and Boadway to Hunter. Spence and myself with gas landed at Hottah Lake and cleaned up Kelly's cache. McMillan and Boadway with provision. Left 400 lbs. at Rae. Coal, steel, B.S. tools, radio and glass. Landed at Hunter 12:30 PM.

Charlton, who was already at the base camp, reported to J.P. that the stories of a mineral showing at Great Bear Lake had been much exaggerated by Murdoff and Boland. J.P. went directly to the showing with the prospectors: Charlton, Bellachey and Sloan. They set up a tent near their showing called the Star claims. This tent would be their home for the next three weeks. The Star claim was frozen solid and the only way to examine the mineralization in the bedrock was to build a fire to thaw out the ground for digging the trenches.

Mar. 27: Built fire and started trenching on boulder showing about centre of Star claim. Ground frozen solid and hard digging.

Each day for the next week, with large fires, the trench was excavated by hand, about 20 feet long and several feet wide. After one week they were down about 8 feet and on Apr. 5 the three planes arrived from Reliance with John Rogers and Hanna. John Rogers was the chief geologist of Dominion Explorers and would direct all geological decisions on the expedition.

The planes also brought gas and radio equipment for the radio shack which would be the expedition headquarters for the next several months. (The shack was still there in 1964.) J.P. was located primarily at the Dominion Explorers' main camp (radio shack) at the mouth of the Sloan River which ran into Hunter Bay. The area was well suited for their camp as the terrain was not as rough and mountainous as some other places along the eastern shores of Great Bear Lake. It was very suitable for aircraft to land on the ice and load and offload supplies, samples and prospectors. It was also near the copper showings located in that area.

Some of the competition mining outfits also used this location because of its central location and accessibility. Their prospectors therefore visited the Dominion Explorers' camp regularly.

The building was log -cabin style, made from small, windblown trees that grew around the lake and were hauled to the cabin by dogs. The treeline was from Great Bear Lake south. To the north was the northern barrens. During the next ten days the walls of the first floor of the radio cabin were constructed and chinked with moss. The radio pole was erected and they put up an antenna. The weather was very windy and from 15 to 38 deg. below zero F.

1930 radio shack, Great Bear Lake, with Harry Norrie in 1962

On Apr. 17: (-38 deg. F) the two-storey wireless station was mostly completed and they had radio connections to Bathurst Inlet, Fort James, Churchill and Resolution. They could now communicate with the Dominion Explorers' head office in Toronto. Water grounds on the transmitter made the signal much louder than it would have been otherwise. The next several days were spent finishing the roof on the radio building and more chinking of the cracks with moss. The temperature was still hovering at 20 deg. below zero. The pilots were still moving supplies from Fort Norman and Resolution to Bear Lake headquarters.

By May 1, with the radio shack livable, they began exploration on several new locations. Sloan and Charlton went several miles south, while Rogers, Bellachey and J.P. stayed near camp. During the remainder of

April and early May the Star and Vega Trenches, near the headquarters, were gradually deepened to about 12 feet, exposing some bedrock and mineralization. Some samples contained in excess of 20% copper.

Supplies were prepared for the Coppermine camp and other prospecting areas, such as Burnt Creek, Father's Lake and Dismal Lake. McMillan took Folker, Thompson and Harling to the mouth of the Coppermine River. McMillan also took McLaren and Gilleran to Dismal Lake and McLellan to Burnt Creek.

During the last two weeks of May, McMillan's plane was used several times for scouting areas around the main camp for good prospecting possibilities.

May 16: Went northeast of Hornby Bay and country around Hunter Bay in plane with McMillan, Rogers, Charlton and Sloan. Landed near break northeast of Hornby Bay and examined surrounding country. Most snow gone on the barrens. Travelling good. Weather clear and mild. Ice getting soft on lakes.

Baker Lake in summer

McMillan's plane was still equipped with skis. Boadway's plane was located at the main camp and its skis were changed to pontoons to use where the ice had melted on the lakes and rivers and along the shores of Bear Lake. Although he was a pilot Boadway was also a mining engineer and was very versatile. He helped J.P. with his prospecting duties out of the main camp when the pontoon plane was not needed. They became very close friends and Boadway would work for J.P. years later.

At the Dominion Explorers' camp they finished the cabin and drained the melting snow away from the cabin floor. J.P. also fixed up the boats as they were now needed where the lakes and rivers were opening up and along the edges of Bear Lake. The ice was still one foot thick on Hunter Bay but the ducks were now returning from the south.

Murray and Bellanger of N.A.M.E. visited the camp. The N.A.M.E. plane flew them to Resolution and J.P. sent the camp mail with them.

The Discovery of Pitchblende

The next section of J.P.'s diary concerns the activities of the Dominion Explorers' party during the period when history has recorded that pitchblende was discovered at Great Bear Lake. J.P.'s diary shows specifically where he and his party were located and it also shows some of their contact with Labine's party.

The diary does not include many of the details history would like to know. However, the possibility or the probability was there that J.P. and his party were involved in the discovery of pitchblende. The following is J.P.'s story of that discovery, which has been known to the family for many years.

The family story has been confirmed by many of J.P.'s colleagues and friends, such as Ed Boadway and Jack Workman, who were there when J.P. identified the pitchblende.

The first sample was found on claims that Labine had staked and J.P. told Labine that he believed it was pitchblende. J.P.'s story was that Boadway was the person who found the sample and described it as "sketchy". He described himself as the second person to see it.

J.P. identified the pitchblende sitting around the campfire with expedition geologists, prospectors and pilots.

J.P. told Labine to send the sample to Ottawa to confirm his suspicion and that, if the material turned out to be pitchblende, he would expect some founders' shares, even though the sample was on Labine's claims and J.P. had no legal rights. This was a normal expectation in geological finds of those days, but J.P. did not receive any shares or any recognition that he had played any part in the discovery.

J.P. helped Labine and St. Paul load the samples on an airplane to be sent to Ottawa for analysis.

May 19: McMillan with Rogers, Charlton, Bellachey, Sloan and myself to Conjuror Bay, then east about 20 miles, and scouted country back to camp. Many deep bays to east of McTavish Bay. Good water routes for prospecting. Saw no dykes or other outstanding features. Rock mostly syenite. In air two hours. Went over Vega showing with Rogers and Sloan in PM. Fine and warm.

"May 19, 1930… Conjuror Bay, then east about 20 miles, and scouted country back to camp." Twenty miles east of Conjuror Bay is the area where Labine was staking his claims on May 18 and May 19, which is now called Labine Point. Eldorado Uranium Mine and Echo Bay Silver Mine would eventually be established at Labine Point.

The diary does not record, specifically, that they inspected the Labine claims or talked to Labine that day, but they were definitely there and could have seen Labine. Men staking claims could be seen easily from McMillan's plane, and prospectors always inspected each other's claims.

This is probably the day that J.P. saw what he later described as "the first sketchy surface discovery" that turned out to be pitchblende.

J.P.'s family story relates that it was Boadway who found the sample and that J.P. identified it as pitchblende. J.P. in his 1945 letter says that he was "the second person to see the pitchblende."

Labine's account was that the first sample was the size of a plum and was discovered on May 16, 1930.

In the Labine account of May 18 and May 19 he did not tell his snow-blind partner anything about his find, pitchblende or silver. Further, they did not stake any more than the first two claims and moved to the claims they had staked in 1929 along the Sloan River, where Dominion Explorers and other companies were staking for copper. This would suggest there was no certainty of pitchblende at Labine Point; possibly that there was no knowledge of pitchblende at all. J.P. may have told Labine then or later about his suspicions.

Mining people were much aware of uranium in 1930. The only use known for radioactive materials in those days was for the treatment of cancer. It was not suspected to be present at Great Bear Lake, nor were they anxious to use up their claim allotment for this mineral. Copper was the reason they were up there; pitchblende was completely unexpected. This is probably why Labine and the Dominion Explorers' party were reluctant to stake the area.

Amos, Quebec
August 7th, 1945

Col. J. M. Thomson
112 Chaplin Crescent
Toronto, Ont.

Dear Mr. Thomson:

It was very kind of you to send me the fine book ends which arrived here today. Thanks very much!

The new atomic bomb should help in shortening the Japaneese war. Too bad we did not have a few to drop on the Germans before they quit. I happened to be the second person to see the pitchblende deposit at Great Bear Lake and was there all through the first summer's work and am naturally very much interested to know that the first sketchy surface discovery has such potential power possibilities.

Thank you for rewarding me so handsomely.

With best wishes

Yours very truly

J. P. Norrie

JPN:H

The weather until the end of May remained warm and clear. Most of the snow was gone on the hills and the rivers and Hunter Bay channel was opening up fast.

J.P. and the main camp crew continued trenching on the Vega claims. They also continued preparing supplies for prospecting parties heading out from the main camp.

McMillan & Wheeler flew Charlton and Sloan to examine a quartz vein in the bay ten miles from camp. Charlton, Sloan and Bellachey then took the dogs and provisions ten miles up the MacAlpine channel to prospect that area from the ground. The dogs, which had been used when snow was on the ground, were now becoming unnecessary, but they could still be used on lake ice.

May 23: John Rogers, the head geologist, arrived at Dominion Explorers' camp from the outside. Many of the other mining companies continued to visit the Dominion Explorers' camp. Ken Murray of N.A.M.E. visited for two evenings and the Consolidated men arrived and left to explore the channel area where the Dominion Explorers' party was working.

May 29: Brunton surveyed Bannerman and Ginn claims with Rogers. Rained at 1 PM. Came home wet. Labine and St. Paul arrived here 4:30 PM, with big load on sled. Helped them transport material with dogs and boat.

This seems to be where J.P. helped Labine and St. Paul load samples to be sent to Ottawa for assay. This entry is not specific enough, however, and these may or may not have been the samples sent out for assay. If these were the samples, J.P. had advised Labine of the possibility of pitchblende before this date.

Some historical reports say the samples were sent out approximately July 21, 1930, but the diary does not support this date.

The area near Dominion Explorers' exploration camp was the area of primary interest for all the mining companies. It was the Labine dyke and the Sloan dyke, located 15 miles northeast of Hornby Bay, the most prospected area. All the companies were in and out of the Dominion Explorers' camp constantly because it was so well located near the mineralized area. The mineralization in this area was copper.

Labine, who staked only part of his allotted claims near Labine Point, moved from Labine Point to the Dominion Explorers' area after staking on May 18 and 19. J.P.'s diary records that Labine and St. Paul visited the Dominion Explorers' main camp at that time.

May 30: Labine and St. Paul here in evening.
May 31: Labine and St. Paul here in evening.

During the first two weeks of June J.P. made several trips from the main camp prospecting the Labine and Sloan dykes. Consolidated had already staked some of this area. J.P. staked a section for Dominion Explorers in the same area.

Sloan, Charlton and Bellachey returned from south of Hunter Bay June 14, having walked overland 35 miles. They found two new quartz dykes. J.P. now joined them exploring the small bays and islands along Hunter Bay and MacAlpine channel, ten miles southwest of the main camp. Eight claims were staked for their evidence of copper.

June 30: Started for bay south of Hunter. Had to camp just before turning the point into this bay, due to heavy drift ice from lake. Looked over Sloan dyke on the islands where the ice blocked us. This may be a low-grade copper deposit of considerable size. Wind changed from west to southeast at 5 PM. Camp already made so decided to move early in morning if wind and ice favourable. Some rain and thunder today, very warm.

At the beginning of July, Sloan, Charlton, Bellachey and J.P. moved further southwest along the coast. For the next six weeks they explored the area from Stevens Island to Richardson Island, including Conjuror Bay, Echo Bay and the Labine claims. They travelled, as usual, by plane and by boat moving inland where the shore was accessible.

For the first few days of July the party searched around Lindsley Bay and Stevens Island, and inland for short distances. This was the best prospecting area to date.

July 9: Examined Labine claims, found some bloom and cobaltite but nothing of much value. Plane with Messrs Lindsley, Hardy and Rogers arrived at noon. Looking over Labine showing along cliff at water level with Messrs Lindsley, Hardy and Rogers, found nicolite and uraninite in place. Had intended going here with boat after breaking camp. In PM Keith and Loan accompanied party by plane to south Hunter Bay. They seemed to be favorably impressed with showing. Plane left at 5 PM for Hunter.

This is an interesting entry in the diary as it shows that uraninite was inspected by Lindsley, Hardy, Rogers and J.P. on July 9. This shows the Dominion Explorers' party knew uranium existed on the Labine claims long before the assay report of uranium was confirmed on August 12, 1930.

Some reports from the Labine discovery state that the samples for assay did not leave the north until the Consolidated plane took them out July 21. Further, when the confirmation of pitchblende was received on August 12, the Labine party hurried to stake more claims near the ones they had already staked May 18 and May 19.

It is also interesting that Dominion Explorers did not stake claims for themselves at any time, even though Rogers and Lindsley inspected the showing identified as uraninite. Nor did Labine stake more claims for himself until after the confirmation of August 12.

July 10: Moved to northeast arm of Conjuror travelling from 8 AM - 9 PM. 20 miles southwest down coast to Richardson Island, 10 miles southeast to mouth of northeast arm, then northeast 20 miles to rapids on river. Camped here...

Charlton, Sloan, Bellachey and J.P. went in separate directions for the next several days but found no minerals of importance. The party then moved to all other inlets along Conjuror Bay but found no copper but lots of stain. On July 14 they moved their camp to the second Inlet of Conjuror Bay. They continued prospecting along the shores and short distances inland making geological observations. On July 18 Lindsley and Rogers arrived again to inspect the Conjuror Bay area.

July 18: ...Plane arrived at 11 AM on small lake then flew over hills to camp... showed samples to Lindsley and Rogers (Baslat trap)... I went overland and took boat back to camp. Plane left about 2 PM, returned to bush 2:30 PM. Plane - Spence, Lorrie (pilot and mechanic) with Lindsley and Rogers.

J.P., Keith and Bellachey looked at copper stringers about four miles south of camp and found a vein in place 1,200 feet long and twelve feet wide with considerable chalcopyrite with specularite and bornite. J.P. sent Keith for more prospectors and staked four claims on the showing in the name of A.G. Bellachey. McLellan and Charlton arrived from Conjuror Bay and staked six more claims. They needed the boat to work the country as the Camsell River was too big. Bellachey was treed by a bear with cubs. The mosquitoes were very bad.

July 21: Heavy rain all day. Washed mended clothes. Made maps, etc. McLelland and Charlton arrived at 8 PM.

Some historical reports record that the samples of uranium ore were sent out for analysis from Labine Point by a Consolidated Mining and Smelting plane July 21. J.P.'s story says that he helped Labine and St. Paul load these samples. This, of course, does not seem possible as he was 20 miles away at Conjuror Bay and had no recorded contact with Labine. J.P. had been prospecting near Labine Point for several weeks after June 15. If he helped load the samples this would have had to be at Labine Point, but the area around Labine Point was much more difficult of access. Most planes came to the Dominion Explorers' camp because of its accessibility. The July 21 date does not seem to be verified by any specific historical facts.

July 26: ...Moved from camp on 3rd Arm of Bay to 12 miles up river of unknown name, running northeast then northwest. Camped at first rapids. River 300 feet wide and water fairly swift. This looks good prospecting country. Examined quartz dyke in Conjuror Bay north of river mouth.

On the last several days of July the party prospected up the Camsell River. They needed to portage two sets of rapids to access the country. They met the plane with Rogers, Boadway, Spence and Lorrie en route. Rogers examined the showing and liked it. Instructions were to work carefully from the rapids to the mouth of the river, then along the south shore to the south entrance of Richardson Island.

They examined the south shore of Richardson Island for cobalt bloom and the south shore of Conjuror Bay for galena. They expected to meet Stevens with a boat somewhere on the south shore. The plane was to return in one and a half weeks.

For the first two weeks of August, J.P. and his party continued to prospect inlets of Conjuror Bay, working primarily off the Camsell River. They also prospected the inlets around Richardson Island. They did not find any further impressive areas of minerialization.

Aug. 7: Moved camp to east shore of Richardson Island. Went 2 miles north with boat… Red cranite and syenite hills presenting a novel setting with the sun just above the horizon. Blueberries, raspberries, and gooseberries plentiful. Keith mapping islands etc. McDonough landed near Gossan at camp where Lindsley and Rogers had last landed.

Aug. 12: Moved southwest through south entrance to Richardson Island and up the shore to Labine claims… Camped at Labine claims at 6 PM. Keith and McLellan went on to Hunter Bay for provisions and canoe.

Aug. 12 is the date that assay reports confirmed uranium on Labine claims at Labine Point. Labine is reported to have immediately staked two more claims adjacent to those already staked. Dominion Explorers obviously were still not interested in uranium.

Aug. 13: Bellachey and myself examined Labine claims again. The north vein was looked at on a previous occasion. It seemed to be on original examination a flow rock covered with a black coating.

On second examination we found some narrow stringers of uraninite. The vein seems to be about three feet wide.

McLellan and Keith arrived back and Bellachey left with McMillan in CFAO for Hunter Bay.

Aug. 16: Left with all equipment for Hunter Bay. Bay very rough and motor running poorly. Camped on west side of bay for 3 hours due to rain and snow. Twelve noon moved into Echo Bay. Camped on shore due to heavy rain and snow... until 6 PM. Started for Hunter. Picked up 300 lbs. gasoline and oil at Labine Point but could not carry on due to rough water.

Aug. 17: Left Labine Point at 3 AM... Water in lake smooth now. Arrived Hunter at 11 AM. Hold up 1 hour with motor. Dirt in gas line... McMillan and Rogers to Coppermine for MacLaren, Folker and Thompson. Very cold this morning, wind and one-quarter inch of ice.

Back at Dominion Explorers' headquarters J.P. had to catch up on recording of all claims. Maps, sketches and samples had to be prepared for all work done while he was prospecting Conjuror Bay, etc. This included those of all parties on Coppermine, south of Hunter Bay, the islands in Great Bear Lake and north of headquarters.

The prospecting parties began arriving back at Hunter in preparation for leaving for Waterways. The parties who arrived early spent a few days prospecting areas nearer camp.

Rogers, Gillingham and McMillan examined Achook and Hogarth Islands noting a large body of low grade copper. McMillan and Milne also took photos of islands in Hunter Bay from 11,000 feet.

Aug. 26: Surveying and geologizing Star claim Hunter Bay with Rogers and Gilleran.

Aug. 27: Examined Jim claims and country to north with Messrs Labine and Rogers.

Stevens, Workman, Byles and Priest arrived from prospecting south of Hunter Bay. McMillan brought Don McLaren, Thompson and Harling from McLaren Lake, and Charlton, Gordon, McLaren and McLellan from Coppermine.

The winter conditions were beginning to set in. There was six inches of snow and it was very cold and windy in Coppermine. This meant ice covering the lakes and flying would become much more dangerous as they now had pontoons on their planes.

Aug. 31: Gilleran, McLaren and Charlton tried to take off with McMillan at 1 PM but could not make it. Upon return to Hunter Bay base plane (AKY) almost sank due to hole in right pontoon. Pulled plane out on back and repaired hole. Took off for Conjuror at 5:30 with gang.

All prospecting companies, N.A.M.E., Labine, Consolidated etc., were now removing their prospectors, samples and remaining supplies south to Waterways.

Spence, Boadway, Wheeler and Lorrie arrived from Waterways and three N.A.M.E. planes arrived from Coppermine. Spence went to Coppermine and returned with Blanchet and three priests and continued on to Resolution. Sept. 3 Boadway and McMillian flew to Waterways with McMillan, Workman, Stevens, Finlayson, Harling and Byles.

Sept. 4: Moore from Waterways W.C. (Western Canada Airways). Two Air Force planes from Fitzgerald.

On Sept. 9 the last party left Hunter Bay with Rogers, J.P., Spence and Wheeler and full load at 7 PM. They landed at Rae, Resolution and Fort Smith and had lunch at the Mackenzie Hotel in Fort Smith. J.P. then flew from Smith to Fitzgerald with McDonough and had a narrow escape, the plane crashing taking off from Fitzgerald. Arrived Waterways at 7 PM.

Sept. 10: Went to McMurray in AM...

Sept. 11: Left Waterways for Edmonton 2:30 PM.

Sept. 12: ...Got clothes from Royal George Hotel. Left Edmonton for Toronto 5:05 PM.

Margaret Norrie McCain recalls the deep distress her father felt at his role in the discovery of uranium deposits, which later provided fuel for the atomic bomb:

On August 6, 1945, I was sitting alone with my father in our living room, quietly playing on the floor beside his chair. He was listening intently to the news report of the horrendous destruction which resulted when the world's first atomic bomb was dropped on Hiroshima. Speaking out loud, more to himself than to me, he said, "I discovered the main ingredient that went into that bomb!"

Intuitively I knew, though I said not a word, that my father was deeply distressed by the magnitude of the human catastrophe and his part in it. For some reason, perhaps because I sensed his distress, I did not repeat what he had said to me. Two months later, my father died. In the years following his death, two members of the Dominion Explorers expedition to Great Bear Lake came to visit my mother and they told her about the uranium discovery. Only then did I repeat the statement made to me by my father on Aug. 6, 1945.

Letter from Sammie Boadway, Ed Boadway's widow, to Margaret Norrie McCain

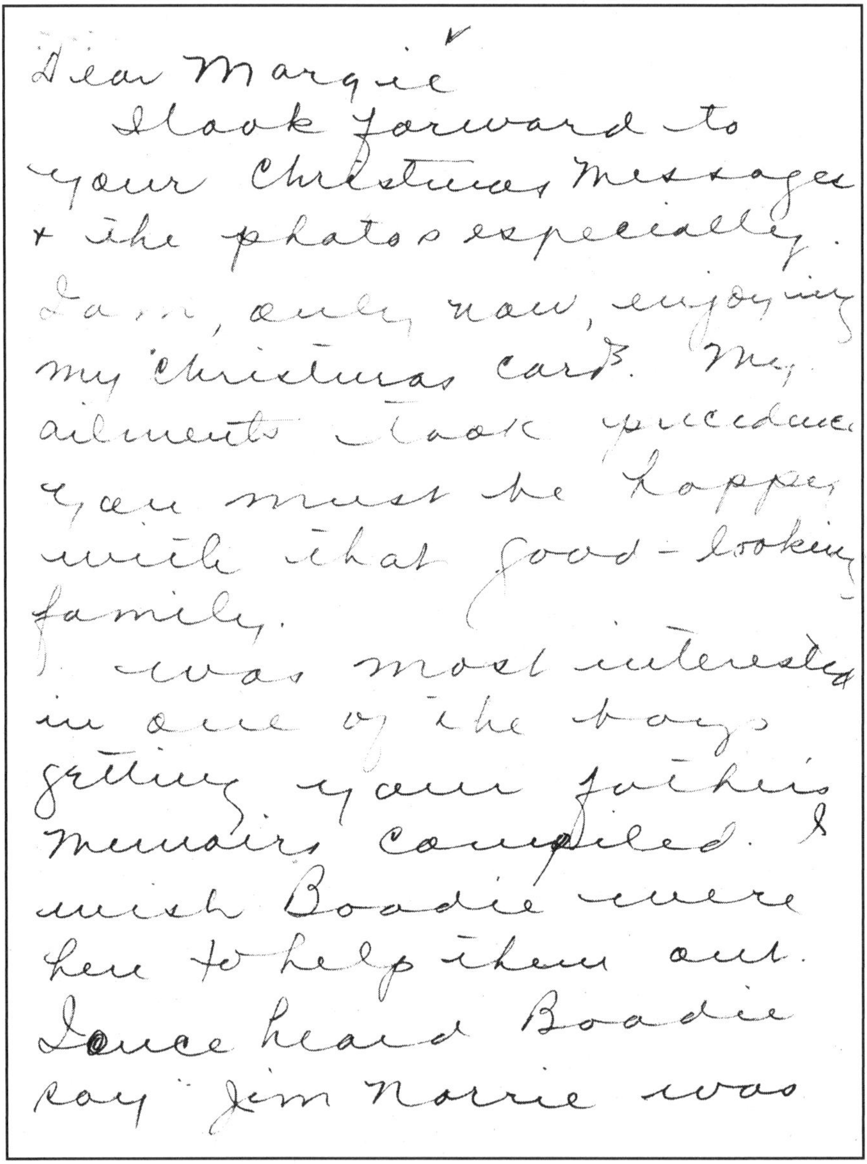

Dear Margie
I look forward to
your Christmas Messages
& the photos especially.
I am, only now, enjoying
my Christmas cards. My
ailments took precedence
You must be happy
with that good-looking
family.
I was most interested
in one of the boys
getting your father's
memoirs compiled. I
wish Boadie were
here to help them out.
I once heard Boadie
say "Jim Norrie was

an honest mining
engineer!" I think
that was a great
compliment because
there were many dis-
honest ones.

The name Rupert Bain
may not be in your
memory. He was a
"successful?" mining man.
He had been the Boy Scout
appointed to represent the
Scouts at Queen Victoria's
Jubilee! Boodie worked
for him a short while
under a written agreement
that B would never be
called on to "tidy up" one
of his reports. The day
came when Mr Bain

asked him to join him &
a Pittsburg contingent at
lunch at the King Edward.
B. declined & Mr Bain wanted
him to go & describe the "Wa Wa
prospect (in west Ontario).
B said "you have my
report Mr Bain & the property
is worthless". Walked
out. Returned the next morning
with his written resignation.
Mr Bain said "I know
what you are going to say
Boadie" & then proceeded
to apologize & beg him to
stay. Having failed that, he
had one more request -
that he not work for Jim
Norrie, which was exactly
what he did! Went right to

work for your father. I do wish B had left some written memoirs. I do remember his telling about one of the prospectors or at least, a field hand Boadie seemed to use _ "Spud Arsenault" & he came from nova scotia. One night he & Boadie were in their eider downs out on the frozen snow. Your dad was in yellow Knife that night - B. heard a noise & a bear had Spud's entire head in his mouth B. threw caution away & managed to pull the bear jaws apart & free Spud. He flew him to Edmonton & he lived, pretty scarred, for a long time. Died on the West Coast I believe with your Dad taking care of him always.

I hope you can read this. My hands don't always obey.

Win Johnston came in late September having decided that what I needed was a trip to Nova Scotia. She drove very sensibly, and I had a happy vacation going in & out of the coves. I wish I had realized that one of your brothers was still on the farm. We could have visited him. Was he the "plump" one?

Most fondly
Sammie B.

Albert Trueman, once president of the University of Manitoba, later of the University of New Brunsiwick, married an aunt of Margaret (Fawcett) Norrie. When he had published an article giving credit to Gilbert Labine for the discovery of uranium at Great Bear Lake, Margaret Norrie wrote the following letter to him but never mailed it.

Truro N.S.
RR # 6
Ap 1/63

Dear Albert:-

You, no doubt, are surprised to hear from me. I have a story to tell contradicting the story about Gilbert LaBine's first sentence, on page 2 in the Star Weekly magazine, March 9, 1963, Namely: "Gilbert LaBine became a mining legend when he made (the first find of pitchblende) in North America".

This may sound very "far fetched" to you but my husband, James P Norrie gave these samples of pitchblende to Gilbert LaBine & recommended to him that it be sent to Ottawa to be assayed because he was sure it was a find of pitchblende. Gilbert never saw the formation or disregarded it. My husband also said if it proves as I say let ~~him~~ me have some of the credit. It was pitchblende but Gilbert LaBine never gave Jim any credit, by telling the truth or in money.

2.

It seems to me you, ~~on~~ the National Film Board, had a film on northern Canada mentioning this fact about La Bine.

For history's sake I feel Jim should be given credit for showing La Bine where it was.

This is telling the tale really backwards.

Around 1930 Jim Norrie headed an expedition called Dominion Explorers. & Three or four planes went to Great Bear Lake to prospect, I believe one side of the Lake. They spent 18 months there & Gilbert La Bine was on the other side of the Lake. Jim wandered further afield into Gilbert's area & noticed this strange showing. Three men hovered over a camp fire one night with samples & decided this was surely pitchblende. As Jim was in charge of the expedition he decided to show ~~it~~ the find to Gilbert.

These three men could have said nothing & returned to stake claims

3.

later but they did not & told Gilbert about it expecting by Gilberts promise to get some credit for it.

Gilbert La Bine never said a word. He took all the honour & still has.

There is only one man left (one of the three) who was at the scene - a mining engineer, pilot and a wonderful chap Lieut Col Edward Boadway, Brighton, Thetford Mines, Quebec.

History is history & facts are facts.

Should this be left alone & let Gilbert La Bine carry off all the honor when it is not true. I've heard the story so many times & from many people.

I'd like Ed Boadway to tell you the story too because I may have some points wrong.

Kindest regards to all the family.

Sincerely

Margaret Norris (Mrs J.P.)
(nee Margaret Fawcett)

P.S. I feel very bold even writing about it but — I must.

Chapter 6

J.P. Returns to Quebec

Treadwell Yukon Gold Mines 1931 (Bussiers)

by George Norrie

In the fall of 1930 J.P. returned from his Arctic adventure at Great Bear Lake to the mining area he knew best. He had staked claims south of Amos with Lee Coffin and others and generally knew most prospectors and mining men in this part of the country. Georges Bussiers and Charles Massicotte were two of these prospectors. They had staked properties near the Perron Highgrade Mine, south of Senneterre, Que. And J.P. knew just what to do with them.

Joe Errington was founder and general superintendent of the Errington iron mine owned by Treadwell Yukon. He had previously been associated with General Hogarth, Thayer Lindsley and J.P. at Malartic Gold Mines, owned by Ventures, in 1928.

Now that Errington was General Manager of Treadwell Yukon, J.P. approached them with the Bussiers claims and made a deal. He also took the job of mine manager to open up the Bussiers mine under Victor Clauson, who was superintendent of Bussiers and J.P.'s immediate boss.

Prospectors in the early days of mining were independent operators. They either financed their own explorations or were grub-staked by a mining company or mining men like J.P.

When they found something promising they staked their claims. However, this was only the first step; they then needed to make a deal with some company to develop the claims. Most prospectors did not have the ability to do this and J.P. was the man many of them turned to for help in northwestern Quebec.

When a deal was reached with a company such as Treadwell, a subsidiary company such as Bussiers was set up and the prospector's claims would be the company assets. This was usually a 3,000,000 share public company.

The prospectors were the ones who incurred the initial exploration costs and were called vendors. When the claims were sold the vendors received a small amount of cash, usually about $3,000, and 300,000 shares in the new company. This was split in various amounts among the vendors.

At least 1,000,000 shares were retained in the company treasury and 1,000,000 could be made available to the public or held in the treasury. The vendors received 300,000 and the remaining 700,000 shares could be used for the stock brokers to sell to their favourite clients and the general public.

The Treadwell Yukon was owned by the Bradley family of San Francisco: Fred and his son Phil. The family was very wealthy because of their vast mining interests in North America: their Nevada and California Division, the Alaska and Northern British Columbia Division, the Wernicke Mines in the Yukon Territory and the Errington Mines in northeastern Ontario.

When they took over Bussiers, Treadwell committed to further explore, diamond drill and sink a shaft to explore underground. If gold ore bodies were found, they would naturally produce gold, and all shares (vendors', public and treasury) would become valuable. Joe Errington also made a promise to the vendors that he would attempt to make a public market (distribution) for the shares on the Toronto or Montreal stock exchanges. This would prove very difficult, primarily because of the low price of gold and the Depression, which was at its peak.

The Bussiers was approximately 25 miles south of Senneterre, not far from the village of Pascalis. A road passed through these bush villages but could only be used at certain times of the year and under good weather conditions. The roads were useless during freeze-up (Nov – Dec) and break-up (April – May). Delivery of supplies had to work around these unpredictable periods.

The water routes supplied most materials and machinery, but they were useless as well during the periods of freeze-up and thaw. The CNR passed through Senneterre but the connector to Noranda via Val d'Or and

Malartic would come later with more mines. Teams of horses were the most reliable means of transporting the goods over land, summer or winter.

The first surface work, of course, was carried out by the stakers or vendors, Bussiers and Massicotte. But now that Treadwell Yukon had taken over the surface exploration, a diamond drill program was begun under the direction of J.P. and Vic Clauson. The result of this exploration was a decision to sink an exploration shaft and to construct a mill to recover the gold from the ore. These were completed in September, 1932. The shaft was located on a hill where all their mine buildings were constructed. The camp to house the men and their families was a short distance away with fire breaks cut to protect the townsite.

The mill began operation in September, 1932. The only ore available came from four crosscuts that were part of the mine development program. No ore was "in sight" which meant a steady supply of gold ore could not be provided to the mill until a stope or gold ore body could be opened up or developed in the mine. The mill would have to depend on any gold recovery from the drifts and crosscuts. Return on the money spent to date would have to depend on this unpredictable supply of ore.

The development and financial circumstances at the time of startup of the mill were explained in a letter to Vic Clauson from Fred Bradley, the owner:

> In view of the unexpected, disappointing and discouraging results underground, all future expenditures must be confined to a simple underground prospecting program, and all work or expenditures not directly connected with the program must stop. The single exception however can be the completion of the flotation unit and its accessory equipment in the mill. This will leave the mill prepared to make a recovery from ore that may be found.
>
> Some construction work such as the refinery and the heating boiler installations should be stopped at some good place. i.e. nothing should be left in such shape that it would be destroyed or wasted through exposure or theft. No equipment should be ordered or installed.
>
> Norrie should write a 5 day letter (in addition to that already sent) describing in language what

has been accomplished during the period. This should be written regularly and promptly so the information that reaches my hands will be current and up to date. The letter can be addressed to you and a carbon copy sent to me.

In general see that money is spent for two things 1) hunting for ore 2) milling at a profit such ore that is mined in the course of development.

Obstacles were obviously many and difficult. However, each month little by little the gold recovery from the supply of ore increased to the point where in May, 1933 the revenue was able to just about cover the operating expenses. The future of the mine seemed more secure as more stopes and veins were opened up insuring a higher grade ore for the mill to process.

The mine steadily progressed through the summer and in September this newspaper article gave the public the details of the mine's progress:

An achievement, 'From Bush to Bullion', in an incredibly short period of time can be particularly applicable to the Treadwell Yukon. They have shown what can be done in Quebec by the way of opening up a property. This performance has not been duplicated in the mining north, not only in speed of opening up the property but also in recovery.

The Bussiers operation to date has been an achievement well worth special mention. The Treadwell people have done a man sized job in a manner reflecting great credit on them. This can readily be understood when one considers that the mill was started only 8 months ago with just 4 crosscuts. As the first rock from the drifts and crosscuts went to the mill, and ore development was started in different places, better grade ore was treated and mill heads gradually increased month by month. Obstacles were many, but as stopes were opened up and mill heads improved, production mounted likewise.

It is of interest to note that first impressions of the Bussiers led to the belief that most veins were small ore shoots, and flat lying. However they have just opened up 2 veins that strikingly resemble

Siscoe in structure and ore conditions. One of these veins has a dip of 20 degrees, which naturally afforded the operators some concern. The problem was successfully overcome when scrapers to remove the ore were put into practice. The other vein dips at 45 degrees and the ore runs out itself.

In conjunction with underground operations, surface exploration is going forward on some interesting veins. This material is taken directly to the mill and has been running better than ten dollars per ton. One exposure that outcropped for 10 feet has now been extended for 125 feet and making ore all the way.

Current production is around $30,000 per month, while current expenses are around $24,000. This is indeed exceptional when it is noted that for the last 2 months plenty of construction work has been done, adding to expenses. However, recovery from the mill has more than covered operating expenses and it can also be said that costs per ton are relatively low.

The operation is now complete with a 175 ton per day mill and it is quite probable that an increase is in sight. 160 men are employed on 2 shifts. A town site has been cleared for ½ mile in all directions from the shaft and 30 houses have been erected for the miners and their families. Enthusiasm for the future of the mine is evident with all employees and management.

The responsibility for the development and operation of a marginal gold mine was tremendous, especially during the Depression. As long as the mine broke even, it would stay open. It was a month by month decision. The Bradleys made it clear they would not put any more into the mine. There were 160 men and 30 families depending on this for their livelihood. The alternative was welfare and family handouts. J.P. carried this burden for the men as well as the security of his own family. All J.P.'s wealth was also at stake in Bussiers as all the vendors' shares in the mine would be worthless if the mine shut down.

J.P. by this time had many friends and contacts, such as prospectors and mine managers. He had brought some of these contacts together in the Bussiers as we have seen.

While he was employed at Bussiers he was constantly searching for more opportunities like the Bussiers claims which he had promoted to Treadwell Yukon. While at Malartic Gold Mines in 1928 he and Lee Coffin had staked some claims just east of the mine and in 1931 most claims around there had come open, and they added some of these to their stake for a total of 1,700 acres in Fournière township (future East Malartic Gold Mine). They also staked a block of claims north of the Bussiers claims and adjacent to the Perron (future Senore Gold Mines), which totalled 1,200 acres. In December, 1931, the Coffin Mining Co. was formed to develop these two blocks of claims and a deal was made whereby the Treadwell Yukon agreed to develop these properties as well.

J.P. had put the Malartic Gold Mine on the map, but with the price of gold at $20 it could not survive. With the price of gold in 1932 at $32 per oz, everything looked at least twice as bright. The Malartic Gold Mine was reopened as Canadian Malartic, and Treadwell now had two excellent properties under Coffin Mining to develop, one of which was directly adjacent to Canadian Malartic.

These two properties had tremendous potential but only a limited amount had been spent trenching on the claims. Any profits from Bussiers could be used to develop all the properties together.

As difficult as the management of the mine operation itself was, J.P. spent all his free time trying to build up a market for the Bussiers shares. Now that the mine was showing better potential, it was time to sell Bussiers shares if possible. This would supply money for the treasury to develop the mine, and his own shares would appreciate which he needed badly as well.

Remember Joe Errington had promised the vendors to make a market for the shares, but with the price of gold at $20, the Depression in full swing and the Bussiers in development stages, the market was not demanding Bussiers stock. Now that was all changing and J.P. wrote Errington to remind him:

The vendors of the property insist that you promised them that the Bussiers stock would be listed, and I see no reason why it should not be. I trust you will see it is listed early.

Conditions here are looking better every day, and I feel sure the proposition will work into one of quite some proportions. Having to feed the mill [with high grade ore] holds up the development [for other ore bodies]. However, I think we can keep it going, pay expenses and make a mine.

I am glad you are willing to trade 5,000 non pooled shares of Little Long Lac for 5,000 shares of Bussiers. You see everything I have in the world is tied up in this venture. I played the same way on Malartic and lost, and this time am anxious to split my bet, as I have a family to support which is gradually becoming more expensive. I am sending 5,000 shares to Mr. Morlock to have stock made in your name. You can send me the 5,000 shares at your convenience. I trust you will come and see me soon, when I am sure you will be surprised with the picture.

J.P. was almost broke again. Errington had not fulfilled his promise to the vendors so J.P. began to contact all his own associates. He needed a broker to help promote the shares as well.

A bonus for the broker selling the company stock usually came out of the allotment of 700,000 shares and could be as high as 300,000 shares depending on how much stock they sold.

The job of the brokers was to sell as much stock as possible to as many investors as possible. This was called making a market or distribution. When the required number of investors was reached, the stock exchange, such as the Standard Stock Exchange in Toronto, would list the stock. Doherty Roadhouse, Montreal and F. Crawford & Co., Toronto were the brokers for the Bussiers stock.

The vendors' shares usually were held in escrow or pooled for one year or more. This meant the vendors could not sell their stock or dump it on the market at a price below what the broker was trying to get. The shares were usually held by a trust company. Oddly enough, Treadwell Yukon did not demanded an escrow arrangement with J.P. and the other vendors. This was unusual and allowed the vendors to sell at any time, which was a real problem for Doug Wright of Doherty Roadhouse.

The only way these free shares could be controlled was through the efforts of J.P. While Wright was trying to make a market, any block offered below the asking price would destroy the broker's efforts. This was inevitable.

J.P. replied to Wright:

> I cannot understand who is offering the stock you refer to. I am absolutely sure none of the holders outlined in the list I sent you are offering any, and all loose stock is covered in it. i.e. full 10% of the capitalization. I am absolutely sure no stock can be delivered. Your letter is being sent today to the other holders of the stock with my advice to follow your suggestions. I will advise you further when I hear from them.

It could be that vendors and associates of J.P. could not resist selling the Bussiers shares by themselves, or it may have been a fabrication of the broker, an excuse for his inability to sell the stock.

After several months J.P. was disappointed in the results from Doherty Roadhouse and in May, 1933, contacted Crawford & Co., Toronto who had been recommended by the Treadwell executive to make a distribution of the stock. By August J.P. was angered when he was told of Fred Crawford's efforts:

> Recently Mr. V.C. Clauson, one of my associates, was advised by officials in your office, that due to the rush of business on the big board, sale of our shares had been forgotten, and that to date no sales had been made.
>
> At the time of offering these shares to you, we had offers from other brokers for distribution. However we had faith in your house and thus gave you the option. Frankly we are disappointed with the treatment we have received at your hands to date, and would be glad to have you make an early effort to distribute the shares so that the stock can be listed early on the Standard Exchange. The mine is

looking better each month, and we all feel sure that shortly another substantial producer will be added to Canada's profitable gold mines.

The vendors in the Bussiers deal received very little cash and have been depending on the market value of the shares to recompense them for the property. We would be glad to have you make an early concentrated effort to distribute the shares. If we can be of aid to you please advise us.

He also offered stock to all other brokers he knew in Montreal, Toronto and New York. The times were depressed and it was almost impossible to sell stock unless the mine was a sure bet. Bussiers was just marginal at that time and J.P. needed money to be able to develop his own properties. His own properties at that time were the Tibelmont Island claims and Malartic Goldfields claims.

The East Malartic and Senore claims were in the Coffin Mining Company, which Bussiers were committed to develop. As his efforts to develop both the Bussiers mine and stock continued, a severe blow to the Bussiers operation occurred with the sudden death of Fred Bradley in August of 1933. It became clear that Bradley's wife, who inherited the estate, was going to dispose of all Canadian holdings immediately, including the Coffin Mining Co.

The Coffin Mining Co., which had been taken up by Treadwell, was now taken back by J.P., Lee Coffin and Charles Massicotte at a time when they didn't have any money, even for their basic daily needs. They were not prepared for this, and money to hold these properties, even for basic claim requirements, would be hard to find.

The Bussiers itself was also offered for sale. Quebec Gold Mining Corporation, who already held a substantial block of shares, took Mrs. Bradley's stock and control of the mine. It would be several months before this takeover would show its effect on the mine.

As J.P. told Joe Errington, everything he had in the world was tied up in the Bussiers and he needed to sell his shares to support his family. He also wanted to sell his Bussiers shares to be able to hold his other interests, and now the Coffin Mining Co. was added to that burden.

Tiblemont

The Tiblemont Island Mining Co. would come to J.P.'s rescue just when he needed it most.

This property was on an island in Lake Tiblemont just east of Bussiers, and along with his other claims had been costing J.P. money for some time. (This property was tangled up with legalities and claim jumpers and will be explained later on.)

Tiblemont also brought J.P. into contact with Ron Macaulay. Cominco was interested in an option on the Tiblemont claims and Macaulay was their representative and negotiator. The negotiations went on during the summer of 1933 and finally J.P. got fed up with Cominco, which was arguing over many of the vendors' demands. He sold the options on the Tiblemont claims to an American group who accepted all the vendor demands. This upset Cominco, and particularly Ron Macaulay, who had felt sure that J.P. would option the claims to them.

Just after J.P. signed the deal with the American group, Quebec Gold Mining took over operation of Bussiers. Ron Macaulay became their vice president in charge of the Bussiers. Of course he became J.P.'s boss and naturally this would become a problem. This became clear after January 1, 1934, when Macaulay wrote J.P.:

It is the beginning of a new year and the change of management has now been in effect for two months. During this period I have been content to allow things to shape themselves, but it seems to me now that we should have some distinct understanding. This mine is crying out for management at the present moment and while I am not afraid of the outlook, I am quite worried about some of your viewpoints. I have nothing to say about some of your viewpoints. I have nothing to say about previous operations and no criticism whatever to offer because, starting with no information, mistakes were bound to happen.

The outstanding thing at the moment is that the development work has not been so planned as to give the maximum information. Attempt has been made to maintain production beyond the capacity of the stopes opened up, and we have not the direct supervision underground for economical work here.

The effects from this are that some ore has been mined and fed to the mill that was too low grade, and this has brought the average grade of the mill feed below the economical limit. It is quite clear to me that there must be, at all times, a big excess, perhaps 50% more working faces opened up than are being mined. Otherwise, when any section gets too low a grade, it will be mined just the same in order to keep the mill going.

The remedy for these things will probably take some time to perfect, but the immediate plan to work on is, first to pick out the most likely area and do more intensive development work there. But in addition to this, we must have real sound practical bosses that will insure return from labour and guard against using unnecessary supplies, and I might add, wasting air and all other economies.

I expect your plan of operation will anticipate these improvements.

Macaulay didn't let up. J.P. was asked to explain his activities regarding the marketing of the Bussiers' shares and Macaulay was very dissatisfied with the answer and accused J.P. of not stating the facts. J.P. answered Macaulay in no uncertain terms. He definitely was not taking this from Macaulay:

As a matter of fact my statement in my letter is an absolute fact and can be easily proven and most of those in your letter of the 17 are absolutely without foundation. I have no object in misleading anyone in regard to this stock. Thus I think your information should have been more substantiated before the remarks in your letter were written and I was instrumental in making the deal with Treadwell Yukon and, although I do not remember it, Bussiers and Massicotte contend that a verbal promise was made to them at the time that an effort would be made to list the stock and both Mr. Clauson and myself tried to have this fulfilled. Your statements re: F.W. Bradley, P.R. Bradley and Mr. Morlock are correct down to the approval of Mr. Morlock and he did approve but no action was taken. So with Mr. Clauson's approval, in fact he remained at the mine while I went to Toronto to investigate the situation, I first went to Mr. Errington and he advised me that there was no objection to listing but advised me to see Mr. Morlock which I did. Mr. Morlock said there was no objection to listing other than one raised by the stock exchange and advised me to see Fred Crawford president of the Standard Stock Exchange.

Mr. Crawford advised me that the only objection to listing was with limited number of shareholders and agreed to distribute 100,000 shares providing the other vendors were willing. The necessary arrangements were made, however, Crawford did not do anything with the stock. Arrangements were later made with D.G.H. Wright of Doherty Roadhouse to do so. Anyway before Doherty Roadhouse would proceed with distribution they wanted all stock placed in escrow which was done except a few thousand shares which could not be located.

The whole situation is not at my instigation but in accord with the wishes of all vendors. I was appointed to make arrangements at all times. This stock of course is cheap but did cost me considerable. I spent many months with quite heavy personal expenses making the deal with Treadwell Yukon and previously with others without results. All arrangements were made before Quebec Mining Corpn took over the property.

Macaulay was satisfied with this explanation. However, he didn't let up on J.P. concerning the operations of the mine.

As mentioned earlier J.P. had signed a good deal with the Armstrong group from New York. The deal had come just in time as J.P. was just about broke. The deal was in fact a sweetheart deal which will be explained in the next section.

J.P. didn't need Macaulay's bickering and when Alex Perron came along with another good offer, including the Nesbitt Thomson financial house, it fit J.P.'s personal agenda perfectly.

Chapter 7

Life in a Mining Camp - Summer 1933

by Harry Norrie

J.P. Norrie was a unique mixture of good humour and physical strength. His brother Joe boasted of his own strength but workmen would say J.P. was much stronger but never intentionally displayed it. One guy said, "He could lift anything loose on one end." When you are charged with bringing a northern mining property from tree roots to production, you need these characteristics. It is written somewhere that those who look back will die from remorse. J.P. had little for which to be remorseful.

During two years of development at Treadwell Yukon a small village of log cabins was erected by workmen to house their families. A short distance from the village, at the shaft site, a large building served as a warehouse and mine office. Through some official sanction, employees' wives would go every Friday afternoon to a window at the mine office to pick up their husbands' pay.

The summer I turned thirteen, Dad gave me my first job: warehouseman rated at .25 per hour, to pass out nuts, bolts and screws to construction workers in need of supplies. From my post in the warehouse, I could see Dad through an opening in the wall when he was at his desk in the office. One day, in the middle of the afternoon, he got up and went outside. It immediately struck me that he was off to investigate some part of the mine that needed his attention, leaving me free to abandon my post, go out and get in the way of some workmen while keeping my eye on anyone approaching the warehouse.

I opened the door to find that Dad had simply stepped out and was sitting on the office steps with his head down in deep concentration. Too late now to turn back! I thought I could quietly shadow past him, duck around the corner and be away from his concern. Suddenly at full arms' length he reached out, grabbed the collar of my shirt and ripped it full away. With another yank he removed my pants and underwear. In the beat of a butterfly's wings I was newborn bare naked, standing in nothing but ragged running shoes.

For a century I hopped up and down, clenching both hands over modest evidence of things to come, otherwise fully exposed, not twenty feet away, to the laughing, lascivious eyes of a long line of workmen's wives there to pick up their husbands' wages.

I have often wondered why Dad's fun fountain suddenly erupted just at that moment. I don't know how long he continued sitting there, reverted to his thoughts, but I bolted back into the building to a pair of oversized pants offered by Louis Rocque Morceau, a friendly office worker.

If it puzzles you about what got into Dad to do such a thing, think what it has done to my fragile mental processes over a period of seventy years. You will not need to ask how easily I remember that fractured Friday afternoon.

I don't think J.P. got a lot of fun out of this example of his humour but the ladies certainly did.

On an evening at the end of June, 1933, Virginia and I were on our way to northern Quebec, leaving the farm for the last time. Pauline had completed her year at Mt. Allison and had gone to northern Quebec in May. Jim was attending Royal Military College in Kingston, Ontario.

Virginia and I had made this trip several times before, leaving little of interest along a familiar route. Just for something to do, we decided to get off the train in Sackville to see a little of the town where Pauline had spent her school year, but by the time the train reached Sackville it was already dark, so we decided not to bother and stayed on the train.

The train was still at the station, with Virginia sitting in one seat and I in the opposite, when suddenly a voice at my right shoulder said, "Are you the Norrie children?" Virginia, facing the voice said, "Yes, we are." The voice said, "I'm Margaret Fawcett and I'm going to northern Quebec with you. Here, I have a telegram from your sister Pauline." She handed the paper to Virginia and sat down to join us. I found the remainder of the trip and the rest of my life a lot more interesting.

Dad had come to the conclusion that Pauline, nineteen years old, in the bush with about two hundred guys, would be lonely for female company. I am not in a position to challenge Dad's conclusion on this but I can't see where loneliness was anywhere involved. Under normal conditions she probably would have fought off any female that showed her head over the horizon. But anyway, when she was in the mine office with Dad one evening, he suggested that she get on the phone and try to coax one of her girlfriends at Mt. A. to come and be with her for the summer, rather than just sit around with a mine full of lusty workmen gazing at her, out of reach. The call was made; arrangements were made, but Virginia and I were not told what to expect when we arrived at Sackville.

The day following our meeting with Marg I was sitting with her and awkwardly asked what she worked at. She said that she was a school teacher, and at Mount A. had been an instructor to Pauline. With respect for all teachers having been thumped into me, I started calling her Miss Fawcett even though she insisted that I just call her Marg. Who would have guessed that in a few months I would be calling her mother, never having remembered another?

We arrived in Levis the next morning, walked to the ferry and crossed over to Quebec City, to spend the day on the boardwalk in front of the Chateau Frontenac, sitting in one or another park and going to a movie before boarding the train in lower Quebec at eight o'clock that evening.

We were met at the station in Amos by Dad and Pauline with pleasant gestures but no great fanfare. Dad was introduced to Marg. I very much regret not ever having discussed with Marg what her impression must have been to arrive at that frontier town of semi-civilization.

We left Amos by car for Lac la Motte and from there by "pointer" to the boat landing at Treadwell.

After that, about halfway to the mine site on the "Muskeg Limited", we were stopped by a forest fire burning on both sides of the track. J.P. stopped to appraise the situation but quickly decided to go on through. I think he found that since the track ties were not burning it was probably safe to keep going. Jim and I and the assistant mine manager were not so sure. We got off and walked around the fire to meet the others on the far side. Here again one would wonder what Marg must have thought about how she had gotten herself into this mess.

The fire was not an accident. It had been set on purpose. Four Finnish workmen had been hired to cut four foot lengths of wood and pile it beside the tracks. From there it would be taken to the mine to fire the boilers during the coming winter. But they had come to the end of their work and would soon be laid off. There was only one thing to do, make it necessary to cut more wood. They were fired anyway. But the Finnish spirit is seldom defeated. These four built a shack in the bush and with a few potatoes and two thousand rabbits, survived the winter.

The summer months were drab for the most of us, but for Marg, in such an unusual environment, it must have been quite exciting, especially as time went on and a formidable masculine presence was starting to pay attention to her many assets.

Early on she had made it known that she would have to be back in Sackville at a certain date to attend the wedding of a girl friend, but by then the decision had been made, she would be coming back.

In the north country during summer months it is quite common to see what is called heat lightning in the clouded western sky. But with the lightning there is no rain and little if any sound, just a bluish flash of light. One evening, in the last light of the day, I happened to look out a side window of the manager's house. (It was never called home, just the manager's house.) Marg was usually in the company of Pauline or Virginia or both. This evening, with a background flash of lightning, I noticed Marg and Dad walking very close together on the road back from the mine office. Right then it occurred to me that the wheels had started to mesh!

The first power at Treadwell was supplied by wood-fired boilers but when a power source was found at Rapid Two on the headwaters of the Ottawa River, a dam and power plant were built and power brought eleven miles through the bush to Treadwell.

The mine buildings: engineering office, warehouse, hoistroom, assay office, etc. were all located, of necessity, close to the shaft and mill. The office staff lived in rooms above the main office. Dr. Eugene Audette had a small room there also, for his dentistry, but would later get a small building for himself. All the other buildings: bunkhouse, cookery, store and manager's house were all located in a rough square about a quarter of a mile from the mine workings. These buildings were widely spaced for protection in the event of a forest fire, even though a wide firebreak had been cut around the entire property.

Mr. Leger, a diminutive Frenchman, ran the store, but other fingers were probably also in the pot. Some of his customers took exception to his high prices but he maintained that he only made one per cent profit; if he bought his stuff for one dollar, he sold it for two dollars: one per cent.

In the construction of the mill a great deal of blasting had to be done to provide the three levels for the machinery. The night shift drillers would blast the results of their work at seven o'clock in the morning. With the office staff living close by in the office building, the blasting was a wake up call. One morning Charlie Brazel was lying in bed counting the blasts as they detonated. One loud bang. And in a few seconds a very large chunk of rock crashed into the rafters just over Charlie's head and quivered there with nothing but the hand of God and a small sliver of wood holding it - with the rest of Charlie's lifeline in question. Charlie moved.

During construction of the mill, there was much concern in the family about J.P.'s health; he worked day and night. At one point, in the middle of the night a family member went looking for him. They found him at the mill pushing a cement wagon.

Extracting the Gold

The ore from underground was raised to a large bin in the headframe. From there it went by upward inclined conveyor belt to a bin in the mill and from there to what was called the picking belt where chunks of rock that were obviously waste material that had been blasted from the walls of the stopes and did not make any ore were removed by hand.

The rock went from there to the primary and secondary crushing units. The secondary crusher was of particular interest. It consisted of two wheels about ten feet in diameter with a wide flange on the outer rim of each wheel. The wheels were placed with the flanges facing each other and closely spaced, and the rocks from the primary crusher fell between the flanges. The spacing between the flanges was determined by what is called the angle of nip.

From the crushing units the ore was conveyed to two ball mills. These were in the shape of a drum about five feet in diameter, slightly inclined and partially filled with five-inch diameter steel balls for fine grinding. The driving mechanism was a geared wheel fitted to the top end of the drum. The fine material was gradually washed through a small opening at the end of the drum.

From the ball mill the ore went to the classifiers where the fine particles were carried away to the blanket tables and the coarse ore returned to the ball mill for further grinding. The blanket tables are of particular interest because of their purpose and because that's where I worked. Three abreast, the tables were about four feet wide by ten feet long raised by about twelve inches at the top end and with a low strip along each side to direct the flow of water.

The tables were overlaid with a particular type of blanket of corduroy design with bristles pointed against the direction of flow. They were secured to the tables by wood strips along the sides. The light, barren rock particles washed away to the dump outside while the gold settled against the corduroy ridges. Every few hours the blankets were picked up and washed out in the amalgam barrel. When the blankets became worn they were thrown out. The amalgam barrel was about four feet long by two feet in diameter and partially filled with two-inch diameter steel balls. After a period of build up from the blanket tables, the barrel was charged with a measure of mercury. It was then rotated for eight hours allowing the mercury to pick up and unite with the gold, forming amalgam.

After the required grinding period the amalgam barrel was emptied onto the amalgam pan. This was a slightly concave steel dish about five feet in diameter, that was agitated by a concentric wheel that forced the material into the centre of the pan where a few small steel balls finished the amalgamation process. Any remaining waste material was washed away by hand with water from a short length of garden hose.

(Roy McKay, who had taken a year off from his studies at Mt. A., was the pan operator while I was his flunkie.)

At the end of the process, the amalgam, along with the free mercury, was returned to the flask to be used again.

In the mining and milling process samples of ore are taken at various stages and assayed for gold content. This is to evaluate the amount of gold going into the mill and the amount coming out the other end. If there is a wide variation, someone is in trouble. The average of the amounts going into the mill is called the heads and the waste going out the tails or tailings. In an ideal situation the tails would assay nil but that never happens. Not all gold is recovered with the milling, and some gold is lost in the mill, but there is a relative balance. A question often asked is "How are the heads?"

One bright August day, Murray Fullerton, who worked in the mill, returning from the outhouse, happened to notice a pile of worn-out blankets lying on the waste dump. An inquisitive little bug entered his head at eye level and wandered around looking for a home. Murray found an old carbide can (carbide was used in the miners' lamps before the electric ones were invented) gathered up all those old blankets, now all completely dry, and burned them. The ash from the blankets he took back into the mill and threw into the milling circuit. The following day when the heads were tested and were far higher than expected, the immediate question was where the hell did all that extra gold come from? The pointed finger eventually rested on Murray with "He did it." No one had thought that a lot of gold would be held by the weave of the fabric.

Two views of Treadwell Yukon gold mine

Amalgam pan on gold processing line

Picking belt on gold processing line

The Story of Pascalis Township

If a man is very thirsty he will walk many miles for a drink, and if the drink he's looking for is alcohol he'll walk faster. It was known throughout the north where the name Jim Norrie was recognized that booze of any kind would not be allowed on any property that he managed. In the early stages at Treadwell none of the miners had their wives with them, a condition that existed at most of the other camps in the area also, with the inevitable result that a little town called Pascalis was built up for the sole purpose of supplying forbidden fruits to base men: those in need of a drink and the company of women of equal desires. It was ten miles from Treadwell to Pascalis but a well worn path was soon tramped out to the town, and dragged out on the way back. J.P. even went to the extent of hiring a uniformed policeman to intercept any drunks returning over a little bridge that led to the Treadwell property, and to confiscate any bottles that might tempt other workers. The policeman's uniform was his only connection to any authorized police agency.

One weekend, two of the workmen at Treadwell decided to find a little relaxation at Pascalis. They soon found that their money had run out, and rather than return to work they decided to rob what was advertised as a jewelry store but was, in fact, a front for liquid commerce of illegal nature.

The police were summoned, but before the thieves could be apprehended they took off through the bush toward the nearest railhead. The police followed. The thieves came to a lake, found a prospector's canoe and climbed aboard. Slumped down as low as possible in the bottom of the canoe to present as small a target as possible, they paddled away.

The police arrived when the canoe was some distance out on the lake. With no indication that the canoeists were about to return on command, the police fired several warning shots into the air and finally three into the canoe. The three shots hit one bandit. One shot entered under the chin and exited at the top of his head, giving him a mild headache and a six-week stay in the hospital. Further information about this man's activities is unknown; he was not encouraged to return to Treadwell.

J.P.'s Chinese Aristocracy

Somewhere in his early life J.P. came to recognize the superior intelligence of the Chinese people that led him to declare them to be the "aristocracy of the human race" and I have found no persuasive argument against that opinion. When three Chinese men requested employment at Treadwell, he set them up in a small building to establish a laundry. Workmen coming up from underground would enter a room called the dry where soggy clothes were hung on hooks to dry out for the next shift. It might also, with reason, have been called the wet because showers too were provided for the miners.

As with other industries in remote areas, miners' work clothing was worn until it disintegrated and fell off. A story is told about one guy who finally took a bath and found three suits of underwear. Top side clothing was more carefully attended to, so the Chinese laundry was of considerable benefit to them.

Along with the laundry the Chinese eventually took over the cook house and dining room, but with a moment of dissention when, through no fault of their own, the water supply to the kitchen became mildly contaminated causing extended waiting periods at the outhouses. Not long after the effect of the cathartic water was recognized, some bright mind remembered that one of the diamond drill holes at the top of the hill was artesian, overflowing. In other words, an endless supply of crystal clear water was running continuously from the hole. It was a simple matter to lay a pipe to the kitchen and forfend the action of some yet unrecognized shit disturber as well as protect the health of the employees.

In any work force, big or small, there is always at least one shit disturber. He is a little man looking for problems where he can encourage other, weaker men to join him in a revolt. He proclaims himself the leader, and others must follow his direction while he hides away from the final action. Napoleon stood high on a hill watching through field glasses his thousands of men die in the valley below. Hitler never went to war, never saw the inside of Russia.

The situation we had at Treadwell was not nearly so grand but it was explosive. When our little S.D. found himself in a lineup at the outhouse, he called around him his weak-minded cohorts and set up a plan to dynamite the Chinese. Explosives were not hard to come by. One dark night they took a small amount of dynamite and blew away the front steps of the Chinese laundry. I can imagine the chief coward hiding in the bushes while he directed his flunkies with signals. The damage was not extensive, the steps were gone and there was a small hole in the ground, but the Chinese were back in their kitchen in the morning under protection of a very strong man. The S.D. is now probably sitting in a beer parlour somewhere bragging about his glorious moment of rebellion.

Charlie Jinn in his native clothes.

When J.P. left Treadwell the Chinese followed. He took their collected earnings and enlarged the amount by shrewd investment to a point where they were able to enjoy the profits of a small hotel, at the time the only hotel in Malartic. The chief proprietor was Charlie Jinn so the building was called Charlie's Hotel. It would not have been proper to call it the Gin Hotel although that would not have been too misleading.

For a time I shared a room at the hotel with Charlie George (working at Malartic Goldfields). One morning a friend came to pick me up to go to work. He asked of the Chinese clerk, "What room is Harry Norrie in?" "Allie Nollie loom flive, dlunk again."

There were some wonderful parties at Charlie's, night long, usually ending in the morning with everyone in a fist fight. I missed one really good party where a handsome Cape Bretoner was so pounded I failed to recognize him when I met him on the street later.

Forest Fire Scene

The hair on a dog's back: it was on one of our early trips to the north that Dad took Jim and me to see one of our first movies. I can't remember if it was in Montreal, Quebec City or Toronto, but wherever it was it was a rare treat. It would be many years before motion pictures came to Amos, let alone any of the mining districts.

In those days the main feature was usually preceded by a cartoon comedy of some sort: Bugs Bunny, Elmer Fudd, Daffy Duck, whatever, and this was no exception. On this occasion the principal character was a bug. It was not until later in the action that you realized that this bug was really a flea. The story line showed the bug on its hind legs, with a pack on its back, trudging through a landscape of what appeared to be trees, but trees without branches. The bug, obviously tiring, finally throws off its backpack, knocks down a couple of trees, splits some wood and builds a fire in preparation for cooking his dinner.

The scene changes to show a large dog sleeping on a comfortable rug in a well-furnished room. Then a curl of smoke starts rising from the dog's back. The dog's eyes pop open, then, yapping furiously he starts whirling around the room snapping at the offending fire. It was by any measure the funniest cartoon that I have ever seen. And many years later, while staking claims or prospecting for mineralized outcrops, struggling through the north bush with its blanket of tall, thin black spruce, at ground level also almost limbless, and in proportional retrospect much like the hair on the dog's back, my belly would still ripple with the recall of that scene. Somewhere, somehow that clever, artful cartoonist must have led the life of a prospector, not to suggest - well never mind.

For protection against forest fires, at Treadwell a wide fire break had been cut around the entire property. The tree trunks were used to fire the boilers that supplied power for the mine, but the slash (tree limbs, windfalls, underbrush) was left in place and with a strong wind was still dangerous. Eventually a power dam was built at Rapid Two, eleven miles from the mine on the headwaters of the Ottawa River, that greatly reduced the necessity of wood-fired boilers.

One day a thick cloud of smoke built up until it became obvious that a forest fire was headed directly toward the mine. An urgent call went out for all unoccupied hands to turn out at the powder magazine. About five hundred cases of dynamite had to be buried quickly. A large exploration trench was used, with the dynamite covered over by a thick layer of overburden. In the evening the wind changed and the danger passed.

J.P.'s First Gold Brick

My introduction to gold bricks came with two incidents, one at Treadwell and one at Perron. The first occurred when J.P. came to where I was working with Roy McKay at the amalgam pan. This time, with the rattle of machinery in the building, he didn't even say, "Come with me." It was just a motion with a crooked finger that beckoned me. I followed him to the office where the steel box containing the accumulated blocks of amalgam were stored. With a long length of 2 by 4 lumber through the handle we carried the heavy box to the gold refinery.

The refinery was a small building erected just for the purpose of turning the amalgam into a gold brick. There were two furnaces, one a retort to drive off the mercury, the other to heat the gold to a molten state so it could be poured into a mould.

The retort was a metal tube about twelve inches in diameter by two feet long with a small pipe extending from the rear end into a container of water. The open front end was equipped with bolts so that an airtight cover could be fitted in place to prevent any deadly gases from escaping. Semi-circular dishes into which the amalgam was placed were inserted into the tube. This arrangement was bricked in to provide a wood burning fireplace below the tube.

When J.P. had completed the transfer of the amalgam from the metal box to the retort, he started the fire in the furnace and, when he was about to leave the building, he instructed me to keep the fire stoked until he returned to complete the process. Most emphatically he told me not to touch the bolts securing the door at the front of the tube. He said that if I did the gases in the tube would come out and knock all my teeth off at the gum line; and I believed him. But a knowledgeable chemistry professor later told me that if I had opened that door I wouldn't have had time to get my teeth knocked out; I would have already been dead from mercury poisoning. The simple facts were, the wood fire delivered sufficient heat to vaporize the mercury permitting it to escape through the pipe at the rear of the furnace into the water where it was transformed into liquid mercury, ready to go back to the mill to pick up more gold.

After J.P. returned to the refinery and the retort was allowed to cool down, the gold in the trays was transferred to the other furnace. This device was a barrel-shaped affair resting on bearings in order that it could be inverted when the time came to pour the gold into brick form. The heat in the retort furnace was not sufficient to melt the gold, only to leave it in semi-granular form.

The second furnace was heated with oil propelled by a strong draft. Fluxing chemicals were added to slag off any impurities remaining with the gold. When the gold became molten J.P. placed a small steel mould, about a foot long by four inches wide and deep, on the floor below the furnace and instructed me to turn a wheel that would invert the furnace and allow the gold, with a small amount of slag, to flow into the mould. This done we sat around until everything had cooled down.

J. P. and the first gold brick produced at Bussiers mine of the Treadwell Yukon.

J.P. then took the cooled brick to a table to chip away any remaining slag, while I, happening to pass the now cool furnace, glanced in and noticed some little blobs of gold, some as big as a dime, clinging to the ceramic lining of the furnace. When J.P. finished his work at the table he turned and found me half inside the furnace. His immediate comment was, "What the hell are you doing there?" I showed him the results of my labour, a small handful of gold nuggets of which he quickly deprived me. I wonder in which depository in the world that little handful of gold is now resting, or is it in a gold band around a slender finger?

I'm sure this was J.P.'s first gold brick because he went to the trouble of having a picture taken of himself with it. Instead of having the gold brick propped up on an old dynamite box and himself in disarray, he should have had the brick resting on bright green velvet and himself in formal wear.

At another of J.P.'s mines I was present when a mill manager purposely misplaced a mould allowing a trickle of gold to flow out onto the cement floor and form into numerous little buttons. I inquired of the manager, "What will happen now?" He replied that he would sweep it all up and place it along with the gold brick in the mine vault. I imagine that spilled gold is now also resting somewhere in a world depository, but only after an improper transfer of paper money, probably several times.

Note: I doubt that J.P. ever read Dante and he certainly wasn't religious, but looking upon his son with the love which the one and the other eternally bear at the fount, the primal and ineffable power made everything which evolves through the mind or through space with such order that he who contemplates it cannot be without taste of him.

Chapter 8

A Mine of His Own

Tiblemont Island Mining Co. 1933

by George Norrie

J.P. was manager of Bussiers but it was not his own. Tiblemont Island would be his own. When suddenly Fred Bradley died, J.P. had the Coffin Mining Company back on his hands with no money to do anything about it. There didn't seem to be any way out.

As luck would have it, during the fall of 1932 the Tiblemont Island fiasco appeared out of nowhere. It was this property that would, in short order, solve J.P.'s money problem. In the fall of that year a miner at Bussiers asked J.P. to become a partner in his claims. There was a problem, however, regarding the ownership of the claims and claim jumping about which J.P. wrote A.O. Dufresne, Director of the Quebec Bureau of Mines:

On Nov. 1, 1932, I was approached by a prospector, Rodney Nadeau, to join him in ownership in certain claims staked on the big island of Tiblemont Lake near Senneterre, Que. After investigation as to title I secured one half interest and have paid the taxes and done the assessment work on these claims.

Mr. Nadeau had also, according to a legal agreement, a one quarter interest in claims with Bill White and two Russian prospectors, each holding ¼ interest. Mr. Nadeau had already disposed of ½ of his interest so I purchased one half of the balance which gives me 1/8 interest.

I understand one of the Russian partners, Mike Zuwan, was mixed up in a May Day communist celebration in Rouyn in May, 1932, and was later deported to

Russia. The other Russian, Peter Repanda, disregards any agreement with Nadeau and claims full rights to Nadeau's claims, but Repanda's name is clearly signed to the agreement before the lawyer who drew up the agreement. Both Nadeau and myself have correspondence to show that we paid our share of the taxes and assessment work. As soon as the staking was done they proceeded to get him out. This action coupled with the favourable location is the reason I became interested.

Knowing the actions of Russians in regard to the Beaufor claim two years ago, I sent Rodney Nadeau accompanied by an experienced prospector C.E. Metcalf, to check up on all points and tags on these claims. They spent four days working on the lines etc. and both report all posts and tags correct according to the plan of the Dept. of Mines. At the time these men reported to me that the Russians had built a cabin on the claims and on personal investigation, I found that they had built a cabin on the claims on which we had a joint interest.

On recent personal investigation, I found that the Russian prospectors had just made a discovery near the cabin. Posts had been either destroyed or removed, so that the south boundary is now south of the discovery. I have a Quebec land surveyor on the property making a survey of the claims. He has taken an affidavit from Nadeau and Metcalf and a sketch showing the location of some of the posts with remarks as they are now found.

A full report with a plan of the survey from the land surveyor Mr. C.M. Dechenes will be sent shortly. Pending a decision from your Department I would like to have an order issued prohibiting further activities on the property.

I am anxious to have this matter cleared up as early as possible, as plans were made previous to this trouble to do extensive work during the open season.

J.P. Norrie

P.S. I would like to have the condition of the posts on these claims inspected by Mr. Tachereau, and am willing to pay the expense. Your Dept. will have some idea then as to the trouble these foreign prospectors can make for the Canadians.

The matter of the ownership in the claims of Mike Zuwan, who had been deported, had to be settled by a judgement from the court in Amos. This would delay any deal for two more months.

The Russians also attempted to jump the claims by moving the claim posts and this was settled by a decision of the Quebec Dept. of Mines.

A further dispute over the claims developed six months after the original staking by Nadeau and the Russians. This was explained in a letter to A.A. McGee, legal counsel for Consolidated Mining and Smelting, who were trying to option these claims from J.P. and his legitimate partners:

I have been advised by Mr. R. Macaulay of Consolidated Smelters to advise you details re the staking of claims on Tiblemont Island on which Mr. Macaulay is taking an option. Claims staked by Pete Repanda have been transferred to J.P. Norrie.

It was the intention of the stakers to cover all the island and were assured by the mining recorder that such was the case. Fifteen claims were staked, covering the entire island. The plan showed the dividing lines to be straight and the irregular shore line of the island to be the outside boundaries.

Approximately 6 months later water claims were staked by Ventures covering the water adjoining the land. The writer was assured in the presence of several witnesses by the staker of the water claims that he was to have only the water and the shore line was the boundary. When survey was made to locate posts on the shore of the lake it became known that some land extended beyond a line between posts and the stakers of water claims expected to have this area.

I am enclosing copy of a letter from Dept. of Mines verifying the assurance of the mining dept. recorded at the time land claims were staked that all land belonged to the original stakers. And that the shoreline would be the boundary.

According to agreement I have full authority to sign all papers and am obliged to form a limited company to operate the claims. This is now being done in the name of Tiblemont Island Mining Co. 3,000,000 shares N.P.V., 1,000,000 to go to the vendors and 2,000,000 shares to finance the operation. I am verbally bound to have the shares listed immediately on the Montreal curb, and as soon as possible on the Standard Stock and Mining exchange.

Shortly the certificates will be issued to the various holders when a receipt in full for each interest will be secured except that of Mike Zuwan, who has disappeared from the country. A judgement is being secured against him for nonfulfillment of agreement. This action should be complete shortly.

J.P. Norrie

In a letter to Hector Authier as late as June, 1933, J.P. stated that a further complication had arisen. Kellorans was jumping Ventures' claims:

Kellorans of Senneterre tried to jump two of Ventures' water claims in which we had the dispute. In other words our dispute was either with Kellorans or Ventures. Kellorans claimed that Ventures had no tags on their posts and that their claims were too big - which is not a good reason. I know Ventures posts were on the ground and were tied in by the surveyor, although I did not see their tags.

With this tangled web of ownership, claim jumping and claim disputes beginning to sort itself out, J.P. began to plan how he was going to finance this property. Cominco was interested in an option on the claims as explained to Authier. But J.P. seemed to favor the same method he tried at Bussiers, making a market for the shares with a broker. This had failed before but this property had such a spectacular free gold showing, everyone wanted a piece of it. He first wrote his old associates Lindsley, Hogarth and Anderson. Further he contacted a financial house to feel out their interest. He wrote Fred Crawford in May:

Within the last month an important gold discovery has been made on ten claims located on an island in Tiblemont Lake, 10 miles south of Senneterre, Que. I personally own 7/16 interests in the claims and have power of attorney to dispose of the property in the best possible manner.

The claims are ideally situated in that the motor road from the south ends right at the edge of the lake. Any size scows can be used for transportation to the find. Much money otherwise used in road construction etc. will be saved.

The Treadwell Yukon Co. would take over this show, but think they have a substantial mine in the Bussiers and wish to confine all activities to it in the meantime. In fact, I have been advised by Mr. Clauson, general superintendent of Treadwell, to take the matter up with you. Several attractive offers have been made by various mining organizations. However it is the wish of my associates and myself to form our own company and develop the show with capital raised from sale of shares.

My associates are poor prospectors who need money. I have taken care of them by buying a portion of their interests for cash.

I would like to handle this show along similar lines to the Little Long Lac, and already have the assurance of some prominent men who are willing to buy an interest for cash and act as directors.

I am very busy here with Bussiers operation and wish to turn this show over to someone who can look over the formation of a company etc.

If you are interested in handling this proposition, please advise me by return mail as to the best procedure, as I want the operation to be clean and well handled from the start. I might say that the find has created a good deal of interest in the north, and from the very limited amount of work done a lot of coarse visible gold has been found.

The formation and gold occurrences is very similar to Siscoe with a much better surface show. I have promised my associates to get things moving early and would be glad to hear from you at your earliest convenience.

J.P. Norrie

J.P. asked Hector Authier to be president and explained the financial option that was being considered by Cominco:

I would be glad to have you act as permanent president of the company and would like to have the temporary directors act in the meantime.

The property has been carefully examined by two engineers of Consolidated Mining and Smelting and they have made the following offer [to option shares].

Option agreement: the option offered by Consolidated Mining was the usual method used by mining companies to finance prospective mines. Consolidated offered to buy 500,000 shares @ .15 per share or $75,000 on a specific date. The proceeds would be used to develop the mine. If after the first $75,000 was spent and a prospective ore body was located they would choose if they wanted to proceed with the second stage of the option: 500,000 shares @ .22 per share or $110,000. This procedure was followed for the four specific dates of the option.

500,000 shares @ .15
500,000 shares @ .22
500,000 shares @ .28
500,000 shares @ .35

They must pay fully for each block before stock from this particular block is taken down. They have asked to have the 1,000,000 vendors' shares pooled for a period of at least 6 months. I have refused this, as Smelters would not have to raise funds from the public, and thus the pooling of shares would be of no benefit to them.

They are considering taking up the property without pooling the shares and have agreed to advise me in a couple of days. If they are not interested on that basis, then we will try to get a broker to finance, which would suit me better. Consolidated have agreed to pay all expenses incurred to date including formation of the company.

Another nice gold show has been found in recent work some distance from our discoveries. This helped impress the Consolidated engineers.

Mr. Clauson of Bussiers and myself are trying hard to make a successful mine out of Bussiers, and having our names as directors on another co. (Tiblemont), people might think our interests are divided. We are not anxious to have this thought developed, and for this reason would not like our names on the board of directors. Mr. Key of Stabell and N.A.M.E. has requested that he be given a chance to finance Tiblemont.

J.P. Norrie

During the month of July, 1933, Ron Macaulay of Cominco and J.P. bickered back and forth by wire and letter. J.P. and the vendors were asking for all cash expenses spent on the property which amounted to over $4,000:

The judgement being secured in court against Zuwan should be completed in about 25 days from date. The total claim is 1/16 interest or 62,500 shares. So I do not see why more than this number of shares should be held up.

Regarding expenses: Messrs Clauson and Hershey [Bussiers consultant] recommended the property to Treadwell Yukon on a deal similar to yours, whereby all expenses were to be paid by the new company. Mr. Bradley's illness and death prevented the deal going through. When we were discussing the deal a figure of $3,000 was mentioned for expenses. As a matter of fact it will be closer to $4,000 all of which I have paid in cash. In case I am not to be reimbursed I am paying well for my shares in the company without anything for the time I have spent in straightening the matter out. I am willing to accept the $3,000 but I must pay everything above this amount as I cannot ask the other vendors to pay at this time.

When the Bussiers deal was made they paid all to date for the deal.

From our conversations I understood the treasury stock was to be taken down in blocks of 500,000 which Authier advises is not Magee's views on this.

I think everything can be straightened out to our satisfaction when we meet in Amos. However I think Consolidated should reimburse me for expenses of $3,000.

J.P. Norrie

Cominco were not only not going to pay all initial expenses, they now wanted to take down the stock in blocks less than 500,000 or however they wanted. This was entirely unacceptable and J.P. advised Macaulay Aug. 11, 1933:

I regret that it became necessary to sign a deal on Tiblemont Island with American people.

Mr. Clauson and the other associates were not in favour of signing the deal unless all expenses were paid and the stock taken up in blocks. The American people were dug up by other associates who called me to Montreal to complete the deal. They agreed to pay $5,000 cash and take up the stock in blocks of 500,000. Thus I had no other course than to sign, which I did two days before receiving your wire in Amos.

I would have liked working with you on this property, as I think it has a good chance.

J.P. Norrie

This was nothing less than a sweetheart deal for J.P. and the vendors. The American group headed by Armstrong would put up their first money when they took the first block of 500,000 shares. They seemed in no hurry to take the first block and J.P. had almost 400,000 which were free shares (not in escrow or pooled) in one of the hottest properties in the north country at that time. J.P. began offering stock to many of his friends, immediately, at prices much below the market. This made a distribution for the shares and also gave his friends a share in this spectacular show of free gold.

Walter Warden was a close friend of J.P. from his days in Toronto and he got some cheap Tiblemont stock from J.P. Walter had just sold some Tiblemont stock. His profit was obviously remarkable considering he paid approximately .10 per share for it:

There is no need to rush the Tiblemont stock certificates to me, I am in no hurry to receive them. The Tiblemont Island declined with the rest of the stocks during the last two weeks of depression in the market but seem to be back around a dollar. Since very little stock changed hands perhaps the prices quoted are not very significant, but I had the satisfaction of selling some from $1.55 to $1.60. Hope this has not ruined my perspective.

This article appeared in the *Quebec Mining News*, Aug. 17, 1933

It seems only a few short months ago that the little mining town of Senneterre was buried in gloom over the retirement from the Woods property of the big Hollinger. Dire pessimism supplanted jubilant optimism and it appeared for the time being as if the story had been told.

Optimism is again in control thanks to the fearless courage of such figures as Jim Norrie, Pete Repanda, Jake Smith, Bob Potter, Sweet and Martyn, Ed Storie and a myriad of other tried and true prospectors operating in the Senneterre section of the Quebec gold belt.

Headed by the spectacular Norrie Repanda discovery owned by Tiblemont Island Mining Co. new finds were reported in quick succession in all sections of Tiblemont. The town of Senneterre is a seething mass of excitement and prospectors are daily holding weighty conferences on the street corners discussing the relative merits of Jake Smith's big find on the south Tiblemont property or the Tiblemont Island show.

The Norrie Repanda find added impetus to the field with the result that prospectors were spurred on to greater achievements in their relentless search for the yellow metal.

Work has already started on the Tiblemont Island property, with important American interests providing the capital. It is interesting to note that officials of Tiblemont Island have been besieged with offers for control of the property from all points on the compass. There were a few exciting moments in the deal when one big Canadian mining company, a leader in the industry, made a last minute attempt by telegram to enter the picture on very attractive terms. It was indeed a very thrilling week for Senneterre.

Over one year after the Tiblemont deal was finalized J.P. explained the deal to P. Thomson, pointing out that his deal, which was so advantageous to the vendors, was possible with other mining options such as the Arrowhead Mines Ltd.:

In the deal I made with the Brady interests on Tiblemont Island the 1,000,000 vendors' shares were free, with no cash payment except a small amount to cover expenses, and thus took on immediate sale value as no treasury shares were available to Brady's until they had spent on the property $75,000 equivalent to 500,000 shares. The rest of the shares were also to be taken up in blocks of 500,000 each. Average .25 each.

I may say that Brady's have yet secured no treasury shares of Tiblemont. The vendors' shares have sold close to $1 without effort.

In Jan., 1934, J.P. resigned as vice-president of Tiblemont because of Macaulay's criticism but stayed on as a director for another year.

J.P. stayed on managing the Bussiers mainly because of his loyalty to the employees. He obviously made enough from the free vendors' shares from Tiblemont that he didn't need to take Macaulay's abuse any more. By the summer of 1934 he had avoided bankruptcy again, and when Alex Perron offered him the position at Perron, he would be able to spend half of his time on his dream of founding Inspiration Mining and Development.

Chapter 9

Perron Gold Mine, 1934

by Harry Norrie

J.P. seemed to disregard or have no conception of dimension; one mile or ten miles, one hour or ten hours. If things didn't change by themselves, he would change them and have fun doing it. Time had no measurement. If he carried a watch it was for decoration only. Day or night, weariness was just an irritation, not a call for rest. I was with him one night in the car between Senneterre and Amos, one o'clock in the morning. He fell asleep at the wheel and went into the ditch. He cursed the car and the road, not the time of day or the one hundred miles yet to go. Getting the car out of the ditch was a minor problem.

He was with Bellachey at Great Bear Lake and an angry bear interrupted their progress. The bear was there and had a right to be there but J.P. was there as well. Where is the nearest climbable tree? There is no sense in trying to fit J.P. to any of the usual standards.

J.P. was never a man to throw words away just to hear himself talk and as his daily work load increased he became ever more reticent. One day in August, 1934, he turned to me and said, "Get your stuff; we're leaving."

It was never wise to ask J.P. "Why?"

My personal stuff filled the bottom of a small haversack.

With J.P. in the lead we took off through the bush for Perron about five miles away. It was clear that he had walked this trail several times before since he had gone to the trouble of having the low swampy areas bridged over with pole walkways. I did not know it at the time but I think, on this trip, he was leaving Treadwell for the last time, and we were beginning my one and only visit to Perron. He had taken on the management of Perron, and all his attention, for a while, would be directed to that endeavour. But since I was following behind him on the trail, I can feel I was the last in the family to leave Treadwell. Not that that means a hell of a lot.

Before J.P became manager, Perron had been developed to a point where a small fifty-ton per day test mill had been set up and had, just then, poured its first small gold brick. The purpose of the test mill was to determine the best method to treat that particular type of ore.

When we arrived that day at Perron, J.P. led me directly to the staff house and told me to leave my stuff in one of the bedrooms where he and I would be sleeping in the same bed. For the next three days I saw him only when we were in bed together; the rest of the time he was either underground, in the mill or in the mine office, his attention occupied full time. With his natural reticence and his active mind he had little need to banter with me.

At supper on the third day he told me to get my pack and meet him at the headframe at seven o'clock. When the time came I went to our bedroom to get my haversack and was a little surprised to find it on the bed rather than under the bed where I had left it. I went to pick it up, but it remained on the bed: its weight had been greatly increased. My immediate thought was that some of the office staff were playing a joke and had put a bunch of rocks in it, but when I opened it to remove the rocks, I found a small wooden box. This was my second experience with a gold brick. On the wooden box was printed in very legible English: The Royal Mint Ottawa. The haversack was now quite heavy but I put it over my shoulder and went to see Dad at the headframe.

This was new territory for me but I followed J.P. down the trail and through the bush to wherever he was going. In a little over an hour we came to a car parked at the end of a gravel road that was under construction from Senneterre to Perron. As we got in to drive away Dad told me to throw my pack onto the back seat, but when he turned the switch the motor was dead. He had left a light burning when he left the car there several days before. Stowing his reticence for a moment he told me that we would have to walk for another hour down the road to a guy he knew who might have a battery or a boost. I asked him, "What about the gold brick?" He said, "Leave it. It's insured." I guess he figured it was unlikely that anyone would be driving to the end of an unfinished road at that hour of the day. And it was insured; a post office had been established at Perron by the Royal Canadian Postal Service.

J.P. found his man and, with compensation established, we were driven back to the dead car. A spark from one to the other and we were on our way again.

All in all it was a mean day for a reticent man. Driving along in deep gravel ruts was not a time to encounter a skunk occupying one of the ruts. The skunk found it difficult to extricate itself. Persistent honking and cursing only tended to raise its tail higher, revealing the muzzle of a fleshy opening only too anxious to fire. Eventually it relented and with an extra effort climbed out and returned to the bush, a more normal habitat.

In Senneterre we drove directly to the hardware store where the owner, a particularly good friend of J.P.'s, was versed in the field of politics. For a long time they fought the issues while I slept in a straight-backed chair. Eventually with "Come along" we were back in the car on our way to Amos and home, a hundred miles away, at one in the morning. Not long after leaving Senneterre a physical condition not recognized by J.P., fatigue, caught up with him and he fell asleep at the wheel and drove off into the ditch - but I've already told you about this.

I don't think Perron needs a lot of attention other than that J.P. brought it into production and that it was in granite where you're not supposed to find gold. The main feature was that it opened the door to Nesbitt Thomson and funding for East and Goldfields. Rob Kennedy said that Alex Perron was a blasphemous character and while making the deal with Nesbitt Thomson and operating the mine with him it was difficult to contain him since Mr. Nesbitt was a Bible thumper, very much averse to blasphemy.

While money is the fuel of industry, mental energy is the fuel of conception. When J.P. entered the Nesbitt Thomson circle his future funding requirements were pretty well assured and he had ample mental energy to look after that part of it.

Only a few minutes with J.P. would reveal his strengths: his physical and mental energy, solid character, high sense of honour and dedication to his work.

After the '29 Depression had crippled most industries worldwide, the American President, F.D. Roosevelt, raised the price of gold to $35 per oz. With an open offer to the world, it was seen as a pretty good idea since it would allow the U.S. to print and circulate double the amount of paper money then in circulation. That, along with funding from Nesbitt Thomson, opened the door wide for J.P. He was in the gold processing business at a time when money was backed by gold, and had a good idea where more gold could be found.

Previously, Nesbitt Thomson were power (electrical) people and one of the largest brokerage firms in Montreal, at 355 St. James Street. But with the new price they funded the development of a mining company: the O'Brien Gold Mine in Cadillac township, a little more than ten miles to the west of Malartic. The O'Brien ore body was quite small but very rich and it made a lot of money for shareholders, particularly Nesbitt Thomson. It was only a short time before J.P. was introduced to them.

Through Nesbitt Thomson J.P. met Andy Davis from Newmarket, Ontario, who, at the time, was a director of the Bank of Toronto and who, later, would lend assistance in funding Malartic Goldfields. Eventually two of Mr. Davis' sons came to work for J.P.; one at the Amos office and one at Perron. The one in Amos, Aub Davis, built a house on the lot adjacent to J.P.'s, making J.P's home look like a doghouse.

It was J.P.'s habit to share his knowledge of promising gold prospects and the stock market with friends and business associates around the towns and mines, some of whom made substantial profits. J.P.'s business office in Amos was in a long narrow ugly building, equally dark on the outside as on the inside. It was adjacent to a large vacant corner lot. The vacant lot was owned by a businessman who operated a store on the other side of the lot. One day J.P. came to work to find large earth moving equipment in the centre of the vacant lot. He went to the owner to find our what was going on. He was told that a new motion picture theatre, the first in Amos, was to be erected but that J.P. had nothing to worry about; his building would not be affected. But the hole for the foundations of the building grew larger and larger and daily closer to J.P's building until finally it started to fall into the pit. J.P. abandoned his office and moved to another location sooner than he had planned. He never again spoke to the owner of the lot who had very much gained from his advice. Fortunately he found a large, two storey building close by that had been the home of a large family and was more suitable for his growing staff.

Many years later at the airport in Montreal I ran into the son of the owner of the store and spent a few moments talking with him. He seemed a very nice fellow but I found him a bit obsequious. Perhaps he was thinking back to other times.

As mines developed in northern Quebec a strange condition arose that, I think, happened nowhere else in Canada: even before a mine went into production it became necessary to lay out a townsite for workmen to build homes for their families, and for businessmen to locate, and of course there had to be a bank. And there had to be a mayor, police chief, fire chief, all of which costs money. So now come taxes to support all of this organization.

But there is always an objector to organization and taxes. This imaginative fellow decides not to follow the norm. He finds a likely-looking spot just beyond the town limits and taxation, and with millions of trees, all free. He knocks down a few dozen, stacks up four walls and has for himself and family a snug little shack. Some of his fellow workmen appreciate his tax dodge and follow his lead.

All the while there is another log shack hidden away behind a thick clump of trees, also off town limits. This building is known to a few of the single workmen, and well, yes, perhaps some of the married men too. Further description of the business carried on in this building in the midnight hours is unnecessary, except to say that a secondary commodity for sale here will tend a man to drunkenness.

By now quite a large number of tax dodgers have established homes off limits but, inevitably, one stout fellow who has always wanted to be mayor of something convinces enough of his neighbours that this community too should become incorporated; and that is why in northern Quebec and Ontario you have: Bourlamaque-Val D'Or, Malartic-Paris Valley, Cadallic-Little Italy, Rouyn-Noranda, Kirkland Lake-Chaput Heights, Timmins-Porcupine-South Porcupine. Another thing, perhaps not so strange, the add-on towns become the main centres of commerce as with Val D'Or and Noranda, while the original towns remain sleepy little communities. Some of the would-be mayors never fully succeeded in incorporating their domains but still had vigorous commerce in fluids and flesh.

J.P. was much too busy with Perron to go out and stake claims for himself. It was not necessary in any event since there were many men in the district who were only too anxious, under daily wage or contract, to prospect or stake claims for other people. Sometimes these men would ask for and be given a small carried interest in the claims they were hired to stake.

It was reported that one of these men, Rube McEwen, was hired by J.P. to stake the claims that became East Malartic. But no one seems sure of that.

Rube McEwen , prospector and town bum.

Every settlement, village and town has at least one character, male or female, inoffensive, loners, non-entities. Rube McEwen was one. No one knew where Rube came from; he'd probably forgotten himself. It might have been Haileybury. He looked like a person from Ontario. He showed up intermittently on the streets of Amos. His disappearances could probably be traced to his interesting prospecting.

Besides his unkempt clothing, his failure to bathe unless he fell into a river somewhere, his disregard for shaving (his whiskers never seemed more nor less than a quarter of an inch long), when in town he was usually drunk, out in the street making a nuisance of himself. He very well knew that J.P. hated whiskey and he tried, on occasion, to shove his intemperance into J.P.'s face. One day he went too far and stumbled into the office building clutching a half empty bottle. J.P. was on the ground floor talking with one or another of his staff: Bill Lipsett, Bill Todd, Aub Davis, Jack Workman, Jasper Irving. Rube shoved the bottle into J.P.'s face and said "Look at that, Jim Norrie." J.P. grabbed the bottle and poured the contents down the sink, then returned to his private office on the top floor, whereupon the members of the staff grabbed Rube and sat him down in a chair. First they poured a bottle of white paper glue over his head and slicked his hair down in a way it hadn't been since childhood, if then. Then they took a bottle of red ink and poured that over his head and turned him loose in the street. It made not a whit of difference to Rube as he wandered along looking as though he'd been hit by a broad axe or caught in a buzz saw. He scared the hell out of a number of women, a few of whom fainted dead away and had to be revived by kindly town folk.

Many of you will feel it was a cruel and inhuman thing to do and I may agree with you, but I can tell you it worried Rube not at all. With the passage of time, the additives wore off in a natural way.

After the town of Malartic built up a bit, Rube moved there from Amos. One day I met him on the street and it suddenly came to me that it might be interesting to learn a little of his background. What made him tick. Not well thought out, since most drunks live only from one bottle to the next. A few people wondered how he could afford an endless supply of whiskey and it may be, as some people said, that East Malartic paid him a monthly stipend for having been one of the original stakers. Anyway, in my room in the jail, I prodded his memory as gently as I could but I found out nothing. He was either evasive or he simply would not answer my questions.

About that "jail" thing - I knew it would grab your attention. Canadian Malartic had found it necessary to erect a building for just that purpose, for men who needed to be restrained for some reason. But after the town was established with a nice new jail of its own, our jail was no longer needed. It was refurbished to house a couple of Canadian Malartic employees, rent free. I lived there for a while with another guy who had been caught in an underground dynamite blast that took out one of his eyes. I enjoyed sharing living quarters with him except when his glass eye started rotating on its own.

One day I heard that Rube had died and that's all I know about it. Probably some religious figure tracked down where he lived, mumbled a prayer and planted him somewhere under a tree.

Every life is a story. A book. Some long - some short. Some are sagas - some not, but all are worth listening to. Even a newly born child has up to nine paragraphs. The beginning of the first one, among other things, is probably the most exciting, considering the diverse locations where it starts - a sandy beach, a mountain retreat, a stairwell, a snow drift... Don't argue with me!

Making a Mine out of Perron

Perron Gold Mine, 1934

by George Norrie

The last days of the Bussiers were not at all pleasant for J.P. or many of his loyal friends and employees. From the time Macaulay and Quebec Gold Mining had begun to direct operations of the mine in Jan., 1934, the mine began to flounder, production dropped and morale was low. J.P. knew that the future of Bussiers was hopeless under Macaulay. The families who depended on J.P.'s knowledge, judgement and tenacity knew, as well, that they would soon be out of work, in the middle of the Depression.

As for J.P. personally he had struck it rich with Tiblemont Island and could now turn to his own agenda, and he had quite an agenda planned. He felt that the East Malartic and Inspiration were his number one priority. He wanted money out of his Bussiers shares to develop these companies. But this never materialized. He also felt that money from the Tiblemont shares would be a better bet than holding on to the Tiblemont property, even with its show of free gold.

Nevertheless, seeing the Bussiers go down was depressing. In the past two years J.P. had spent endless hours trying to find financing for both Bussiers and Tiblemont and had not been satisfied, particularly with brokers Fred Crawford and Doug Wright.

In June of 1934, financing was to find him, right out of the blue, but not for Bussiers. Alex Perron was operating a small high grade mine just two miles north of Bussiers through the bush. Perron had been trying to find money to pay for further development of his high grade operation. Somehow Nesbitt Thomson, who had an investment west of Perron, decided to make Perron an offer.

J.P. later wrote the details of the offer to his lawyer, Mr. Michener:

On June 13, 1934, when I was employed as manager of the Bussiers mine of the Treadwell Yukon Co., I was approached by Mr. Perron and an engineer by the name of R. Howe to take charge of the Perron mine.

In the first conversation a proposition was tentatively discussed, however I did not like conditions at the Perron and made no decision to go there.

Sometime during the latter part of June, 1934, Mr. Perron visited me at the Bussiers property and again discussed my going to the Perron. At this time Mr. Perron advised me that I could have all of his shares that I wanted at the same price that certain of the Perron shares were optioned and sold to Nesbitt Thomson Co. of Montreal. One of the conditions of the Nesbitt Thomson option being that I be secured as general manager of the Perron.

When, eventually, J.P. accepted the Perron offer he had to answer to Nesbitt Thomson. He made sure they were advised of every detail. He explained his plan to Mel Thomson for the Perron operation:

You will notice that your letter was a long time reaching me. This is due to the distance from the steel [railroad] i.e. 75 miles by slow and irregular boat service, thus mail is delayed both coming and going.

According to my agreement with the Perron company, made with Mr. Perron in the presence of Mr. Davis, I am to spend one half my time on the property and believe the mine can be efficiently developed under such an arrangement during the next year at least, as I do not think that large salaries should be imposed on the property at this time. When the mine is ready for it, and I hope that eventually it will be, I am prepared to spend all my time there. During the next year all available money should be spent on the underground work.

The Perron organization is as follows: all surface work, construction etc. is under a competent foreman who works under my direction. The mill is being supervised by the former mill supt. of the Treadwell Yukon. He is a graduate engineer and a competent man. Several changes had to be made in the mill in order to get better recovery and tonnage treatment.

The mechanical staff is being supervised by a very good mechanic who has direct charge of all men in the mine, and is responsible for engaging and dismissing them, while I direct him in the various working places.

Although my resignation has been in the hands of the Bussiers Co. for some time, I have not yet been replaced, and it was impossible for me to walk out and leave the mine unmanaged. My successor is to arrive here on the 15th instant when I will definitely be eased out. In any case I have been spending more than one half my time on the Perron since I took over the management there, and aside from the setback due to the fire everything is progressing favourably.

The Perron mine is located 2 miles from the Bussiers i.e. 40 minutes walk, so that I have been able to keep the work there well in hand. On the evening of the fire I left there at 7 PM. leaving Mr. Perron at the mine. Arrangements were made to step up the available steam power by the use of very dry wood. The extra heat in the boiler was responsible for setting the fire to the building, as they were busy in the discharge of their duties, and as the fire caught on the outside it had quite a start before being discovered. When it was observed the men became panicky. It is possible that the building could have been saved by the proper application of water. However, it is hard to size up a situation of this kind, not having been there at the time. The plant has been completely repaired and is now in full operation. Installation of a new boiler is under way and should be soon in operation.

Mr. Perron takes a very keen interest in the operation and spends approximately one half of his time on the property. Personally I think there is nothing to fear in the Perron organization. My advice has been followed in every case to date, and when it is not (or if things are done which are not for the company's good I will advise you).

If the setup I have described is not according to your wishes, please advise me, and I will see that they are changed according to your wishes.

I believe that one half of my time on the Perron during the next year will bring satisfactory results, and from present ore indications I believe the property is a real good bet.

J.P. Norrie

The manager's house at the Perron mine, showing the firebreak. This was J. P. and Margaret's first home.

J.P. also gave Nesbitt Thomson a complete geological report on the Perron property as soon as he took over the operation:

The Perron mine of 553 acres is situated in Pascalis and Senneville townships, Quebec, near the eastern contact of granodiorite batholith with older Precambrian rocks. The Siscoe mine bears the same relation to the contact on its extreme western extremity. Other mines on the contact are the Green Stabell and Sullivan, now reputed to be in profitable production, and the Sigma, Lamaque, Newroy, Bussiers and Beaufor, all of which have excellent mine making possibilities.

The original discovery on the Perron property was followed by extensive surface trenching and diamond drilling. The trenching exposed several important veins of good length, width and grade. In 22 diamond drill holes ore of various widths was indicated, assaying from a trace to $60 per ton, development work to date being confined to a small portion of the property. Diamond drilling was followed by sinking of Number Two shaft to a depth of 345 feet. On the 175 foot level, 1,750 feet of exploration work has been done, and on the 325 foot level 200 feet of crosscutting.

The underground work has opened seven of the ore sections disclosed in surface drilling. The amount of work on any one of these sections has been limited to date, but an intensive development program is to be undertaken immediately. Up to the present sufficient has been done to demonstrate geological structure similar to Siscoe with ore width up to 15 feet and a grade of $10 per ton.

A mill of 50 ton capacity has recently been placed in operation to test the ore from the various underground disclosures. The first test of 200 tons gave an extraction of $11 per ton, with gold at $20. Tailings assayed $7 per ton. Further additions to the mill will reduce the tailings to .20 per ton.

I believe the Perron mine is a prospect of more than average merit, and now with ample funds with which to push development of the ore sections already disclosed, it is reasonable to expect a profitable producing mine will be developed.

J.P. Norrie

Along with J.P.'s ability to supervise the mining of gold at Perron, he also brought with him the usual list of contacts and friends. However, Mel Thomson began to wonder:

In spite of the wonderful property that you have up there I have not received any orders from Amos; I thought they would be looking for at least 50,000 or 100,000 shares in your vicinity.

When J.P. was manager at Malartic Gold Mines the local citizens had invested heavily in the venture. This was just prior to the stock market crash and everyone was optimistic. All shareholders, J.P. included, would lose everything when the mine closed. Now five years later and in the middle of the Depression, investors didn't have a great deal of money for investment. J.P. replied:

It is hard to sell stock of this class in the north country. Most of those interested invest their money in small prospecting syndicates and in penny issues, generally on the strength of so called inside information.

I do not believe they make any money in this way, however, you can sell .15 stock up here. There is little demand for the higher priced shares.

At the Perron mine the class of employees is different from O'Brien. The O'Brien is an operation of long standing and most of the employees there are English speaking, while at the Perron the majority are French and foreigners who are not interested in higher priced stocks.

Following is a list of the Perron men who might be interested: Cameron Yule, John Black, H. Duffet, N. Murray, R. Bryson, D. Livingston, G. McLean, John Riley, Isadore Cleman, Frank Blais Sr., George Blais, Hector Authier MPP, Pierre Beauchemin, C. Massicotte, L. Massicotte, Edmond Carriere, David Gourd, Rosario Gagnon, Frank Dorval.

I trust that some of these will prove good prospects.

J.P. was able, however, to suggest mining investors whom he had been associated with and who might be interested in placing money in the Perron venture:

The friend I mentioned in Toronto is C.D. Kaeding, mining engineer.

Mr. Kaeding has been connected with several big mining operations in Canada and U.S. He was at one time in charge of one of the nickel mines in Sudbury and was later manager of the Dome. He is now consulting engineer of Sherritt Gordon, Little Long Lac, God's Lake and some other properties. He handled much of the financing of God's Lake.

Mr. Kaeding wrote to me some time ago inquiring of Perron and the chance of his acquiring a block of shares. You can write him and I will do the same.

Re: Selling shares in Amos. I suggest that you get in touch with J.M. Blanchard, local agent for Burke Dansereau who is well known there. Mr. Blanchard advised me today that his other connection will not interfere with his handling Perron stock.

J.P. also contacted Thayer Lindsley who always took some stock on J.P.'s advice. Lindsley was interested in Perron because Ventures owned the Pascalis adjoining Perron to the east:

We have been interested in Perron Gold Mines for a couple of years and are holders of a small block of stock. I suggested to Mr. Neelands that while he was in the north, he drop over and pay you a visit.

Any assistance you can give him in studying the ore deposits at your property will be much appreciated.

Even Nesbitt Thomson with all their contacts and financial force had trouble with market breaks. It was still the Depression and money was still difficult to find.

Thomson discussed the first blocks of Perron shares taken down and marketed:

Under our option we have to date taken up 30,000 shares of stock and paid to Montreal Trust for your account a total of $15,000. These will become due and payable on the 15th delivery and payment for a further 30,000 shares at $15,000.

We started our men to work yesterday on placing these shares but with very poor results. The break in the market, particularly Hollinger, and the proposed new taxation by the Province of Quebec, is not helping.

Mel Thomson, right

In April of 1935 J.P. advised Mel Thomson of a very interesting development:

I am enclosing an up to date map of the 175 foot level of the Perron showing the present relation of developed veins to the boundary of Beaufor and Pascalis Gold. You will notice that vein 190 is striking fairly into Pascalis ground and will reach there by advancing 307 feet. Work since this map was prepared has reduced this figure and the vein is still going strong. This vein 190 is a very important one and could easily make a mine in itself and I fully believe it will. Recently we have been running a mill test on material from this vein alone with the mill recovery of $14.70. This is much better than indicated in the first drift on the vein and is as I expected.

This vein 190 dips 60 deg south into Beaufor at a vertical depth of 659 feet. You will notice a drill section worked in red just south of the Perron-

Beaufor boundary. This section was secured in the Beaufor drilling just previous to the closing down. We saw the Beaufor plans and located it on our own. You will notice that it is high grade. This section is the downward extension of our vein 190 at the point where it dips into Beaufor.

You will notice that vein 192 is also pretty close to Beaufor. This is a very important vein and promises to be equally as good as 190 in both width and grade. This vein is also dipping approximately 60 deg to the south.

I believe some immediate action should be taken to acquire both Beaufor and Pascalis Gold and take advantage of the information we have secured before others learn of our favourable developments close to the boundaries of these properties. I do not know how many shares have been issued by Pascalis Gold but I am sure most of the shares remain in the treasury as they have done very little work. My suggestion is that you take an option on their treasury at low prices which I feel certain can be done as some of the big shareholders have already approached me on this. In this way you could secure control of the property and at the same time develop it. This could be done by making an agreement with Perron to extend their 190 drift into Pascalis ground. Should the vein continue and I think it will, you can see that Pascalis Gold shares could quickly be made worth a substantial sum. Your profits in placing the Perron in its present shape have been small compared with developments and I think you can repay yourself by taking on this venture which is about the most legitimate one I know of at the present. After the Pascalis property is developed and definite information secured on its ore value an amalgamation with Perron could be consummated on more favorable terms than today.

Developments on Perron are so good at the moment that I am sure I can substantially help you in financing the properties by selling large blocks of shares to inquirers who will readily see the importance of the property and to back up my recommendation I am willing to take down a substantial block of shares myself.

I believe the same mode of attack should apply to Beaufor. Of a 3,000,000 share capitalization only 900,000 shares remain in the treasury and they are very anxious to do something with the property. I have been talking to Beauchemin, a large local shareholder. He told me they would be prepared to put back in the treasury enough shares to give control to any responsible people and they were sure other large shareholders such as Timmins and Lindsley would do the same. I know Lindsley is very anxious to have someone take over both Beaufor and Pascalis.

The Beaufor could be opened up by extending Perron workings and in this way could be made high figures quickly. Then when development is completed an amalgamation could be made with Perron making a real big mine of the three ventures. I believe this is the only equitable way to attack the properties. You could make substantial sums in sponsoring the issues and later in the amalgamation.

I am sure I can raise the money for both of these ventures myself and am willing to back my recommendations with my own money. I know most of those interested in both companies and I could be of material assistance in rounding them up for you in any arrangement desired.

Until something is done I am not going to take anyone underground, at least, the 190 section of the Perron mine or show anyone the plans. In fact, I have the plans locked up.

J.P. Norrie

Mel Thomson immediately began work on acquiring control of both the Beaufor and Pascalis mines. The Beaufor deal developed quickly and will be discussed later. Lindsley, Timmins and Beauchemin all wanted the deal as the Beaufor could not develop further on its own merit.

The Pascalis would wait a few years but eventually J.P. would work on the further development of the Pascalis together with the Perron.

It became clearer that more veins were running over onto Beaufor and that the mill serving both mines should go on the Beaufor property.

J.P. wrote to Bill Dunn, the designer of the mill:

We propose to build the mill on Beaufor property on the hillside south of the Perron shaft as arrangements have recently been made by Nesbitt Thomson to finance Beaufor.

At this site Perron veins will be close to the shaft at 1,000 feet and Beaufor veins will be right there, i.e. if they go to depth.

I believe this site will be much better for all concerned and much better than the one we had selected on Perron property.

We are now driving a crosscut out from Perron bottom level and will raise a shaft adjacent to the mill.

Recent developments on the Perron bottom level have been excellent and much ore has been added to the first level since you visited the mine.

Any report on mine business was a very important consideration for any mine. Gossip and uncontrolled reports were not tolerated. When a report appeared in the *Quebec Mining News* with no known source and some erroneous information, it was disturbing to mine directors. In this case J.P. had a personal interest. The article suggested that Andrew Waltz would be the engineer in charge. J.P. had already seen the damage that a consultant engineer could do to a mine when Pope Yeatman recommended the shutdown of Malartic Gold Mines contrary to J.P.'s advice. (Waltz later took credit for the development of Malartic Gold Mines, now Canadian Malartic, under Waltz's direction.)

J.P. was not going to lose either credit for Perron's success or control of its future as he explained to Andy Davis, new director of Perron. He had worked much too hard for his success:

I'm enclosing a clipping from the *Quebec Miner* which speaks for itself regarding the operations at the Beaufor.

I understand this article was given to the press by Beaufor people and I'm sure will not be in the best interests of the property.

Practically all the ore on the Malartic property was developed by me and this can be verified by those now there, while Waltz comes along and checks up on some of the ore intersections and claims credit for making the mine.

In the Pascalis area I have had four hard years work and much intensive study of the district in general and as a result of this I have without assistance surely put the Perron on the map. I would like to do the same on Beaufor and get full credit for my work.

Waltz has had no experience in the Pascalis area, so why should he come in and take credit for the work I have done. I frankly do not like the situation. Credit for one's work in mining is worth much more than all the salaries in the world and in this game we find many such men horning in on the work of others.

It is fine for Waltz to come in and check up and make reports on my work but not when it comes to consulting.

However, if this is necessary then I wish to consult someone who will be of some use to me. I feel that having Waltz as consultant engineer on the Beaufor is to suit Beauchemin.

All I ask is to be left alone and I think that Perron successes can be quickly attained and I surely hate to be classed as an associate of Waltz in ore development in the Pascalis district. In this case I would be glad if you would correct this statement in the *Quebec Miner* or else give me permission.

The new vein on the bottom of the Perron surely looks important. We are just getting into the production zone now and I think news from the bottom level should come rapidly now.

J.P. Norrie

Andy Davis, associate of Mel Thomson, was quickly becoming J.P.'s best ally:

I received a telephone call yesterday from Thomson and today received a letter from him enclosing copy of a letter you had written him under date of June 15, 1935, with reference to the item which appeared in last week's *Quebec Miner* with regard to the operations on the Beaufor property.

Mr. Thomson was considerably worried, naturally, that this interview had been given out without their being consulted in the matter, and also statements that were put in regarding Mr. Waltz.

As far as I can understand, the feeling of the Nesbitt Thomson Co. and I have discussed this angle of the situation with them on several occasions, not only with Thomson, but also with P.A. Thomson, and also with Mr. Nesbitt, and they have been satisfied 100% with the efforts and ability that you have shown in handling the Perron situation, and they have not at any time, to me at least, intimated that there should be a consulting engineer tied up in the proposition. But what they do say, and what they do claim is that in order to present the facts properly to their clients, they must have a qualified engineer go in periodically to the property and check the work that has been done, and also the various angles of the situation, and report to them. This, I understand, you entirely agree with, and I think you can appreciate the necessity of it as far as the general public is concerned.

I think the whole trouble in this case is that someone, bubbled over with enthusiasm because the Beaufor property was again financed, probably ran into a representative of the *Quebec Miner* and felt it was necessary to spread himself before the public.

I agree with you absolutely that it is very unfortunate and I also agree that a correction should be made, but I think the proper people to make the correction are Nesbitt Thomson people, and I don't think there is any doubt but what they will make it. Of course you can appreciate the fact that until the new board of directors is organized on the Beaufor proposition and they have finally got control of the board, they naturally have to be a little careful in the way they move.

Mr. Thomson asked me in his letter which I received today if I would become a director on the Beaufor proposition, and I have written him today to accept this position, and you can rest assured, for your own information, that if I am accepted on the board you will have one director that is 100% behind you, and I am quite sure that Thomson is just as enthusiastic for you as I am. I don't think there is anybody that is closer to M.A. in this proposition than I am, because we have spent a great many hours in a variety of different places covering this whole situation from all angles.

I don't want you to feel that I am trying to butt into any of your affairs, but I just wanted to let you know how the situation looked to me, and assure you of my endorsement.

A.J. Davis

As we have already seen, J.P. hired on with Perron on a handshake. The offer and promise to buy shares at a specified price was a spoken commitment which Alex Perron would fail to carry out. Eventually J.P. was forced to take the matter to Mr. Michener for legal action. This was after several written requests that Alex Perron make good on his commitment and after almost two years making a mine out of Perron Gold.

I understand that you were recently successful in a court case against Alex Perron of Kirkland Lake. I believe I have a very similar case against Mr. Perron. The case and some of the evidence is as follows.

On June 13, 1934, when I was employed as manager of the Bussiers mine of the Treadwell Yukon Co. I was approached by Mr. Perron and an engineer by the name of R. Howe to take charge of the Perron mine.

In the first conversation a proposition was tentatively discussed, however I did not like conditions at the Perron and made no decision to go there.

Sometime during the latter part of June, 1934, Mr. Perron visited me at the Bussiers property and again discussed my going to this Perron. At this time Mr. Perron advised me that I could have all of his shares that I wanted at the same price that certain of the Perron shares were optioned and sold to Nesbitt Thomson Co. of Montreal, one of the conditions of the Nesbitt Thomson option being that I be secured as general manager of the Perron.

The date of this conversation with Mr. Perron is being ascertained, however two days later I definitely promised Mr. Perron to take charge of the Perron mine. A few days following this I placed my resignation with the Bussiers company.

When Mr. Perron found that I had tendered my resignation he then put off the matter of shares to me. He promised to settle with me many times until about three weeks ago he definitely refused to do so and said he never made any promise to me.

I have written him about these shares on two or three occasions to which I had no reply.

In Mr. Perron's dealings with Nesbitt Thomson Co. he promised Messrs Nesbitt and Thomson and Mr. A.J. Davis another director of the company to fix up which he never did and now refuses to do so.

It is my intention now to take this matter to court to get 25,000 shares of Perron at the price that Nesbitt Thomson paid for their shares i.e. .55 per share.

My arrangements with Mr. Perron were made in Quebec, however, I think it would be better to have the case brought up in Kirkland Lake, which would suit all concerned better.

Would it be possible to present the evidence of Messrs Nesbitt, Thomson and Davis at their respective places of business as I do not want to have them appear in person due to loss of their time. However I believe some of them would do it if necessary.

I would be glad to have your comments on the situation giving me some idea what charges would be for handling this case along with some idea of complete cost.

J.P. Norrie

Both Mel Thomson and Andy Davis did their best to make up to J.P. what Alex Perron would not. Eventually the matter was taken care of to the satisfaction of J.P. as described in this letter from Andy Davis:

As you know Alex Perron lost his suit with the Miss Oakes' estate and has to pay the balance of a note amounting to $5,000 plus interest and other court expenses.

I had dinner with Alex in Toronto the night of the day he lost the case and he told me then that he had to raise some money immediately to take care of this situation. After discussing the matter back and forwards he agreed to sell me 10,000 shares of Perron Gold Mines stock at .75 per share. He said he did not mind giving me a bargain but he would not give it to Nesbitt Thomson. I had to get Nesbitt Thomson's consent to allow him to sell it because you know he is tied up by contract to offer them any stock he wants to sell first. However, Mel was quite agreeable to me taking the stock and mentioned that his other directors had agreed with him to turn you over 5,000 share of this 10,000 that Alex is supposed to deliver to me at the same price as I am paying for it, .75 which would give you 10,000 shares of the stock sold up to $1.20 in the last couple of days.

Do not think I am trying to force this stock on you but am only trying to do you a good turn if you want it. Let me know if you want to take advantage of this 5,000 shares or not.

A.J. Davis

These matters did not interfere with mine problems, however. All possible effort was dedicated to development of the Perron mine and the financial statements show that clearly.

Mel Thomson thought the mine townsite was not proceeding quickly enough. J.P. responded to Thomson:

> I have your letter of the 3rd re townsite on the Perron. My experience in this regard is to leave a townsite until the property comes into substantial production, and you will find this to be standard throughout the district.
>
> In the Perron I feel we have a very important property and any money spent on a subdivision would not be lost. It was my intention to get after this just as soon as the mill gets into production for two reasons. First, we are all very busy now on the mill etc. and secondly I did not want to spend a cent on anything not absolutely necessary until we get our present program in the clear. It is going to take all available funds to do it. When the surveyor arrives I will see that he gets all help possible.

During the year 1935 an airplane landing strip was cleared on the property and considerable bush cut adjacent to the plant for the usual fire break. An excellent road was built from Senneterre to the mine, a distance of about 35 miles. As the mill neared production, good mine bosses were in demand. J.P. wanted young engineers about 30 years of age and contacted several mines in Quebec and Ontario searching out the best men.

A competent man to take charge of the cyanide process in the mill was also essential. The process was very delicate and he eventually found a very competent man in Frank McLellan.

Mel Thomson was not always the man with the questions. This time A.J. Nesbitt wanted to know:

> I have just received the announcement that the first fifteen days of May production was $33,500 which is rather disappointing.
>
> I am very much disturbed about the Perron situation, as I notice ever since the mill went

into operation there has been a reduction monthly in the value of production.

I would appreciate very much if you would send me a personal letter telling me why the mill heads are steadily going down in value and are much below (if I remember correctly) what you originally estimated. This reduction in value is a bit disconcerting when there is no definite information as to the reasons for same.

Nesbitt was not as well informed as Mel Thomson but J.P. explained the lower production levels:

When the original development work was done on Perron the average indicated width of veins was approximately 4 ½ feet with an average grade of $12 per ton. The first 150-ton unit was built to suit the above figures. However, when stoping got under way it was found that many high-grade stringers went off from the main part of the vein into the walls, which if mined considerably lowered the grade. At the meeting of the Board of Directors in the spring of 1936 it was decided to increase the mill to 350 tons to accommodate this larger low-grade tonnage, as it would be poor business to leave low-grade ore in the walls of the veins when the development work had already been done.

The mill had during 1936 averaged $10.33 per ton on a 150-ton milling basis. During that period we needed money badly and little attempt was made to mine the lower grade marginal ore.

During the first period of January 1937 the mill was brought up to 315 tons per day. Production figures since January first are as follows:

1937	Grade	Tons Per Day
Jan 1-15	10.16	315
Jan 16 - 31	7.99	321.5
Feb 1-15	8.43	338
Feb 16 -28	8.36	336

Mar 1 - 15	8.68	342
Mar 16 - 31	8.36	326
Apr 1 - 15	7.17	332
Apr 15 - 30	7.18	335
May 1 - 15	6.63	343
May 16 - 31	7.10	334

In the average shrinkage stope mine, where large tonnages of ore are broken ahead, an even grade can be secured by selecting various grades from different places. At Perron we are forced to take the ore as it comes and if we happen to run into a lower grade at any particular time it must be milled as it comes. This is what happened in April and the first period of May.

J.P wanted to work with the Pascalis interests to develop both mines together. However, the Perron Board didn't agree as discussed in this letter from Needlands, of Ventures, June, 1939:

I have your letter in which you inform me that some of the Perron Directors were adverse to the idea of Pascalis developing from the Perron Bottom Levels.

I think it would be well for us to think this question over seriously with respect to the future because it is probable that, if no arrangement for working together can be made, Pascalis will decide to sink its own shaft with the result that the final position would be less favorable to both companies.

Eventually J.P. would develop the Pascalis and the Perron mine together.

The following article was found regarding the Perron mine and is included to demonstrate the community life at Perron in 1938. The author, G.L. Caya, obviously moved to Perron with his family as a young boy. There were only French schools, no English schools. But it was still the Depression and jobs were scarce.

Sergeant G.L. Caya (M.M.)
Algonquin Regiment, "A" Company, 7 Platoon

In 1938 at the tail end of the Depression, my family lived in a small mining town called Perron, 16 miles from Val d'Or, Quebec. There were no roads in those days so we [rode] on a pole track after leaving a ship that had taken us some distance across Lake Superior. A truck had been modified to ride on the poles that were used as rails. This vehicle travelled at a rather slow speed but it was much easier than walking. We saw some wildlife such as a bear with two cubs going up a tree but they were not very close to the pole track and we soon lost sight of them. The truck was loaded with our bedding, a few cooking utensils, very little clothing and little food. After leaving the truck, our goods were taken by a team of horses pulling a stone boat leaving us to walk the last two miles on a bush trail.

When we arrived at Perron Gold Mine, we were met by our father who had arrived there a few months earlier and had built us a house of sorts with square logs. We didn't go to school from 1938 to 1939 so our mother would teach us some religion and many other things she had learned when she was in a convent. I come from a family of 8 children, which consisted of the oldest ones being 3 girls, 3 brothers and the two youngest ones being girls.

When I was fourteen years old I was employed at a movie theatre. I earned very little money but I was allowed to watch the movies for free. In the spring of 1940, I got a job working for the Inspiration Drilling Co. working underground for 5 or 6 months. By then the government made it law that all persons 16 years of age and older had to be registered. Being too young to work and produce a card led to me being laid off. My first and last underground job was over.

In 1941, I was employed by TransCanada Highways cutting and burning the right-of-way. I was offered a cookee's job and took it with pleasure. The contract with Crawley MacCracken would soon come to an end and summer was now over.

In September 1941 I found myself without a job and no money to talk about so I decided to join the army.

G.L. Caya

Perron Gold Mines, Limited

(No Personal Liability)

(Incorporated under the Quebec Mining Companies' Act)

PROFIT AND LOSS ACCOUNT

For the year ended 31st December, 1939

Bullion Production—less Shipping and Mint Charges		$1,455,987.16
Miscellaneous Income—including Rentals, etc.		15,435.56
		1,471,422.72
Mining and Development	$613,775.94	
Milling	112,871.77	
Administration and General Expense	103,100.64	
		829,748.35
		641,674.37
Depreciation on Plant and Equipment	105,422.24	
Provision for Provincial Royalty Tax and Federal Income Tax	60,116.79	
		165,539.03
Profit for Year		$476,135.34

Chapter 10

The House in Amos

by Harry Norrie

When J.P. was at Treadwell Yukon he lived in the "manager's house", a building erected and owned by the company, but when he took on the management of Perron it was to his advantage to buy a house for his family and rent a building for his office in Amos. From Amos it was only an overnight trip by train to Montreal or Toronto when he was called there on business matters, and he could rely on a telephone system that was more than just a fragile wire strung through the trees from the mine sites.

When he jumped off the train in Amos for the first time he found a very small town with a very, very large church with its nearly equally imposing priest's mansion.

The business section of the town was one block long with a rather large post office, considering the size of the population, a hardware store, two drug stores, a ladies' wear store and the Radio Cafe. The Transcontinental Hotel and Isador Clement's men's wear shop were on a short side street.

The house in Amos was quite small and always had more people in it than it could hold. They flowed out onto the lawn. There were friends and business associates of J.P., friends of Marg, friends of friends, friendly passers-by drawn to the warmth of the place, and children. Everyone was young, ranging in age from expecteds to about forty-five. One room in the house, half of the top floor, was sacrosanct, J.P.'s bedroom, out of bounds to all miscreants causing trouble on the first floor.

In the evenings there were card games, board games, treasure hunts and one-on-one games further afield.

J.P. enlarged the house by adding on a sunroom and a small bedroom on the ground floor. Then he stuck on a greenhouse where in the winter months he cultivated roses to be transplanted into

The wedding of Pauline Norrie and Dr. Robert Halet, a geologist took place at the Norrie family home in Amos, Quebec

his summer flower gardens. One of his quests was for a black rose. I don't think he ever found one but a supply house in Holland came close with a very dark purple variety.

There were maids in the house but only during the day. Their homes, of necessity, were in town. Their job was to clean up after the night before and to prepare food for anyone who wound down from overexertion. When hand-held movie cameras became available, J.P. bought one to record the frolic. There was one large easy chair invisibly marked with J.P.'s initials in a corner of the living room, beside the radio. This was before TVs in homes and before motion picture theatres in Amos. At eleven p.m. everyone went down while J.P. listened to news from the rest of the world. When that was over he went to bed and quietness continued. Frequently the revellers moved to another dwelling or out into a field; revellers not having the weight of management on their minds.

Mother Marg and I became very close. My mother died when I was four so I remember very little of her. For a while after Marg and Dad were married I called her Mother. She was a little young for the honour but who cares? I was in dire need of guidance from some source. That may have partly been Dad's intention in the first place.

One day Marg and I were in the living room talking generalities while little Margie played on the floor. After a while Margie came to her mother and asked for a glass of orange juice. Marg told her to go to the kitchen and ask the maid. I don't remember the maid's name; maybe it was Jeanette. In any case, both maids had been told never to give Margie a glass that was cracked or chipped. Margie returned with the juice but in a moment started to whimper and came to her mother's knee. There was a spot of blood on her lip. Marg immediately became very concerned. She examined the glass and found a rather large chip had been broken from the lip. A sharp edge had undoubtedly been the cause of the cut. She called Jeanette and asked if the glass had been broken when she gave it to Margie. Jeanette said, "No. It was okay." We supposed that Margie had knocked the glass against some hard object that caused the chip.

Marg examined the inside of Margie's mouth but could find no other damage. It was not a serious injury and Marg and I forgot about it until about 11:30 that evening when Margie woke up crying with pain in her stomach. This was unusual since she generally slept through the night until natural pressures awakened her in the morning.

Marg immediately panicked, remembering the glass incident of the afternoon. Her suspicion was that the glass that Jeanette had given to Margie was cracked and that the little one had swallowed the shard. She awakened Dad, who had been asleep for about half an hour, and told him of her concern about Margie and the broken glass. They decided to call Dr. Rudd in Noranda and ask his opinion. He said that Margie should be brought to his office immediately, about 100 miles away. After a heavy day Dad was in a state of anger and frustration. The usual gang of people were milling around the place showing anxiety and making suggestions - everything in turmoil! Dad got dressed while Marg arranged blankets to go in the car to keep Margie comfortable.

As usual I had gone down to the Radio Café early in the evening, and just at the height of commotion I arrived back home. A rare moment of cognition parted the mist. I figured that if the glass had been broken while it was in the cabinet, perhaps having been knocked against another glass, the shard might still be there. I went to the cabinet, moved some glasses around and found a small piece of glass. With a little more detective work I found the glass

Amos in the 1930s, a very small town with a very large church

that had been given to Margie. The chip fit perfectly. We called Dad back as he was backing out of the driveway. He was too tired for further comment. Marg called Dr. Rudd. And Jeanette..? I don't recall ever seeing her again.

As a general rule J.P. didn't pay much attention to religious outlets, even his own, but in this case he made an exception. He was never hesitant in expressing his views when witnessing what he regarded as injustice. The huge Amos church, circular in shape, made from beautiful cut stone that must have been brought from a distant quarry probably somewhere in southern Quebec, was out of place and a great burden to the small community. Its copper dome alone was said to be the second largest in the world. Some self-indulgent religious figure seems to have felt that he could not effectively deliver "the message" except from an edifice equal to his own importance, with the result that the local population (mostly French-speaking) was called on to carry a very heavy load. This upset J.P.

Still there is always a counterpoint. As J.P. became increasingly well-known in the town diverse people were drawn to him, and one of these people was a priest by the name of Father Morasse, who was somehow attached to the overly large church but with far broader interests.

Father Morasse was quite tall and thin. His hair looked to have been trimmed with a dull push-pull lawn mower with a couple of blades missing. His face seemed made from spare parts covered with greyish colored skin that could have been the result of being mistakenly locked in a wine cellar for a few weeks. But the eyes gave the man away: clear, bright windows to a mental library with long shelves of philosophy and philosophers but longer shelves of history, economics, comedy and many other just plain interesting things. It is little wonder he was drawn to and welcomed by J.P.

Father Morasse was a stand-out figure in an area where conversation was more likely to be profane than salvation seeking. Except for a very small number, the central interest of all people is the pursuit of money, and I think in this Father Morasse was probably not an exception, although in his case it would be the pursuit rather than the acquisition - perhaps he had hungry nieces and nephews.

In his black clerical robes and on his own schedule, Father Morasse would visit one or another of the mines under J.P.'s management. At Treadwell Yukon he rented a room above Mr. Legere's general store but in the evenings would be in J.P.'s living room before an open fire.

Mr. James Norrie June 28, 1944
Amos, Que.

Dear Mr. Norrie:

I have received the 3,000 shares of Croinor Pershing Mines Ltd. sold to me and for which I thank you very much.

I don't forget your goodness for me and I pray God to bless you, your family and to give you great success in your business.

Yours very truly,

Father J.H. Morasse

Note: the stock sold to Father Morasse would be some of J.P.'s personal holdings. Since he was financing Croinor by himself, without Nesbitt Thomson, he would have in excess of 300,000 shares and he probably sold it to Father Morasse for .25 a share or less.

One day one of the maids came to work with tear-filled eyes to reveal that her husband had died and that she would need $40 (which she didn't have) to have him properly interred in consecrated ground; otherwise he would be planted in just plain mud, like the Protestants. When Mother Marg broached the subject to J.P., he flatly refused to fork over $40, probably with the full realization that Marg would relieve the poor woman's suffering out the back door in the dead of night. But knowing Marg, she probably gave the money to the maid with a flourish right in front of J.P. and with J.P.'s acquiescence. But the point was made and would be carried back to the local father.

J.P. didn't argue with his wife Marg for one reason only. He was aware of the consequences of natural history, or how the Camel got its Hair Lip.

Marg Norrie

The legend of the first cam-u-el: An Arabian apologue

Across the sands of Syria,
or, possibly, Algeria,
or some benighted neighbourhood of barrenness and drought,
there came the prophet Sam-u-el,
upon the only cam-u-el,
a bumpy, grumpy quadruped of discontented mouth.

The atmosphere was glutinous;
the cam-u-el was mutinous;
he dumped the pack from off his back; with horrid grunts and squeals,
he made the desert hideous;
with strategy perfidious,
he tied his neck in curlicues, he kicked his paddy heels.

Then quoth the gentle Sam-u-el,
"You rogue, I ought to lam you well!
Though zealously I've shielded you from every grief and woe,
it seems, to voice a platitude,
you haven't any gratitude,
I'd like to hear what cause you have for doing thus and so."

To him replied the cam-u-el,
"Beg your pardon, Sam-u-el,
I know that I'm a reprobate, I know that I'm a freak,
but, oh! This utter loneliness!
My too-distinguished onliness!
Were there but other cam-u-els I wouldn't be unique."

The prophet beamed beguilingly.
"Aha," he answered, smilingly.
"You feel the need of company? I clearly understand.
We'll speedily create for you
the corresponding mate for you.
Ho! Presto, change-o, dinglebat!" He waved a potent hand.

And lo! From out vacuity
a second incongruity,
to wit, a lady cam-u-el was born through magic art.
Her structure anatomical,
her form and face were comical;
She was, in short, a cam-u-el, the other's counterpart.

As Spaniards gaze on Aragon,
upon that female paragon,
so gazed the prophet's cam-u-el, that primal desert ship.
A connoisseur meticulous,
he found her that ridiculous,
he grinned from ear to auricle until he split his lip.

Because of his temerity
that cam-u-el's posterity
must wear divided upper lips through all their solemn lives!
A prodigy astonishing,
reproachfully admonishing,
those wicked, heartless married men who ridicule their wives.

Arthur Guiterman

Now it may seem strange to see such a light-hearted poem in a serious book like this, but the Norrie family was always full of fun and poetry, particularly Virginia and me, ever since we lived on the farm in Nova Scotia with Annie and Henry. There we had few childhood toys, so we made our own recreation by memorizing magazine articles and poems. Virginia, now in her eighties, can still recite the legend of the camu-el.

Chapter 11

The Excitement of Prospecting

by Harry Norrie

The development of a mine must go through many stages: the search, discovery, the claim staking, surface exploration, subsurface exploration, preliminary underground examination and decision on mining method to be used; the removal of the rock and milling to remove the economic minerals. Like most other professions mining and milling are boring. But the search (prospecting), that is where the excitement lies.

J.P.'s search followed the allure of that elusive, duplicitous goddess, gold. She would make him rich and then destroy him.

A claim is staked by what is called pace and compass. We all know what a compass is, but a pace can vary in length depending on how long your legs are. To make everyone equal, the recommended figure is five feet for a double pace. In double pacing you count only on one leg, but when hacking your way through the bush, it is easy to lose count, so most prospectors, when completing twenty paces, will break off a twig and put it in their mouth. When they get to 1,300 feet and a mouthful of twigs, they know they're almost there. But a prospector is not expected to be a surveyor although he is expected to be reasonable. Some claims have been found to be four or more times larger than recommended. Those stakers must have had very, very long legs. They certainly claimed a far greater acreage than the next guy.

Fifty years ago, to keep your claims in good standing you were required each year to do five man-days work on each claim, which you would then record along with a sketch of your claims at the Department of Mines, and with a small fee your claims would be renewed for another year; work would include trenching and sampling. The guy with the long legs had to do far less work per acre claim than the honest guy.

It is surprising how much work you can do with a few sticks of dynamite. But if you did not do your work commitments by year's end, you could take a chance and try to restake before another guy got there. Claims came open at seven in the morning and if you were in a hot area you might arrive at your #1 post to find a mob of other guys waiting, all hungry, muscular and younger than you. If a prospector has not been successful in his own search, he will rush to where a discovery has been made with the hope that he can claim ground as close to the discovery as possible.

Claim staking regulations have changed since J.P.'s time when a claim was a quarter of a square mile and one man could stake twenty claims in his own name and beyond that he could stake twenty for each of two other people by proxy. That is he could, with permission, acquire an additional forty claims, but the proxy would have to sign the application for registration with the Department of Mines. When a major rush was on, it was not unusual to find guys at the local graveyard where names could be picked from gravestones. Although it has always been frowned upon for a man to sign another man's name, a dead proxy is unlikely to voice an objection.

Meandering through the bush one day I found my name on a stake where I had not placed it. I did not object either, when, on inquiring, I found the staker to be my own father.

Before setting out, a staker should have three things: a pair of boots, sturdy and as near waterproof as possible; a sharp axe; and a reliable compass. The compass should be set at astronomic north. That is, the declination between magnetic north and astronomic north should be accounted for on the compass.

By regulation, the northeast corner post is set first. That post, cut from a nearby tree, should be four feet high with the top end squared to four inches. On the post, the staker will write his name, the time of day, the number of the post, and affix a metal tag supplied by the Department of Mines on which the number of the claim is recorded. From this location he will blaze a line due south one quarter mile to his #2 post where again he will write the required information. From here he goes west to his #3 post, then north to #4 and finally due east back to his #1 post. But this format is rarely, if ever, followed, particularly if the area is hot and the bush is full of prospectors. As soon as possible the staker will run the outside lines of the claim group and complete the inside lines later. With some of the prospectors not being able to spell "mining claim" let alone "declination" it is surprising how little confusion there is. On a rush, the prospectors generally know each other and will get together to agree on which area each is to claim according to the time of their arrival on the scene: first man first, last man last.

In Ontario, Quebec and the Arctic, in the early stages of exploration J.P. spent extended periods of time living in tents. In J.P.'s time tents were made of heavy white cotton canvas or silk. Although a silk tent can be packed into a smaller space than one of canvas, it is more an expression of indolence than authority and reflects the intent of the owner to play rather than work.

A prospector carries his home and supplies on his back. His mode of transport is a canoe over water and leg power over land. That carries with it a wonderful feeling of freedom. In the bush, with no communication to the outside world and what is called civilization, the cares and worries of the world quickly fall away, leaving a person free to enjoy natural surroundings.

The prospector works through the day, and only in the late afternoon must look for a dry level spot beside a stream to pitch his tent and make camp. But to the untrained eye what appears level can be greatly misleading. There are no level places in the bush, unless, with luck, you come to a sandy beach, but then you are cursed with gritty food and sandy bedroll. And what you think is sufficiently level is bound to hide sharp rocks, roots and low rotten slumps that only become evident when you settle down to sleep.

Generally a prospector will find a couple of well-spaced trees that can be limbed, to which he can fasten a ridge pole for his tent but, lacking that, he can easily and quickly fashion a framework with long thin poles.

Once his tent is up, the experienced prospector will partially eliminate those pesky rocks and roots with a thick layer of twigs which he strips from the ends of tree branches. These make a comfortable mattress. If he has a groundsheet or a small tarpaulin, he may cover the twigs with that and over that lay his bedroll.

After the tent is up it is time for dinner. A small space of ground is cleared, dry twigs and fire wood are laid down and lit, and two kettles and a frying pan are unpacked. One kettle or pot is used to heat beans, always beans; the other is for tea, never coffee. If the prospector's meat supply is all used up or gone bad, he has only to toss his hook and line into a stream for a fat fish, which is always better than fresh meat anyhow. In foul weather his dining area may be protected by a tarpaulin, or he may have brought along a collapsible stove which he can set up in the tent.

Now to bed. His eyes close. But just as he is about to drift off, an owl on silent wings comes to settle in a tree almost directly over his head. "Wooo-wooo." The sound does not irritate him, actually it may be welcome - there is a softness to it - wonderment - mystery, a part of the world where the prospector is only on the fringes.

In daylight another bird comes calling, the beautiful Canada jay, sometimes called the camp bird. He is there to peck up scraps of food thrown out when food is prepared or with dish water. With care he can be persuaded to come to your shoulder or outstretched hand for more personal attention.

There is one function of the body where any form of comfort is denied. If the prospector is going to be in camp for only a few days, squatting will take care of it; if longer, he will have to establish a more permanent arrangement. This will require digging a shallow hole below a short length of wood attached to two closely spaced trees or on the other side of a windfall. One day in Chibougamau, Ryder Storn looked up from his windfall directly into the unblinking eyes of a Canada lynx. Ryder's need to return to that position was delayed for a number of days. Even much later, when telling of the event, his face and eyes would still register fear. It really scared the... Well!

Camping in the Arctic is not a lot different than at lower latitudes except for one very important feature - the bodily function just mentioned. With the temperature at forty below and more, you have to be very nimble. You do not dally. You do not bring along reading material.

If conditions look promising in the area being worked, a prospector may stay for a week, a season or several seasons, improving his living conditions as needed. He will have to arrange to have supplies brought to him regularly. More likely he will strike camp and go for them himself, possibly also into welcoming arms. With deep snow on the ground he will not be prospecting during the winter months. He will probably stay in town, but if he has made a discovery he may continue to work by blasting and trenching.

In winter his camp setup will be quite different. He will not continue to enjoy the pleasures of sitting by a campfire in the evening; his tent will be heated by a small stove with wood cut and piled by hand. The stove is what is called an airtight or Quebec heater. It is oval in shape, made of light metal, and sits on four short legs. It has a wide opening on top to admit fire wood and the opening is covered by a lid. Here he will also cook his meals. Now pay close attention; this is very important. The prospector will need a few lengths of stovepipe to expel the smoke from the stove, but he will *not* stick the pipe through the top of the tent. A spark is sure to fall down and set the canvas on fire. Rather he will cut a hole in one wall to which he will attach a metal ring to absorb the heat when the pipe gets red hot. The pipe goes out horizontally through the ring, then a ninety degree elbow is attached and several more lengths of pipe erected vertically.

In the winter months, in deep snow, a prospector will not be prospecting but he may want to stake some claims or, as mentioned, do other work. In this case he will have to be careful to remove all the snow from where his tent is to be set up, down to the underlying moss, so that his bed does not rest on snow; otherwise he is going to have some cold nights and little sleep.

There is a part of bush life that cannot be equalled anywhere. Mostly everyone likes the sound of rain on the roof, but that cannot nearly equal the sound of rain on a tent roof. For some reason, that lends a feeling of peace, security and tranquillity, especially at night in a snug eiderdown bedroll, even with varying degrees of violence. And then in winter when a violent blizzard is raging through the tree tops, at ground level, with the protection of the surrounding wall of trees, the wind is deadened and allows the snow flakes to sink gently to the ground.

In the soft warm rain of summer it is a great pleasure to walk for a while in the woods and become completely drenched. There is a sensuality about it, but let's not...

Gus Schlet worked for a while at Malartic Goldfields and lived with his wife in the director's cabin at the rapids. Among other things, Gus was a casual amateur photographer. Before coming to Goldfields he had had a short-time job in the bush and lived in a tent. One night his sleep was interrupted by the sound of rustling in the tent. In total darkness he waited for a moment to cogitate the situation, then felt for his camera that was always at the ready. He got a picture of a startled, full-grown black bear, in the tent, close beside his bed.

On a claim staking trip to Chibougamau in the early winter of 1951 Jim Hobson, Lorenzo Blondeau and I set up our tent in deep bush about seven miles from the highway. We expected that the work would take only a few days, and when setting up the tent, foolishly stuck the stovepipe directly up through the roof. Early the next morning, while Jim and I were asleep, Lorenzo got up and started a fire in the stove, then went back to bed. After a short time, by some instinct, he opened his eyes and found that a spark had fallen onto the tent and had already burned a large hole in the canvas. He jumped up, and with his upper body through the hole he whipped around and around with his bare hands and was able to extinguish the smouldering fabric. If the smoulder had burst into flame, we would all have been in extreme danger. Fortunately we carried with us a large tarpaulin, which we fixed over the tent. Then we properly installed the stovepipe through the wall.

The bones of all old prospectors should be ground to fine powder, taken to a clear lake in a wilderness area and wafted over calm waters by a soft evening breeze. Or if his life has been of a more rambunctious nature, ashes should be placed in a very small canoe-shaped vessel and launched into the waters of a fast running stream just before it enters an area of white-water rapids.

An exploration program, following staking, involves stripping, trenching and sampling, when chips of rock from mineralized zones are removed for assaying. If the results from this work are satisfactory, a diamond-drilling program is instituted. Diamond drilling is not a search for diamonds. It is so called because the cuffing bit is studded with industrial-grade diamonds. The diamond bit is affixed to a string of rods which are then ground down into the rock. The rods are hollow, permitting a core of rock to be recovered, which is then split in two lengthwise. One half of the core is kept in wooden trays and made available for examination at any time, while the other half is sent to be assayed for mineral content. The diameter of the rods determines the size of the core, which may be from less than an inch to several inches. The characteristics of the rock being drilled determine the diameter of the core to be drilled.

Many professions claim to be the second oldest in the world, the oldest being a toss-up between religion and prostitution, each engaged in cadging cash. Mining's claim to longevity seems pretty sound, however, since it is quite likely that ancient man had to harvest rocks to throw at offending neighbours and wayward animals before he learned to fashion flint and fix it to a stick with which he could then deter his neighbours with accuracy from a distance.

To gain the riches of the earth many mining systems have been developed. Surface mining involves two main types: stripping and open pit. Strip mining is mostly used for the removal of coal, where in some cases mountains are removed to access flat-lying layers of the black fuel. Open pit mining is used to obtain a number of minerals such as copper, gold, zinc, lead, etc. or just for crushed rock as an aggregate for cement and concrete. Here again mountains may be moved.

Following a successful surface exploration program, when an economic deposit of mineral has been indicated underground, then one of the many forms of underground mining is chosen: shrinkage, room and pillar and others. It will be necessary to sink a shaft to further evaluate the deposit.

Exploration shafts are generally rectangular in shape and consist of two or three compartments, each being about seven feet square. One compartment equipped with ladders is used as a man-way when the cage (elevator) is not available.

The depth of the shaft will depend to some degree on where the top of the ore body is located. It may come to surface or be hundreds of feet below surface. The shaft may be vertical or inclined, again depending on the nature of the ore body. It will have an open framework of sturdy timbers put in place as the shaft is lowered, and it will have a signal system to send directions to the hoistman. Vertical guide rails are fitted to the timber framework to keep the cage in place as it is raised or lowered on the end of a braided-wire cable.

As the shaft progresses, rooms called stations are cut at regular intervals in one wall of the shaft. From the stations, tunnels, called drifts, are driven out, usually a short distance. The end of the tunnel is called the face. The floor is called the floor, for obvious reasons, but the roof of the drift is called the back for reasons known only to the guy who first started calling it that. The drift will rise at a low angle, about one half of a degree, as it is driven out from the shaft, to allow ground water to flow back into the shaft, the bottom of which is left open to provide a sump from which the water is pumped to surface. The water will seep back into the ground to create the same problem it did in the first place.

After the drift reaches the end of the ore body and is discontinued, other tunnels called crosscuts are driven at regular intervals into and through the ore body. The walls of the crosscuts are sampled for mineral content. The crosscuts also provide locations for further diamond drilling with the advantage that they can be drilled at any angle, up, down or sideways. The crosscuts also provide locations where vertical or upward-inclined tunnels, called raises, are blasted out. Further sampling can be done by opening up rooms extending from the raises. These rooms are called stopes, and from these, larger samples can be taken to be milled, giving a more accurate evaluation of the deposit.

Broken rock is removed from the underground workings by metal mine cars that are taken by the cage to surface, dumped and returned underground. Until all sampling and test milling is completed the deposit cannot be called an ore body, since the definition of ore is mineralized rock that can be mined at a profit, and that hasn't been determined yet. When all of the sampling, surface and underground, is completed, then you will know whether you have or have not an ore body.

If exploration work is successful, the final step is to bring the mine into production. A large production shaft and a mining system suitable to the nature of the ore body will be needed.

J.P.'s mines used either cut and fill or shrinkage stoping, except in one case where it was recommended by a team of consulting engineers that block caving be used. That was unsuccessful. A large block of ore was lost.

With shrinkage stoping the ore body is completely undercut and removed to a height of ten to twenty feet. A floor of large timbers is then set in place and the remainder of the ore above the floor is blasted out in lateral strips. When rock is blasted, it fills a larger space than when it was in place so that a portion of it must be removed down points in the floor. Each strip of rock is removed by working on the broken rock of the previous strip. When the undercutting reaches the level above or within a few feet of surface, all of the broken rock is removed from the stope. Now, to support the walls of the stope and prevent caving, the stope is filled with gravel (backfill) through openings in the surface rock. With this mining system most of the broken rock is held in the stope until all of the ore has been broken.

Cut and fill mining is much the same as shrinkage except that the ore is removed as soon as it is broken. The open space is partially filled with backfill (gravel) then covered with a timber floor onto which the next strip up is blasted. The advantage of this system is that all of the ore is available to the mill as soon as it is broken.

As soon as J.P.'s mines went into production his interest turned again to other explorations; as I said before, breaking rock is boring, in some situations restricted to felons.

Two miners underground, East Malartic

Chapter 12

Inspiration Mining and Development

by George Norrie

The Perron Gold Mine was developing nicely to say the least and J.P. had to commit only half of his time to its management. He could now put his personal dream into action. The Tiblemont Island had given him the financial resources to see a company incorporated and he named it Inspiration Mining and Development Co. Ltd. Inspiration would be his eyes, ears and nose in all aspects of the mining business. He would be prepared to act on any and all prospects in northern Quebec and Ontario. Exploration, diamond drilling and mine development on each property would provide him with first hand information on which to base decisions to take, increase, hold or delete investment interests on any property. The Inspiration setup was well on the way to incorporation before J.P. left Bussiers. Since he already owned the Coffin Mining Co. he, of course, wanted the new Inspiration to take the Senneville and Fournière township properties as its first investments.

Carl Erickson, one of the intended financial partners, discussed the initial setup of Inspiration:

Since you left it occurred to me that we could probably handle another fellow of a similar type to Jack Workman. You will have met him by now, Ken Carmichael.

I would suggest that we offer him the same terms as we did Workman and have him cover this field, including Sudbury and CNR field to Foleyet from the Sault North to Ota, Matashewan, Temagami, Gowganda, eastern Kirkland, the Quebec belt and Chibougamau.

I have estimated that salary requirements and expenses for each man will be approximately $3,000 for two men, Workman and Carmichael, a total of

$6,000. The estimate is based on a salary of $150 per month and $300 per month for expenses. The expense item is probably high but it is safe in event either of these men need to hire a man or two.

Carl Erickson (left) with J.P. and the Muskeg Limited rail car in the background

In reply, J.P. outlined his thoughts on how financing Inspiration should be structured:

I have your letter of 23 re: Coffin Mining Co. A telegram this AM from Workman says he is definitely tied up with Tech Hughes and cannot accept our offer. So I think it would be wise to send Carmichael out at once to the districts you mention. It may be well to outline a fairly definite course for him in case we want to send someone later on. When the Coffin company was formed we were given [by Bussiers] 100,000 shares for the Senneville group. The Fournière ground was turned in later and no adjustment was made for this. [J.P. wanted 200,000 shares for these claims from Bussiers]. We have considered 200,000 shares to split it up equally among the four new directors. No allowance has been made for a fifth so we will either have to split the 200,000 five ways or else take down another 50,000 shares. Either way is suitable to me. This can be decided by the present directors.

The present share standing for the Inspiration Co. is

Original holding of Coffin & Norrie	300,000
Four new directors @ 50,000 shares	200,000
10,000 shares to each director @ .10	50,000
100,000 shares to be sold @ .10	100,000
5 shares to each director	5
Balance in treasury	2,349,995
Total	3,000,000

Sales made to date of the 100,000 shares to be sold @ .10

Dr. Eugene Audette	10,000 @ .10	$1,000	paid: $500
Arthur Rivest	10,000 @ .10	$1,000	paid: all
Employees Bussiers	10,000 @ .10	$1,000	paid: $200
W.G. Worden	10,000 @ .10	$1,000	paid: all
Isadore Cleman	10,000 @ .10	$1,000	paid: all

One fellow I had counted on for 10,000 shares has temporarily backed down. I have written a few others but have no reply yet. Riley should take 10,000 shares as he made a nice stake in Tiblemont. I wish you would see him. Cleman says he would take 10,000. Please tell him to put the money in the bank...

J.P. Norrie

Erickson (the financial partner) objected to the proposed property acquisitions as well as the basic financing:

...now having your picture on paper where I can study same, I cannot say I agree with the setup. I discussed the matter with Charlie Massicotte and he is of the same opinion, which I will explain.

It is not a matter of setting up an argument but we are entitled to our opinion, namely that we do not care about bringing the Senneville and Fournière groups into the setup. We appreciate that the idea of the picture originated with you, etc.

One reason we are not in favour of the setup is that we cannot see knocking down 300,000 shares for these groups. It may be your opinion that they are worth same to you but we cannot agree, etc.

That is the only reason Parker turned you down, namely that he and his crowd did not want to carry the deadwood with their claims, etc. We are of the same opinion. This is not written in any spirit of argument, but if our representative turns out to find a property worthy of taking up, immediately your load of two groups of claims for which you have already knocked down 300,000 shares becomes a burden.

My suggestion and Charlie agrees with me, that since this organization is partly underway why not start off from scratch with a new sheet. Here is the question. Regardless of what you or Lee thinks about these two groups of claims, they have had an opportunity to make good during the past three years but have failed to do so. That speaks for itself. [Senore and East Malartic!]

Here is our proposal for 3,000,000 shares:

For preliminary finances
100,000 shares @ .05 per share

C.E. Massicotte	20,000	@	.05	=	$1,000
J.P. Norrie	20,000	@	.05	=	$1,000
Carl Erickson	20,000	@	.05	=	$1,000
Lee Coffin	20,000	@	.05	=	$1,000
Frank Dorval	20,000	@	.05	=	$1,000

For exploration finances for two years employing one and probably two engineers for scouting and examinations:

100,000 shares @ .10 per share

Dr. Eugene Audette	10,000 @ .10	$1,000	paid: $500
Arthur Revist	10,000 @ .10	$1,000	paid: all
Employees Bussiers	10,000 @ .10	$1,000	paid: $200
W.G. Worden	10,000 @ .10	$1,000	paid: none
Isadore Cleman	10,000 @ .10	$1,000	paid: none
Fred Reid	10,000 @ .10	$1,000	paid: none

Lee Coffin was an experienced prospector and had learned to trust very few in the mining business. He did trust J.P. but not Carl Erickson. He didn't think much of putting up any cash as Erickson suggested:

Yours of June 16/34 was badly delayed in reaching me in the bush and I had hoped that possibly I might see you in Amos which has further delayed my answer.

I am quite agreeable to your suggestion regarding the Senneville ground [Senore], but am not quite sure that I can comply with Carl's suggestion at this time of putting up $1,000 as a director. I had hoped that you would be able to take the lead in the new Inspiration Co. when I would have given my full slice of financial support.

I still believe in the Fournière ground [East Malartic] but a sizable group should be staked, if any, and we should have our plans well thought out, so…

The fire was very good to me only taking one of my… camps. Have been very busy covering the burned area before someone else came in and found a mine.

Strict gossip has it that you are soon to command the Mathews [Perron] property, so I will be expecting a lot from our Senneville holding with you as a neighbour. Hope all goes well and with kindest regards to you and yours.

Inspiration Mining and Development Co. Ltd. Prospectus

Inspiration Mining and Development Co. Ltd. 3,000,000 shares par value $1. Quebec charter was incorporated Sept. 1934 to take over assets of the Coffin Mining Co. Ltd., a subsidiary of the Treadwell Yukon Co. Ltd., incorporated to operate 600 acres on the granodiorite contact in Senneville Township, Quebec adjoining the Perron Gold Mine and on the strike of its veins. The shares now outstanding are 100,000 for the assets of the Coffin Mining Co. and 100,000 shares to directors at .05 and others at .10 leaving 2,800,000 shares in the treasury. It is proposed to sell further 200,000 shares @ .10 to provide a treasury in excess of $25,000 to cover the prospecting season of 1935.

Purpose: the purpose of this company is to employ two qualified geologists and a number of good prospectors to prospect and explore favourable mineralized sections in northwestern Quebec in particular and other parts of the Pre-Cambrian Shield. To acquire discoveries and work them to a point where the formation of a subsidiary operating unit is justified or dispose of such properties to the large mining companies at a profit to Inspiration. It may be well to note here the Canadian Exploration Co. Ltd. operating from Amos have had great success as during the last two years they have sold the Lamaque property to Teck Hughes and the Sigma property to Dome Mines Ltd. It is further proposed to stake promising ground adjacent to new finds and for this arrangements have been made with local flying concerns to quickly transport Inspiration men to the scene of such discoveries.

Complete equipment including outboard motors, collapsible canoes, etc. are ready at all times for such quick trips to new discoveries. Intensive prospecting during the field season just finished has revealed the presence of many new prospects in the great gold belt extending south of the Canadian National Railways from Porcupine to the Pascalis area of Quebec. The last year's work in the Chibougamau area has demonstrated that it is one of major importance and recent developments of Barbeau Lake, Rose Lake, etc. stamp this area north of the railway as possibly the best prospecting ground in Canada for gold and base metals.

Property: the only property now held by the Inspiration Co. is 600 acres on the granodiorite contact adjoining the Perron mine in Senneville Township Que. It was acquired from the Treadwell Yukon Co. when the Canadian operations were suspended.

Approximately $15,000 has been spent on the ground in trenching, blasting, etc. with the result that an important gold discovery has been made. The ground is heavily covered with overburden which makes surface prospecting difficult. However as the granodiorite contact is sufficiently outlined by a number of outcrops, diamond drilling can be accurately placed. Before further work is done on this property it is intended to await developments on the Perron mine which at this time are distinctly favourable. It may be noted here that aside from the Perron mine, the Bussiers, Sigma, Lamaque, Stabell, Sullivan and Siscoe are all associated with the same granodiorite contact.

Personal: J.P. Norrie, the originator of the Inspiration Co., has many years experience in the operation of mines and prospects in various parts of Canada, more particularly in northwestern Quebec. During the last three years from Aug. 1931 to Aug 1934 he was manager of the Bussiers mine of the Treadwell Yukon Co. Ltd., which position he resigned to become president of Inspiration Co. Ltd. Previously he had much exploration experience in northwestern Quebec, Great Bear Lake, Coppermine

River area, etc. He was the first manager of Stadacona and the Malartic mine in northwestern Quebec and was responsible for the discovery of gold on Tiblemont Island and more recently in the East Malartic property.

Other information: Frank Buckland, recently chief geologist of the Treadwell Yukon interests in Quebec has been retained to take active charge of the Inspiration. Mr. Buckland is a graduate of University of British Columbia in mining engineering and McGill in geology.

The Inspiration company is being promoted by mining men of long experience and excellent reputation. Operations will be confined to the most promising mining areas of Canada. The public are invited to become associated with the Inspiration Co. with the assurance that the sponsors are spending their own money with them in a venture which promises to become highly profitable to all concerned.

J.P. Norrie

With the incorporation of Inspiration out of the way J.P. began immediately to market his shares. The first were the directors and close friends and employees at Bussiers. Then he went to his old boss and colleague Thayer Lindsley at Ventures.

Mr. Lindsley has passed over to me your recent undated letter… accompanying 25,000 shares Inspiration…

Mr. Lindsley advises me that Ventures has undertaken to purchase these shares at .10 per share, and in reading your letter further I note that you are counting on the money for the next year's campaign. In the circumstances, would it be satisfactory for us to make payments as follows:

$1,000 on Feb. 1, 1935
$1,500 on April 1, 1935

Mr. Lindsley has gone over your letter outlining your proposed activities and also your special comments on specific properties. He has not given me any definite word for you with regard to either Barbeau or Rose Lake claims you dealt with but as soon as he returns, I will ask if he wishes to make any statement with regard to these.

Neelands

One property J.P. had been working on personally for a long time was Gwillim Lake in Chibougamau. He was acting as the agent for the vendors at a commission (similar to Bussiers) and wanted to contribute the shares he earned as commission to Inspiration as his personal contribution. There was a problem which he discussed with J.P. Bickle of MacIntyre Porcupine, who took an option on the claims:

Since an arrangement was completed with you, I have had a discussion with the vendors of the Gwillim Lake property as to my commission for handling it.

Due to the fact that substantial concessions had to be made, they do not seem inclined to allow me much. In fact will not agree on anything until I have been in touch with you to see if you will bear part of the expense.

This point should have been brought to meetings, and as it was not, I thought the vendors intended to take care of the matter themselves, otherwise I would have done so.

I have spent considerable on the property in aeroplane fares, time, and other expenses in putting it in shape to make it worthwhile for interested parties to go in there and examine it, and feel that I should have something substantial for my services, and would be glad to have your views on the subject.

Bickle obviously didn't agree:

The suggestion submitted in your letter of Feb. 25, 1935 comes to me as a distinct surprise, for surely all parties concerned fully appreciate that any commission in respect of the deal is solely the obligation of the sellers.

My impression gained in the several conferences we had was so favourable as to character and fairness of all parties that it would be difficult for me to imagine any differences arising between the vendors and yourself that could not be amicably and satisfactorily adjusted between yourselves. I am sure this will be the case.

With kindest regards and best wishes to all.

J.P. again brought up the commission situation with Bickle some time later, but he was really looking for his investment.

I did not answer your letter due to the fact I was unable until now to get a satisfactory commission out of the vendors.

I fully understand that any commission of this sort should come from the sellers and hesitated to write you on this account. However, certain of the vendors requested me to do so before they would settle the matter, the commission has now been set at 5% of the vendors' cash and 5% of the vendors' sales which is satisfactory to me. This commission goes to an exploration company which I have recently organized and know as the Inspiration Mining and Development Co. in which I would like you to become interested.

The company is capitalized at 3,000,000 shares of which 230,000 shares have been issued.

The present assets of the company are as follows:

600 acres adjoining the Perron mine in Pascalis which is developing in a most spectacular way,300 claims close to the Gwillim Lake property on which a gold discovery has been made.

5% of the vendors' share in the Gwillim Lake property.

$10,000 in cash in the bank.

Three diamond drills working on profitable contracts.

Two prospecting parties outfitted and financed now in the Rose Lake Chibougamau areas.

Also a small interest in several other prospects.

I intend sending more men out after the breakup period.

I have done very well in this area by keeping my ear close to the ground and now without salary from Inspiration Co. or immediate recompense in any way I hope to build up an organization and accomplish something worthwhile with it by issuing not more than 500,000 shares and making them worth real money.

Through being acquainted with most of the prospectors in this area I hope to acquire some of the new discoveries made, stake ground adjacent and by various ways and means keep in closest touch with general development in the area for the benefit of the Inspiration Co. I can see more Norandas in the country north of here and hope to be close by when they are found.

I expect this organization is too small for you to bother with. However, if you care to take a small bet at .10 per share I can guarantee that I will make every effort to make your interest profitable.

J.P. Norrie

Nesbitt Thomson was becoming more and more confident in the ability of J.P. as the Perron and Beaufor developed quickly into top junior mining companies. When J.P. incorporated Inspiration and East Malartic as separate and very different companies, they were watching very closely. By March, 1935 J.P. sent Mel Thomson his first shares in the Inspiration venture:

I am enclosing four share certificates of 5,000 shares each... of Inspiration at .05 per share... You are being appointed a director along with Mr. A.J. Davis. Any of the present directors can be replaced at will.

The president, C.E. Massicotte is a well to do Frenchman, a good fellow and a square shooter. This fellow has many local mining connections and was one of the vendors of Bussiers, Leroy, Beaufor, etc. No doubt many valuable mineral discoveries will be made by the French element and with a French president it will be easier for me to acquire them. Massicotte has a very small interest in the company and if you object to him he can be changed quickly. The other directors are Lee Coffin, Geoffrey Porteous and myself. Coffin is a prospector associated with local mining and has been the discoverer of many valuable finds. I appointed Porteous as he can be of some assistance to me in Montreal business. If you object to any of these fellows they can be replaced.

The following is a list of present shareholders:

Geoffrey Porteous	20,000
J.P. Norrie	60,000
Lee K. Coffin	50,000
Walter G. Warden	10,000
Thayer Lindsley	25,000
L.A. Ladouceur	5,000
Dr. Eugene Audette	5,000
Arthur Rivest	10,000
K.D. Clark	2,000
Charlie Jinn	2,000

Making a total of 189,000 shares issued.

All these shareholders, except two, paid .10 per share. Lindsley has nothing to say in our affairs and he is a shareholder at .10. I believe he is regaining prestige in the mining world as he has shaken off all vultures with whom he was connected and some of his ventures such as Falconbridge, Beattie, Canadian Malartic and Canadian Matachewan are coming through. Of the others Warden is a retired civil engineer of Toronto, Audette is a local doctor, and Ladouceur a local lawyer and conservative standard bearer. The present assets of the company are 600 acres adjoining Perron, 1,200 acres adjoining the new McIntyre show at Gwillim Lake, $7,500 and 22,000 shares in the property recently taken over by McIntyre at Gwillim Lake. i.e. 5% of 150,000 plus 5% of 450,000 shares. I was responsible for the deal and turned my commission in to the company. Two complete diamond drill outfits working on profitable contracts and another on order for which a contract has been secured. Two prospecting parties are now in the field and partly financed for the season. Approximately $5,000 in cash is in the bank.

Our share in the MacIntyre property alone makes the stock worth .15 per share, providing the property stands up and I think it will. They are taking down treasury shares at $1.

I think you will agree that this is a nice setup. If there is anything in the setup that you do not like please let me know and I will have it changed to your liking.

You will notice there has been no robbing of the treasury. Coffin and I took 50,000 shares each for the property adjoining Perron. The balance I paid for in cash. My object is to sell up to 500,000 shares when necessary and am sure I can make them into real money. A few more kicks like the MacIntyre alone can do this. Please advise if there is anything further you wish to know.

J.P. Norrie

Andy Davis was Mel Thomson's closest friend and associate at Nesbitt Thomson. It was obvious he was enthusiastic. The whole family wanted stock:

Your letter containing certificates of Inspiration Mining and Development for 5,000 each, or totalling 20,000 shares at .05 per share, I found here at my office on my return yesterday afternoon.

I have also read your letter with a great deal of interest, and have noted the officers of the company, also shareholders, and hasten to assure you that I have no criticism to offer in any way, and think that your ideas with reference to the president are 100% right.

I have pleasure in herewith enclosing you my cheque for $1,000 covering 20,000 shares of Inspiration and have also pleasure in herewith enclosing you Aubrey's cheque for $2,000 for 20,000 shares of Inspiration at .10 a share, also E.J. Davis' cheque for $2,000 for 20,000 shares of Inspiration.

Kindly be good enough to have 20,000 shares of Inspiration made out in the name of Aubrey Davis, and mailed direct to him at Newmarket. Also 20,000 shares in the name of E.J. Davis and mailed to him direct.

I want to take this opportunity of expressing myself with regard to the way you are opening up this company and want to say that I consider you have been very fair indeed, and sincerely trust in the years to come it will turn out to be a very profitable investment, and also the means of my becoming much better acquainted with you personally, and sincerely trust that any time you are personally in Toronto you will let me know so that we may have a chance to arrange a meeting...

A.J. Davis

By the latter part of 1935 Inspiration was well on the way to becoming a successful drilling and exploration company. It was structured very similarly, in its investing and exploration departments, to the larger Ventures company owned by Thayer Lindsley:

May I congratulate you on the successful results attained by the Inspiration Mining & Development Co. during the past season.

They make your prospecting company stand out above the rest.

Thayer Lindsley

Ventures was already J.P.'s prime competitor in northwestern Quebec. Oddly enough, they were also his partner in most of his projects. These words, coming from Thayer Lindsley were certainly very complimentary remarks. Thayer Lindsley was, without doubt, J.P.'s friend and mentor, competitor and partner. Their association was cemented by trust and respect, very rare ingredients indeed.

After Lindsley died in 1976, the *Northern Miner* ran this tribute:

One of the giants of Canadian mine exploration and development - perhaps the greatest of them all - Thayer Lindsley, or "T.L." as he was affectionately known to his scores of associates and hundreds of employees, passed away peacefully last weekend at his New York apartment. The grand old man, still actively pursuing his personal mining interests, was in his 94th year, only three months away from his 95th birthday.

By unanimous consent of people who worked with him and knew him best, Thayer Lindsley was a genius. This is the description that most often comes to the lips of his cohorts when asked their opinion of the man. Some liked him more, and some less but no one disputed his uncanny geological sense and his prodigious capacity for work. Tales of his photographic memory are legend: he had almost total recall down to the last detail of things read, or seen, or heard some 20 or 30 years earlier.

Thayer Lindsley was the man who fathered Falconbridge Nickel Mines and the tens of gold and base metal producers that evolved under the wing of Ventures Ltd., the holding company through which he promulgated his geological ideas and theories. At the peak of his career he was president of 10 mining companies, director of 15 others, and in control of 40 or 50 more. He resigned as president of Ventures in 1955, of Falconbridge in 1956, but he never retired. He formed Westfield Minerals in 1958 and from his New York base was involved in mineral exploration round the world.

Self-effacing to a degree, T.L. shunned publicity like a plague. A few years ago when the *Northern Miner* published some complimentary remarks regarding his career, he thanked us but insisted that "The greatest contribution to the success of various mines I started was due to the stalwarts of Canada with whom I was associated. The gallant crew contributed talent and energy to a degree that far exceeded all others. To them, Canada and myself owe a great debt."

It is hardly surprising that on his death his name had no meaning for the metropolitan Toronto press or radio. They simply had never heard of him. But to Mrs. Mary Millenback, his friend for over 40 years, he was a wonderful man. It was she and her husband Jack who were entrusted with the distribution of $750,000 of the Lindsley fortune to Canadian universities, churches, schools, and hospitals, and who followed his instruction to see that the money was distributed to as many people as possible. He gave away millions both here and in the United States, most of it anonymously, such as the canteen he built and equipped for airmen during the war at RCAF Base Borden. More than that, he directed that all his dividends during the Second Word War be put into a fund to

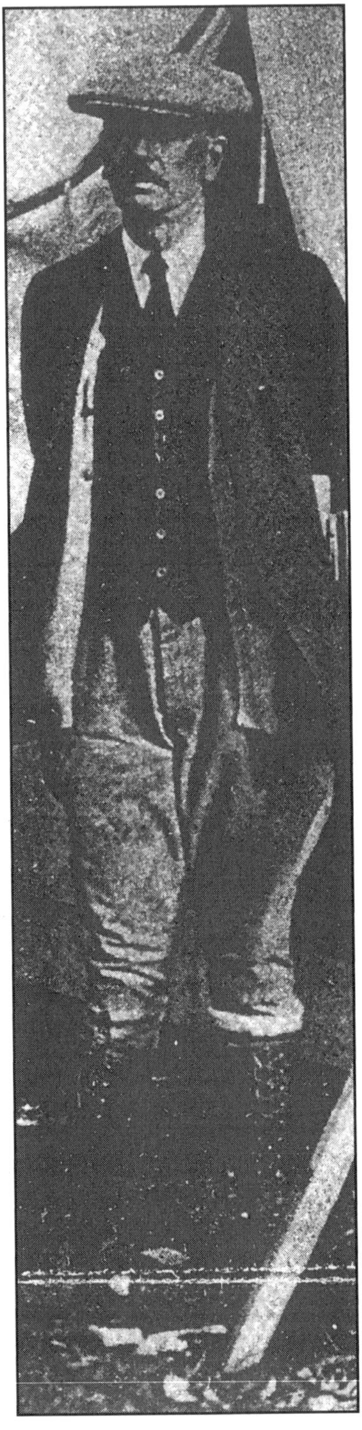

Thayer Lindsley

aid the war effort - a sum that was estimated to have totalled $800,000.

At the age of three score years and ten, Lindsley had more energy in his rangy, frail-looking, six-foot frame than most men half his age. Even then, his prodigious capacity for work used to embarrass his younger executives.

T.L. lived simply and his wants were simple. He neither smoked nor drank, nor did he object to those who did. He never owned a car, and his favourite foods were apples and whole wheat bread. He was so absorbed with his mining and geological projects that he often forgot to carry enough money on his field trips. Friends have had to lend him a few dollars - to tip a porter, buy a breakfast, or get a taxi.

Lindsley was born in Yokohama, Japan, in 1882, where his American father was the representative of the Canadian Pacific Railway. He was educated in the United States and after graduating in civil engineering from Harvard he found his first job with the New York Transit Commission as a rod man at $80 a month. His older brother, Halstead, encouraged an interest in mining and he returned to study geology at Columbia University around 1910. He honed his mining expertise at various operations throughout the western States and Mexico. Preferring to be on his own, says an acquaintance, he rarely took a salaried position. As a consultant he became involved in the Homestead Iron Dyke Mine in Oregon, and after serving overseas in the First World War, returned there, leased the mine from the owners, applied his theories to find more ore, and wound up with a nest egg of close to $30,000.

Lindsley came to Canada around 1926 to examine a property on which McIntyre Mines held a "million dollar option" from the Rouyn prospector and promoter R.C. Gamble. McIntyre had wanted Fred Searls, of Newmont fame, to do the examination, but he recommended Lindsley, then a successful but not widely known geologist in California. Lindsley turned down the Gamble option, but at the same time offered

McIntyre a property he had staked nearby. This later became Osisko Lake Mines. McIntyre spurned the offer but Joseph Errington, then a McIntyre director, went into the deal personally with Thayer, at the same time bringing along Brigadier General D.M. Hogarth for a share. That was the start of the Lindsley epic in Canada.

Lindsley moved quickly into the formation of Sudbury Basin Mines, one of his less successful efforts. Then in 1928 came Sherritt Gordon Mines when Lindsley went into a partnership with R.J. Jowsey that also included Gen. Hogarth, Joe Errington and Col. C.D.H. MacAlpine.

Again it was Lindsley who inspired MacAlpine in the formation of Dominion Explorers and when the MacAlpine party was lost in the Arctic during the fall of 1929, Lindsley put up $250,000 of the $400,000 that was spent on the greatest aerial search that the world had seen up to that time. A more fortunate association was his perseverance in developing the giant Yellowknife Gold Mine at Yellowknife, one of the country's large gold producers.

Thayer Lindsley did not discover the Falconbridge ore body. It had been drilled by Thomas A. Edison in the early 1900s and was subsequently owned by the Longyear interests of Minneapolis. It was from them the property was purchased for $2,500,000. The price was calculated on the basis of .50 a ton for the 5,000,000 tons of nickel ore known to be there.

But it was Lindsley's economic vision that saw the potential in Falconbridge at a time when most people said the idea was stupid; that there were no markets; that he couldn't possibly buck International Nickel. Lindsley, of course, had other ideas and his trump card was the purchase of the Norwegian refinery and metallurgical rights.

Although he would perhaps be surprised by the description, Lindsley was a consummate salesman – of the soft-sell variety. At the time of Ventures Ltd.'s stock issue, he sounded a strong note of warning. He stated that the project, as the name suggested, was entirely speculative, and those who could not afford to lose were advised to look elsewhere. In spite of this warning - or perhaps because of it - the issue was heavily oversubscribed. After surviving the debacle of 1929, the shares eventually reached a high of $67 a share before being absorbed in Falconbridge Nickel.

"A remarkable, unusual man," said Gen. D.M. Hogarth who predeceased him several years ago. "A genius in the true sense of the word. He probably knows more about world geology than any other living man. In my opinion Thayer Lindsley has contributed more to the development of this country than any other man, or group of men."

The same kind of accolade was expressed by the late Col. MacAlpine who termed Lindsley "a brilliant geologist, with the added plus of economic vision, plus energy, plus guts. He has no hobbies apparently. His only hobby is work."

Another voice from the past, the late W.J. McDonough said of him: "The man is a genius. I believe him to be the greatest geological genius this world has ever seen. He is a paradox - very quiet and refined, does not know the meaning of physical fear. I never saw him lose his temper or raise his voice above normal speaking level. He knows no profanity. During the Depression years, he often paid the Ventures salaries himself, rather than see people unpaid. Yet in deals he has been cool, calculating, almost ruthless."

His friends will gather to pay him one last tribute on Thursday, June 3 at 12 noon, when a memorial service will be held at the Unitarian Church, All Souls Chapel 8th Street and Lexington Avenue, New York City.

Inspiration Mining and Development Company Limited

PROFIT AND LOSS ACCOUNT

FOR THE YEAR ENDED 31st DECEMBER, 1937

Profit from operations after deducting expenses incident thereto, Depreciation of $19,453.08 and Provision for Doubtful Accounts		$131,483.09
Add:—		
Revenue from Investments		37,619.10
Net Profit on Sale of Investments		4,543.91
Sundry Revenue		1,096.51
		174,742.61
Deduct:—		
Exploration Expenses	$37,811.11	
Montreal Office Expense	3,323.58	
General Expenses	4,648.63	
Legal Fees	1,670.92	
Directors' Fees	875.00	
		48,329.24
		126,413.37
Deduct:—		
Provision for Income Taxes		17,280.05
Net Profit for the Year		$109,133.32

Chapter 13

Frank Buckland's Reports from the Trenches

by George Norrie

J.P.'s chief geologist and one of the two principal property managers was Frank Buckland (doctor of geology) who has been mentioned earlier. The following is the first contract Frank had with Inspiration. This was a standard deal for Inspiration prospectors:

It is hereby agreed that F.C. Buckland shall receive 10% of all monies, shares or other considerations accruing to Inspiration Mining and Development in the matter of any mining claims or deals in properties, in which the company shall enter and in which F.C. Buckland is the original staker, or in properties which are staked upon the advice of Buckland.

It is further agreed that F.C. Buckland shall receive 5% of all monies earned between the company and parties not employed by the company in which Buckland has acted as agent for the company in locating the property for persons desirable for the company to deal in or with.

He was also allowed and encouraged to invest his own money in Inspiration:

Please be advised that at the next meeting of the Board of Directors of Inspiration Mining and Development a resolution will be passed giving you a call on 25,000 shares of the company's capital stock @ .05 per share. This call to be for a period of 2 years.

Beginning in January of 1935 Frank Buckland, working for Inspiration, began trenching, prospecting and diamond drilling on the East Malartic property. These are Buckland's reports, and the tremendous hardship of this work during the northern winter months is obvious. The trenches were dug to bedrock by hand through overburden of frozen swamp and muskeg. The depth of overburden was sounded with metal bars and when twelve feet or more of overburden was encountered, the trench was considered too deep and discontinued!!

These reports from Buckland talk about payroll, employee problems and supplies. Transportation and communication were very slow and difficult. Everything depended on the horse and the teamster:

Reports to Ken Clark, Inspiration Office Manager

Dec. 20/34 [from the Ford Hotel, Montreal]: I have bought three single beds at $13.95 and one ³/₄ bed at $16.95 for the East Malartic camps, and three pairs of pillows. These will be shipped freight paid but with a draft attached for the purchase price. All are excellent beds and are new. From what I have seen around town yesterday and today they are good value. Second hand cots can be obtained considerably cheaper (approximately $5 cheaper) but are not good and the mattresses look bad.

We also bought furniture for our rooms at the back of the office. This should arrive next week and will be freight paid clear at Amos.

Jan. 13: I have enclosed the payroll which I phoned to you and have added some notes to it. Louis Edmonds was laid off this morning and has 29 ½ days coming to him for December and 12 days in Jan. less $5.90 which he owes John Partanen [prospector] for clothes and supplies. Please keep Edmonds' cheque at Amos. Could you arrange to have a packing or shipping order sent with each load? We have no means of determining what has been sent. The last driver said he had a letter but lost it. Perhaps this was the shipping list.

Will phone you tomorrow to Cleman's or in the evening to Norrie's.

Jan. 14: I will see if I can locate some specimens for Porteous [stock broker] but at present we have only one good one and that is attached to about 75 lbs. of rock!

Jan. 16: Morin has phoned from Cardinal's Corner and says he will not bring his tractor past Canadian Malartic and that Norrie made these arrangements some time ago. I am sending the horse over to meet him but this is unsatisfactory for groceries, as we have no way of keeping them from freezing and the trip takes about an hour, so most likely some will spoil. I do not know what arrangement you made but keep this in mind when the bill comes in.

I have enclosed a copy of the Dec. payroll. If cheques have not been signed perhaps the changes for McKay and Maki could be made.

I am sending four rolls of roofing paper and three boxes of wood screws to be returned by the tractor.

Jan. 25: Our telephone has been out of order for the last three days so I am dropping you this note in case I cannot call you within the next few days. Labelle, the drill runner quit and we want to get another man. Evensen knows of one in Rouyn who is supposed to be good. Call #64 Rouyn. S. Bisant's grocery store, and ask him to send a messenger to Mrs. E. Larson 403 Gamble Street and ask her to get word to Gunnar Roos, tell him he can get work as a drill runner at East Malartic and would like him to catch a plane out. Have him call you back. Evensen is paying .70 an hour for drill runners. If this man comes out send in Charlie Vincent. Do not send him unless we get a drill runner as I can only use him in case we run three shifts on the drill. If Labelle is not paid off, charge him for the amount of the hotel bill and send him in here via snowmobile. He was promised expenses but quit too fast.

We are moving to Sladen's ground tomorrow. We will need more sample bags, also the heavy envelopes I ordered. We need a first aid outfit in a hurry. There have been several cuts already and we have no iodine or bandage of any description.

Martel met our teamster at Malartic and made him take the last load although he had no orders to do this. There were six dozen eggs missing from the case and several dozen were frozen after our teamster took them as there was no way to heat them. Martel hauls to Malartic for .60 according to Neelands. His drivers are dang liars.

We have had considerable trouble and lost a great deal of time. Both pumps are in rotten shape and the creek by the showing dried up.

Please try to get Gunnar Roos as fast as possible. If he cannot come get someone else.

Hope to be able to phone you before this letter reaches Amos.

Feb. 1: Gunnar Roos arrived today and we laid off Alphonse Ladoux. He has no time for February so that his full amount is for 46 hours @ .70 less seven days board at $1.25 and one phone call to Amos. Please put his cheque in the mail, addressed to Alphonse Ladoux, Amos.

We want to get another good man, Albert is far from being the best and will be laid off as soon as we can get another. Evensen believes Desmorest in Rouyn has just finished some work and should be letting out some good runners. Perhaps if you wrote these we might be able to get an experienced man. We are far from making any money now and to do so we must have the very best of crews.

If Ladoux's cheque has been sent here by the time this reaches you I can forward it. I am about fed up with the teamster and will let him out as soon as I can get another horse. Charlie Cardinal seems to have a good horse and I am waiting for him to come and see me.

Feb. 3: I have enclosed the payroll for January and Evensen's expense account from Bussiers to Amos. Evensen started on the 13th. He wants his cheque made out to his wife and forwarded direct to her at Witherbee, New York. I do not know if Norrie understood over the phone about $40 to be sent to Mrs. Evensen. Al has $40 in cash here which his wife needs and he has no way of getting it to Amos at present. He wants a loan of this amount until he can get someone he can trust to take the cash out. Please send him a note confirming all cheques sent to his wife.

Peter Jinn does not seem to demand any wages and will take whatever we offer. I am not sure what cooks are paid at present so have left his rate open.

Martel's men insist that they have orders to take our freight only to Canadian Malartic and are not even bothering to phone me when they arrive there now. Most of the freezable stuff is frozen when we get it here… a single horse with a sleigh suitable only for hauling bags. In this respect, we will need more potatoes within the next couple of weeks as most of the last shipment were frozen and spoiled. Blais comes down three times a week and is met by Cardinal who brings in the Canadian Malartic mail. Perhaps some arrangement could be made to get our stuff in by Blais. Millenback asked that you make a bundle of our mail before giving it to Sealey in order to facilitate sorting at this end. Perhaps we could have a mail bag of our own.

Feb. 6: There is some doubt as to how much time for the teamster should be charged to East Malartic and how much diamond drilling. Every 2 or 3 days the horse works an hour or so bringing wood for the camp. I have not checked this as it is generally done while the horse is not required at the drill.

We will need some more oats very soon. We haven't room to store more than 3 or 4 bags. Please arrange for about 4 bags to be sent in. Please send me 50 three-cent stamps. Some of our mail has been tampered with lately and we only get an odd paper now and again. Millenback (Canadian Malartic) would be glad to let us use the Canadian Malartic bag, I believe I wrote about this before. If you send it by other means please do it all up into a parcel which someone would have to open to get at both letters and papers. Some letters came today in a loose parcel. Evidently someone had ripped the packaging as there were no papers and a hole through which the letters could easily fall. Evensen found a wrapper for one of his papers but no paper.

Feb. 22: Charlie Vincent scalded his knee with a steam hose and raised a bad blister. He thought he should see a doctor but I am keeping him here as the skin is not broken and I don't believe anything can be done but rest it. Is it advisable to let a man go out and who pays their transportation if they do?

P.S.: Al figures on breaking in Larsen as a setter when the new drill arrives. We will therefore need four new runners. Helpers can be picked up here. We will need blankets for 7 men (14 pairs) and six mattresses. Pete says we will need the following: dishes, 2 doz. teaspoons, 1 doz. dessert spoons, ½ dozen cups and saucers, 1 dozen soup bowls, ½ dozen serving bowls.

We will have to get a helper for Pete. There is too much work and one man is coming back from work and doing chores in the afternoon at present. With 7 extra men, we will need a chore boy. Pete will get his brother if you think this is okay. I believe he has landed a laundry job at Malartic to start May 1/35.

Feb. 27: The new horse you sent in looks good but the driver speaks no English. We can get along okay though, I think. The last teamster, Ladoux's son, was lazy and a thief. Please send his cheque in the mail addressed to Henri Ladoux c/o Canadian Malartic Mine. He has 12 days for Inspiration and 3½ days for East Malartic at $3 a day clear…

P.S.: Ladoux took an axe with him. Better hold the price of an axe and handle from his East Malartic cheque.

Mar. 2: John Partanen wants one print of each film enclosed. If possible he wants this charged to his payroll account. Also enclosed is Ross' bill for hay. My check on the bales agrees with his but his addition is out in our favour. This was bought at $24 a ton and is good hay. The first hay you sent was excellent clover hay but the load that arrived with Demontigny had two bales, half of which were straw and chaff; however, that has nothing to do with Ross' account except it was the reason we ran short suddenly. The best thing to do is send his cheque and receipt out here and I will have him sign it. He seems incapable of realizing that a better bill might be a good thing. Let me know how much I owe you for the carrots and the laundry that was left when we came out here. Also anything else you can think of.

If possible to make a load for someone with a team it would be much better to send our freight that way than with Martel. His tractor cannot come in here and the last load he arranged with Malartic to bring over as a special favour to us. Needlands came over the same day, however, and said if we had contracted to have the hardware hauled right to our camp he would bill Martel for it. I still have considerable gas and groceries over there to haul. I have hired a second horse from Ross but the two will be working overtime for the drill as we are hauling water now and this is slow work as we have no large supply to haul from. The Malartic creek is frozen solid and Malartic are hauling water themselves from below Sladen's camps for their cement mixer. Their teams are free every second Sunday but they do not like to work them then… Please send a dollar's worth of .03 cent stamps to Evensen.

Mar. 9: If we are going to put up ice, it must be done quickly. This will have to be hauled from Malartic Lake as the creek here is hardly frozen and the small lake further down would necessitate a road. Let me know as soon as possible about this.

The first operations representative was in here and I gave him the necessary dope. He said that the matter would be concluded as far as we were concerned. This was about the middle of Feb./35.

I am sending a few pieces with gold for Porteous. This is about all I can find now and will have to blast in the old trenches if he wants more. There is none showing.

Mar. 13: Charlie Vincent wants a carton of Winchester cigarettes and a can of old chum and 4 rolls of papers.

Re: my telephone conversation this morning; since phoning I found a case of ketchup and a case of All Bran at Malartic, so if more of these have not already been sent please forget them.

There was one case of plums short when Martel delivered them to the Malartic storekeeper. Martel put 25 bags of potatoes on our freight bill that belonged to Malartic. I hauled these over here and then had to haul them back again and I think he should be billed for this. Other groceries came at the same time but I think a fair estimate would be one day's work for one horse and a man. $4.25 plus horse feed. He told Malartic that he would credit us with the freighting charges.

Cardinal's team worked 5 hours hauling groceries and our horses worked 9 hours. The potatoes were stored in the drafting room and hall of the office and I had to get them moved first night. Millenback did not think you were sending such quantities of perishable goods or he wouldn't have promised to look after them.

If you haven't shipped any meat by the time this letter arrives please get some on the way as we are starting on eggs and ham tomorrow.

Larson says the rats knew when to leave the ship. Anyhow we got two good men out of the Bussiers shut down. With the Lind brothers this should give us a first class crew.

P.S. The 2 in. lumber Al ordered for the new drill shanty is all in 10 foot lengths. Al says he ordered 12 foot lumber. We have used this but we have had to do considerable amount of splicing which means a lot of extra labour and an awkward shack to move.

Mar. 15: You will probably receive a bill from Dominion Skyways for transporting Gus Benson from Rouyn to Malartic. Better pay this and we can charge it against his wages. He is a new helper who was broke when he started. Also Dominion Skyways took a core barrel to Turner's Machine Shop in Rouyn. You will probably receive a bill for this from both Skyways and Turner's.

I have enclosed the payrolls for the first half of March, also have written J.P. about these. We should pay twice a month as it does not seem possible to get the payroll here in less than two weeks after the end of the period, dang near a strike here last night when I phoned.

I do not think we have enough meat here to do until the end of May. A quarter of beef lasts less than a week and we only have 5. Also how about some pigs and sheep.

We will need about 12 more sacks of oats for the breakup. Also need 1/2 dozen plates.

P.S.: Charlie Vincent crushed his finger and the Malartic first aid man advised that he should go out as it is broken. Watch that he doesn't spend much money and think he can charge it to us because of injury. He seems to have this in mind. An Amos doctor should be all that is necessary.

For God's sake Ken don't ship any more stuff to Malartic. They are right on my neck about having to store our stuff in their office which is their only available room. I have no means of handling this freight and have explained in letter to Jim as previously to you by phone.

How about another fifty 3-cent stamps.

Mar. 22: At present the gang is working on the ice house but when that is finished will sound still further west. I could not get ice cheaper than $10 a ton. This price I think is too high. There is about 16 inches to 18 inches of ice on the creek in some places and I am cutting this. It is inferior ice but will serve our purpose just as well. Our freight is causing considerable trouble and expense at this end. Please see if Clark can remedy this. I have phoned him and written him about it but the freight still comes to Malartic. They have no place to store it there except in their drafting room and we have had so much here that they are getting fed up. Our horses are working 11 to 12 hours hauling water and are getting paid for 9. This is rather a sore point with the teamsters. They will not work longer hours hauling freight at night. Today Martel dropped 10 drums of gas about ¼ mile this side of Malartic and told me to come and get them. If Martel is not going to haul them further I will have to get Cardinal to come down and it will cost us around 10 to 12 dollars. Last night 14 cases of eggs came to Malartic and Millenback phoned telling me to get them before they froze as he had to leave them outside. Our team went over but as a result they went to work late this morning and the drill is shut down now.

Re: Charlie Vincent. His accident was unfortunate in more ways than one. Evensen had asked me to make out his time as soon as I could find another helper. He has no interest in this work and is always either day dreaming or inciting the men he works with to dissatisfaction. Our 3 runners have all reported him to both myself and Al and they do not wish to work with him. After his accident he made it known around camp that he intended to get an aeroplane ride out of us. None of our present runners will take him back but perhaps if you could tell him to snap out of it we could give him to one of the new men when the other drill comes in.

Chapter 14

The Battle for Financing

by George Norrie

Although Inspiration had begun operations in the summer of 1934, only two years later in 1936 the company had done so well that it was ready for expansion. To position itself for investment in all sections of Quebec, Ontario and beyond, a major financing input was required. J.P. outlined, briefly, his plans to Andy Davis:

I have been taking up with Mr. Thomson the sale to them of 1,000,000 Inspiration shares @ .50 per share, this to provide our treasury with $500,000.

Right now we have 16 drills working on a very profitable basis and we have been for the last two weeks turning work over to other drilling concerns as we do not wish to further impair our cash position in the purchase of new equipment. Our object in this financing is:

1) to get the stock listed, give it a value and help in our advertising as no other such concern now listed.

2) to provide funds for expansion across Canada and also take advantage of foreign jobs. Our man Evensen is well acquainted with the African situation which provides much profitable work for Canadian and American drillers.

3) to develop a new drill which we think will revolutionize diamond drilling.

4) to provide funds to initially develop good properties one of which we now have in Chibougamau.

5) to provide funds for investment in good stocks made quickly valuable by drill results which we have initial information on.

6) to further expand our exploration activities.

I am very anxious to make this Inspiration into a real achievement and this can more quickly be done by financing at this most opportune time.

I would be glad if you would use your influence with Mr. Thomson to consummate a deal with us on terms reasonable to Inspiration and himself.

In September, 1936, J.P. gave a detailed report on the new company structure and the option granted to Nesbitt Thomson:

Inspiration Mining and Development Co. Ltd. was incorporated under letters patent from the province of Quebec in October, 1934.

It was incorporated primarily for the purpose of diamond drilling, mining exploration, and mining development. Under its charter it has among other provisions the right to invest in other companies having similar objectives, and to invest any surplus funds which it may have from time to time, in other companies as its directors may see fit.

The company's officers are as follows:

Mr. M.A. Thomson	Montreal	President
Mr. J.P. Norrie	Amos	V-P and General Manager
Mr. E.A. Boadway	Toronto	Assistant General Manager
Mr. F.D. Lamont	Montreal	Secretary-Treasurer
Mr. R.F. Martin	Montreal	Assist. Secretary-Treasurer
Mr. W.X. Toole	Amos	Accountant
Mr. A.C. Evensen	Amos	Manager, Drilling Dept.
Mr. W. Proctor	Amos	Assistant, Drilling Dept.

Directors

Mr. M.A. Thomson	Montreal
Mr. P.A. Thomson	Montreal
Mr. J.P. Norrie	Amos
Mr. A.J. Davis	Newmarket
Mr. W.H. Miner	Montreal
Mr. H.J. Carmichael	Oshawa
Mr. L.A. Ladouceur	Amos

The company functioned more or less as a private concern from its incorporation until Sept. 1936, when it was decided to enlarge the scope of its operations by means of purchasing additional drilling equipment and acquiring claims and properties.

With this end in view on Sept. 28, 1936, an option was granted to Nesbitt Thomson to purchase 1,000,000 shares of the company's capital stock at the price of $1 per share. In January, 1937, this option on 1,000,000 shares of the company's capital stock was taken up by Nesbitt Thomson and the $1,000,000 paid into this company's treasury in Montreal. With this new financing completed, the capital set-up of the company then consisted of 2,000,000 shares of capital stock outstanding, each share having a par value of $1 - 1,000,000 shares having previously been issued for cash and various other considerations such as properties etc.

Considerable surplus was left in the treasury after providing the necessary additional drilling equipment… with this end in view an investment committee was appointed…

The company's organization:

Amos: at this point the accounting dept including the cost dept., purchasing dept., keeping of company's records, and such other costs and accounting records are maintained.

The company's general manager J.P. Norrie is located in Amos…

Mr. W.X. Toole is in charge of the accounts

Mr. A. Evensen is in charge of the drilling dept…

The company has part of its equipment in Amos. A machine shop has been erected near the railway station.

Toronto: The company's branch office is located at 80 Richmond St. West, comprising a 2-room office at $75 per month. This office is the charge of the company's Assistant General Manager E. Boadway with W. Rockwell as bookkeeper.

From this office there is maintained the contact with the field staff of explorers with the various mining centres…

There is also an office at Kirkland Lake, which is taken care of through the Toronto office and maintained for the purpose of getting diamond drilling contracts, and as a field contact point for prospectors in the Kirkland Lake district.

The first order of business with any infusion of capital, especially with Nesbitt Thomson, was the pooling of as many existing shares as possible. J.P. called on Walter Warden to pool his shares:

...an agreement has been completed with Nesbitt Thomson to purchase 1,000,000 shares of Inspiration at .50 per share.

To help them in this, one half of the outstanding shares must be pooled. Would you be willing to pool one half of your shares for a period of six months?

If so please send the shares to Chartered Trust Executor Co. Montreal. A definite pooling agreement will be sent you later.

We are getting some nice ore in both shafts at East Malartic. The ore in No. 1 is very spectacular. No. 2 shaft is finished and crosscutting is underway.

During the remainder of 1936 and 1937 Inspiration took several new claim positions due to the input of capital by Nesbitt Thomson. At the beginning of 1938 Inspiration was definitely in a position to benefit:

Diamond drilling: During the calendar year 1937, the diamond drilling department drilled 304,229 feet on 66 contracts. A substantial profit was returned, as shown in detail in the financial report of Dec. 31, 1937, recently submitted. The major part of this drilling was done in northwestern Quebec, but during the past year, the drilling operations have been expanded into the Larder Lake - Kirkland Lake - Matachewan areas.

Considerable shop and drilling equipment has been purchased, and your company has sufficient equipment to undertake the contemplated expansion of the drilling department into other mining areas.

```
Exploration department: In 1937, a field staff
consisting of two mining engineers, a geologist,
and two practical field men, were active in Ontario,
Quebec, Manitoba, and the Maritime provinces. This
staff supervised the exploration and examination of
optioned claims, claims previously acquired by the
company, and claims staked during 1937.
   Three two-man parties in Quebec, and one in western
Ontario, prospected throughout the season; claims
were staked, but no major discoveries were made.
   Many examinations of properties and prospects in
all stages of development were undertaken, and as a
result, interests in seven properties were optioned.
Of these, three were released after further work,
two are being diamond drilled, and on two the full
option payments were made.
   Several parties were employed for short periods
to stake claims. A dip needle survey was made of an
interesting area in Quebec.
   Approximately $54,000 was expended in conducting
the exploration department's activities, and the
details of the expenditures are included in the
financial report previously submitted.
```

When J.P. was courting Nesbitt Thomson he did everything he could to persuade them to take over the financing of Inspiration and East Malartic. This was the middle of the Depression and money was almost impossible to attract to any venture, let alone money for a mining venture. His experience with brokers, as we have seen with the Bussiers, was the major roadblock to development.

As the Perron developed, J.P.'s ability demonstrated itself in no uncertain terms. Profit flowed for everyone involved, particularly the money men, Nesbitt Thomson. After less than two years of developing Inspiration and East Malartic, this fact was even more obvious.

When Nesbitt Thomson placed their money in a project, their work was only beginning and Mel Thomson did his job and he did it well. He monitored every move that J.P. made. It was a hands-on operation for Nesbitt Thomson, Mel Thomson's hands. It was his right and there was no doubt about that. However, there was a point beyond which J.P. would not be pushed. This became obvious when his employees were questioned. He would defend them, without a doubt. He knew they were the reason why he became successful and he never forgot that. He also would not be pushed without a confrontation if he thought Nesbitt Thomson was taking advantage of a situation as we shall see, later, in taking down treasury shares from East Malartic at a price below what J.P. thought was reasonable. This instance involved claims in Louvicourt which had been allowed to lapse in error. Nesbitt Thomson charged that the employees at Inspiration in Amos were at fault:

I have your letter relative to the charges of Riddel, Stead, Graham & Hutchinson, and I will discuss the situation with you on your next visit to Montreal.

I hardly think you realize the position the company was placed in when the Louvicourt claims were allowed to lapse, and I cannot agree with you that the auditor brought this question up to justify extra charges. I think if you will discuss the situation with Mr. Toole he will give you some idea of the grave error that was committed, and I am of the opinion that there is no one to blame for this but your office at Amos.

M.A. Thomson

Mel Thomson continued to question J.P. on the Inspiration drilling costs for quite some time. Finally the morale of the employees was being compromised and J.P. could not let this go on any longer:

When I discussed the Calumet Island drilling this morning with Evensen and told him we would have to further reduce prices there, I found him suffering from a severe case of disappointment and I am sorry to say I feel much the same and have so for some time.

At the time Inspiration became connected with Nesbitt Thomson Co. I felt that we would surely get all the drilling under their control at reasonable prices and thus accomplish all objectives Inspiration had in mind, the leading one being to get advance information on new prospects and pay for such information through drilling profits. Up to now we have been through sheer diligence able to do it, from now on I fear the prospects are not bright.

The Continental Diamond Drilling company gets all of Noranda's drilling both in Noranda Mine and in their flung development projects. Smith and Travers get all of International Nickel work. Sudbury Drilling, all of Errington's work. Ventures goes to Demorest Diamond Drilling Co. in which they are interested. Sigma and Lamaque to Morrisette Drilling Co. at prices much above what we are filling at Perron, Courinor and East Malartic. Outside bids are not called for on any of the above described work and we can get none of it regardless of our bid.

I had considerable trouble convincing Evensen to agree to the link up between Inspiration and Nesbitt Thomson Co. He wanted to go as we were, I told him we would get all the benefit from a growing mining concern and finally convinced him. Today he says he is far from convinced. He says the gears of our interlocking directorships do not mesh and I believe you will agree that this is a fact.

We have been forced to reduce drilling prices on Perron, Courinor and East Malartic to the point where no decent profit can be made. Recent drilling on Malartic Goldfields due to reduction in prices was just barely profitable. Sladen bought their own drill and, we hear, are about to buy another, and now prices on the Calumet are to be reduced to a point where little or nothing will be the result.

Recently Mr. Fontana gave Continental Drilling Co. a 5,000 foot contract at a price much higher than our bids, because one of the directors was related to one of the drilling company directors. I know of many such transactions in this business. If we could capitalize on our connections in this way, we would have more profitable drilling than we could handle.

I surely have given 100% of my efforts to this company and all the others with which I have been connected and frankly I feel disappointed with the support I am receiving from the other directors in trying to make this drilling business a success. In grinding us down in our drilling prices you are asking us to compete with haywire outfits who do not pay their bills, go broke regularly and at the best cannot give good service.

The fact that Noranda, Nickel, Errington, Ventures, etc. let out their work at reasonable prices and help to keep reputable drilling companies on their feet should be sufficient to warrant the support I am asking, i.e. that we be given all our drilling at profitable prices. Anyway be sure you have a much disappointed drilling staff, me included.

A little encouragement at this time would be in the interest of all concerned.

J.P. Norrie

When drilling for customers, the cost measured against the quality of work was always a judgement decision. Gus Rainville was president of several companies and was considering a change in drilling contractors, J.P. wrote:

We have just been advised that our drilling on Jacques Cartier property has been finished. Our contract called for 2,000 feet and the contract was terminated at 2,001 feet in a hole still unfinished and to be finished by another drilling contractor.

I understand there is some hook up in this drilling between the engineer on the property and the drilling contractor and that the holes we have drilled have been in difficult places and generally under adverse conditions while the holes yet to be drilled are easy and generally from bedrock setups.

I have been informed that our drill crew lost only four feet of core in the 2,001 feet drilled which is more or less a record in this type of rock.

Diamond drill rig

I call your attention to this as our engineer has just made an examination of the Fleming cores and he informed me that much of the core in the Fleming vein material has been badly ground up in fact to the point where it is hard to appraise the value of the property. This is born out by the fact that several sludge samples indicate excellent values in places where the core was badly ground up.

Having had wide mining experience our men appreciate the fact that diamond drilling is done to recover vein material which is highly necessary in diamond drilling. It seems that our competitors on the Fleming are chiefly interested in big footage overlooking the highly important point that cores' recovery of vein material is the one important thing.

Please do not think that these remarks are made to secure further drilling. They are simply to advise you that some very bad work has been done on your property to the point where it is difficult to size up the property from the work done to date.

J.P. Norrie

The year 1939 was the beginning of the war and the decrease in profits and cutback in exploration began:

Inspiration Report Dec. 31, 1940

As of Dec. 31, 1940 the company owned and operated 41 diamond drills. During 1940 diamond drilling contracts were undertaken in northwestern Quebec and Ontario. The total footage drilled was 269,270 feet. Exploration activities during 1940 were chiefly confined to the Quebec districts.

Development results on the National Malartic and Senore properties in which the company has large holdings are impressive. On the former, several million tons of commercial ore has been indicated by diamond drilling and extensive underground work will be carried out during 1941. On the Senore property the shaft is being deepened to 500 feet and active underground work will follow. Valuable ore exposures have been outlined on the 200 foot level. Underground work is now proceeding on the Pascalis Gold Mines to open up important ore deposits intersected by earlier diamond drilling. The company acquired 100,000 shares in this mine during the year. One quarter of the financing of the Vicour Gold Mines, where important gold ore occurrences are now being developed, has been undertaken.

The company also owns substantial share interests in Wright-Hargreaves, East Malartic, Malartic Goldfields, Sladen Malartic, Perron Gold, Macassa Mines, Kerr Addison Gold Mines, Canadian Malartic Gold Mines, and several others. In addition, the company holds over 1,500,000 shares in mining projects which are now more or less in the prospect stage. Inspiration also owns 699,995 shares of Insco

Mines a well located and promising copper/gold prospect.

Employees Dec. 31, 1940 - 66

In 1941 it was becoming clear just how restrictive the war regulations were going to be on gold properties, particularly those in the exploration stage of development. J.P. expressed Ottawa's attitude, which is "all the gold possible with the least possible noise."

More restrictions than ever continued in 1942. Inspiration turned its attention to one of the strategic materials, mica:

Inspiration Progress Report Nov. 7, 1942

During the season just closing our entire prospecting energies have been confined to three and at times four prospectors and one field engineer in the Eau Claire Mica field in Mattawa township, Ontario, where large quantities of strategic mica have been found. The result of our work has been the staking of 25 claims near the famous Purdy Muscovite mica discovery on which several important mica showings have been found, the acquiring of 9 claims surrounding the Purdy, and the purchase of the Purdy property itself.

All of these claims have been transferred to a two million share Inspiration-controlled company which will conduct future operations.

This discovery is considered the most outstanding mica deposit in North America.

The property has been equipped with air compressors, rock and other mining equipment. Adequate buildings to carry on operations have been erected.

Rough mica from the mine is trucked twelve miles to a modern mica processing shop in Mattawa where 15 girls are employed. Five tons of strategic mica have been prepared for market. Two tons have been sold to the U.S. govt.; the balance is now ready for shipment. Scarcity of strategic mica has compelled the U.S. govt. to control the entire import of mica into the country, made necessary due to uncertain transportation from India and Brazil, the previous main sources of supply, and greatly increased consumption for war weapons.

It seems that Inspiration has acquired a very profitable property for the duration of the war and one that will operate in peacetime with a fair margin of profit. It gives the officers and employees of the company much satisfaction in being able to materially help the war effort in the production of a strategic mineral.

The company has approximately 20 drills working on a reasonably profitable basis on exploration and development chiefly on strategic minerals in the Maritimes, Quebec and Ontario.

In 1944 the end of the war was in sight and exploration was improving. In anticipation, Sudbury Diamond Drilling was purchased and this greatly increased the potential for exploration, particularly in Ontario:

Inspiration Report 1944

Drilling department: in operation: 15 underground drills and 15 surface drills. Contracts signed but not drilling - 3; contracts pending - 6; contracts completed - 3. Purchased two second-hand drills and have ordered one new drill.

Exploration department: one prospector is covering ground in the Octave River section of Quebec. The other prospector "jumped the job" at the last minute, and has not been replaced yet. Have interested a party in the Vauquelin claims held by Inspiration (Irving H. Hertzberg of N.Y.C.)

Would suggest that something be done with the group purchased from O'Leary Malartic in Mud Lake section.

Due to the Wassa Lake boom, it might be a good time to make a deal on Insco. I would recommend getting as much of Inspiration's Insco sold as possible.

Sudbury diamond drilling have temporarily placed A.J. Davis Jr. in charge. Have transferred Trembly from Inspiration to act as field man for the present. Have negotiated with Al Evensen to take the job of manager of Sudbury, subject to your approval. The arrangement with him is as follows:

Monthly salary of $300 plus 10% of the net profits before taxes. I feel that this method of payment is superior to paying him a straight salary of $600 or $700 a month, because it will give him added incentive to keep the costs down. Also we will have the advantage of his experience and ability for Inspiration.

We have placed an order with Longyear Company for sixteen new drills. Have received the resignations of G.A. Corbett and C. Simmons, effective June 1st.

Even though business was improving toward the end of the war, J.P. had made a personal decision to become a personal mining consultant:

I herewith tender my resignation as General Manager of Inspiration Mining & Development Co. Ltd. to take effect Feb. 1, 1944.

J.P. Norrie

Chapter 15

A Nose for Ore

by George Norrie

East Malartic Gold Mines

In the 1944 Annual of the *Northern Miner*, veteran geologist Dr. C.V. Corless, who was regarded as the dean of Canadian geologists, wrote:

Not merely many, but very many years must necessarily be consumed in discovering, exploring, developing and mining the great tangled networks of veins and vein systems in this single area [Canadian Shield]. For the area extends over a great length. The width, too, is extensive. Besides, there appear to be several outlying vein systems. By comparison, as to difficulty, Sudbury and the Rand of South Africa seem much simpler.

How can the engineer proceed with the unravelling of such a tangle of problems? Nothing can be substituted for close, keen, intelligent observation, backed by accurate knowledge and sound judgement or "horse sense". But amid such complexities, there is necessary besides, an indefinable something - for want of a better term it is sometimes called "a nose for ore". It is a sort of uncanny instinct, which a few men seem to have - Oliver Hall, for example, Eldon Brown, and J.P. Norrie. Some doctors have a natural gift for diagnosis. Some men seem "born detectives".

There are highly skillful mining engineers and geologists aplenty, who have not this instinct. They proceed by the method of elimination, which

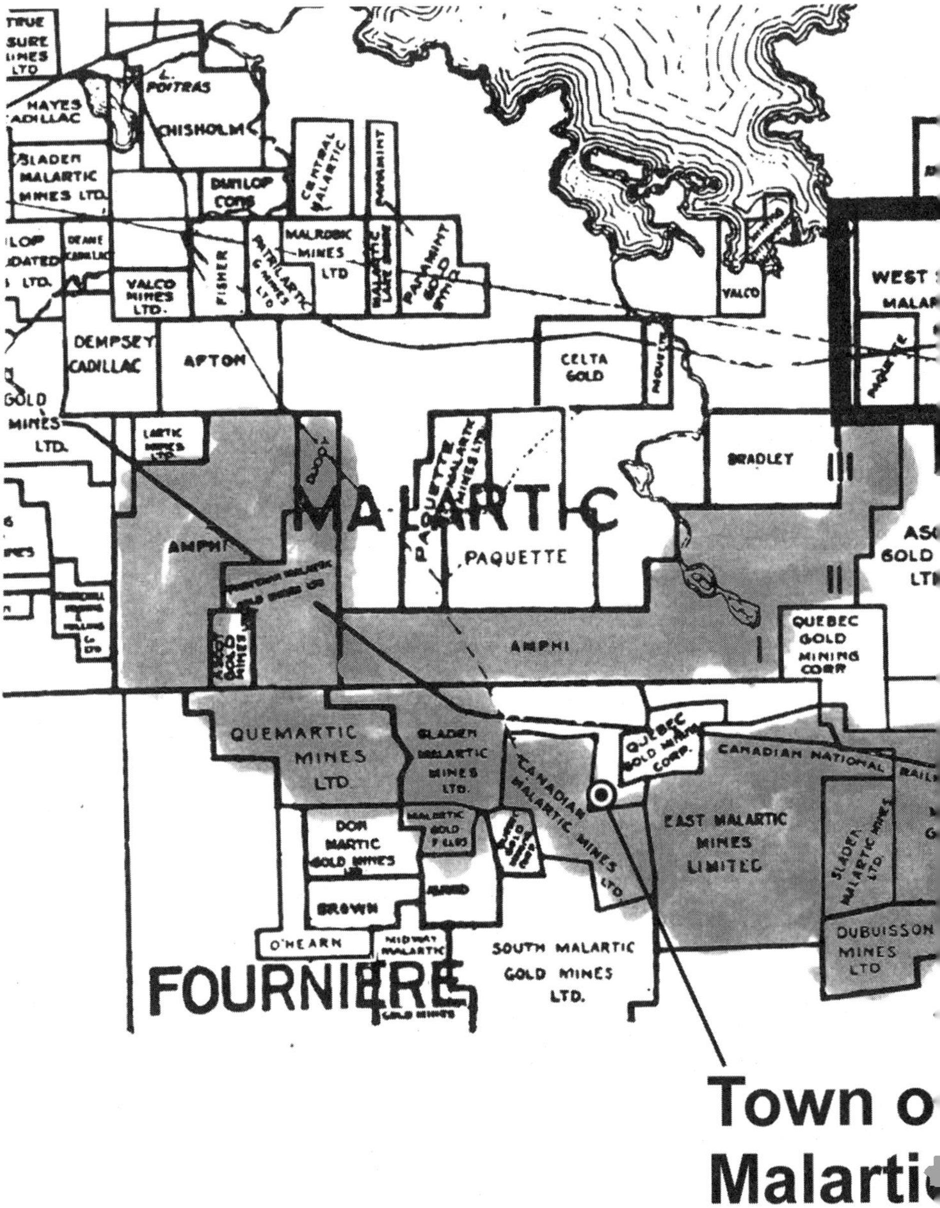
HAYES CADILLAC
L. POITRAS
CHISHOLM
SLADEN MALARTIC MINES LTD.
DUNLOP CORP
CENTRAL MALARTIC
MALROBIC MINES LTD
VALCO MINES LTD.
FISHER
DEMPSEY CADILLAC
AFTON
CELTA GOLD
VALCO
WEST
BRADLEY
MALARTIC
AMPHI
PAQUETTE
AMPHI
QUEBEC GOLD MINING CORP.
QUEMARTIC MINES LTD.
SLADEN MALARTIC MINES LTD.
CANADIAN MALARTIC MINES LTD
QUEBEC GOLD MINING CORP.
CANADIAN NATIONAL
EAST MALARTIC MINES LIMITED
SLADEN MALARTIC MINES LTD.
DON MARTIC GOLD MINES LTD
BROWN
O'HEARN
SOUTH MALARTIC GOLD MINES LTD.
DUBUISSON MINES LTD
FOURNIERE
Town o
Malarti

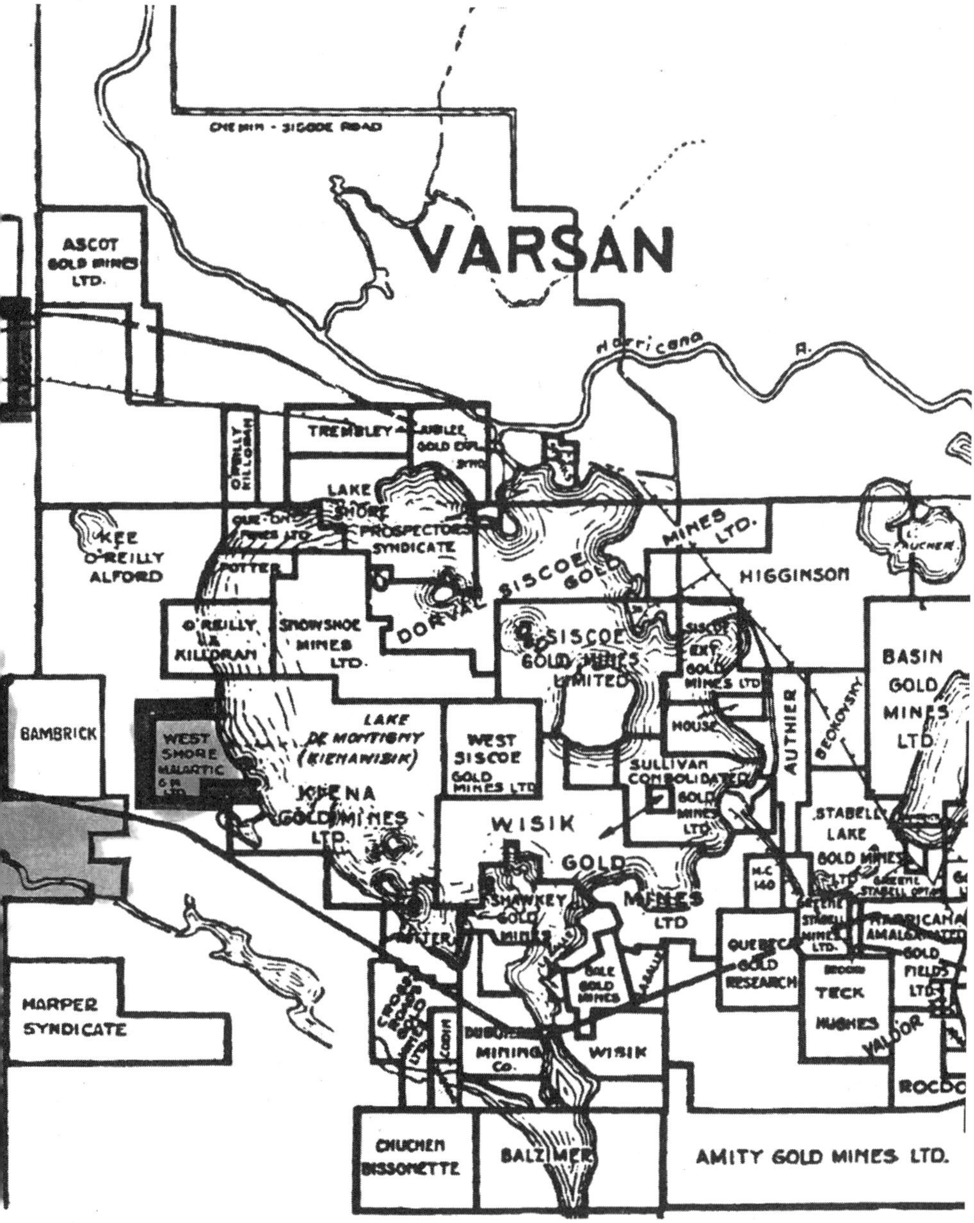

Map showing mining claims along the Cadillac-Malartic geological break, also known as the greenstone band.

will find ore in time but is both slow and costly. The men of whom I speak proceed in the opposite direction. Their shrewd instinct will occasionally find ore with a first or second borehole. After that, of course, the going is easy. Such acute diagnosticians of mining ground are of inestimable value to any exploring enterprise.

J.P. had come to the Canadian Shield over ten years before East Malartic. He had done his apprenticeship as superintendent at Porcupine-Davidson, Sylvanite, Stadacona, Malartic Gold Mines and Bussiers. He had learned his trade well, underground with Porcupine and Sylvanite. From the very beginnings of the Stadacona, Malartic Gold Mine and Bussiers he had made a living, but the primary benefactors were the companies who employed him.

Tiblemont was his own but for some reason he sold his interests for a very handsome profit and moved on. Maybe his "nose for ore" advised him that there would be no gold mine, and there wasn't.

This time, with the East Malartic, that instinct told him that there was gold there and lots of it. It would take many years of hard work, and tenacity, and failed attempts to convince others that in fact the ore was there, that he did in fact have the "nose for ore", but he would do it. It was just a matter of who would be the lucky principal investors and promoters and how much J.P. would be able to profit from his vision and courage.

When the Coffin Mining Company was dumped back into his lap by the untimely death of Fred Bradley of Treadwell Yukon, J.P. was not prepared to develop these claims immediately. Due to his personal shortage of money and financial constraints of the Depression, his dream to incorporate an exploration and development company had to wait, but now was the time and East Malartic would be one of the first properties that this company would explore. Ironically Carl Erickson, one of the original investors in Inspiration, did not want the East Malartic property as an Inspiration investment. It did not look promising to him at that time,and Erickson wanted to begin Inspiration on a clean slate. This left the East Malartic claims for J.P. to promote on his own. It also left J.P. with a greater risk of failure but, if successful, much greater reward. Rather than the vendors' shares being held by Inspiration, they would be held by J.P. personally.

Early History of East Malartic

J.P. wrote the following historical account in East Malartic:

Wilbur D. Gouldie's original discovery in the Malartic area did not bring to him the fame and fortune of some others, as work on his original discovery did not prove an economic ore body. His original claims are now owned by Canadian Malartic Gold Mines.

Gouldie's work in the district was not confined to his original discovery. He later located several accumulations of gold-bearing boulders, individually up to 30 tons in size. These boulders were found on a line nearly three miles long and almost straight along the south boundary of what became the property of East Malartic Mines Ltd. At the time, the significance of these gold-bearing boulders was not recognized and their source was thought hard to locate due to heavy overburden on the country to the north but it is here that the East Malartic story begins.

During the years following the original discovery in 1923 the East Malartic ground was held in small parcels by many stakers who never made any concerted effort to find the source of the boulders. The ground was simply held because it was adjacent to the potentially valuable Canadian Malartic ore bodies.

In the summer of 1932 all this ground was allowed to expire and then was acquired by staking in one block of 3,500 acres by the vendors (J.P. & Lee Coffin) of East Malartic Mines who recognized the importance of the boulders and devised a plan for locating their source.

The property of the East Malartic Mines comprised 45 claims in all. The aggregate area was in excess of 1,800 acres.

The property was located in the northeast section of Fournière township, adjoining the Canadian Malartic Gold Mines to the west, and was bounded on the east by the claims of Sladen Malartic Gold Mines. It covered the main Malartic-Shawkey-Lamaque break for a distance in excess of two miles and was directly on the strike of the veins on these properties.

The Shawkey Gold Mines property (ex-Yankee baseball pitcher Bob Shawkey) on which were spectacular ore bodies, was located approximately seven miles to the east. Siscoe Gold was eight miles east on the line of strike of the formation.

There was a water route from Amos, Que. to the claims, a distance of 45 miles, whereby heavy freight could be delivered to the campsite. Another water route of equal length could be taken by way of Siscoe, whereby freight could be landed by large scow within two miles of the East Malartic property. A road was built early on by the Quebec Dept. of Roads from the Malartic River to the shaft site of the Canadian Malartic mine. Another road to serve the O'Brien and Thomson Cadillac mines was extended to serve the Malartic mines. An aeroplane landing strip on the East Malartic property was served from the bases at Noranda, Amos and Senneterre. A telephone line extended to Amos via the Canadian Malartic property.

The claims of the East Malartic were originally staked by Lee Coffin, a Harvard graduate and veteran prospector, and J.P. Norrie, when he was manager of Canadian Malartic Gold Mines in 1929. In 1932 the claims were sold to the Treadwell Yukon Co. who incorporated the Coffin Mining Co. to operate Canadian Malartic. Following very favourable reports by their engineers and geologists, several thousand dollars were spent in surface trenching and development. A diamond drilling program and other plans had to be discontinued due to the death of the president of the company in 1933. In accordance with the wishes of the late president, all Canadian properties of the company were sold and the assets of the Coffin Mining Co. were acquired by Coffin and J.P. Norrie. In August, 1934, the East Malartic

Mines was incorporated to operate the Fournière claims formerly held by the Coffin company.

Although the ground was mostly covered with light overburden of glacial drift, many outcrops were found, so that geological study and prospecting were not difficult and trenching for hidden ore bodies was easily done by pick and shovel.

The entire property was timbered with jack pine, spruce and birch of fair size in sufficient quantity to provide fuel for initial steam power operation and for general heating purposes. Mine timber of unlimited quantity was available.

The transmission line of Canada Northern Power traversed the country from Noranda to Lamaque approximately four miles north of the East Malartic property. Power to supply a mining operation of any size could be secured from this source.

Sufficient water for milling purposes could be had from the Malartic River which traversed the central part of the property.

The westerly end of the property and that of the Canadian Malartic to the west was underlain by Temiscaming sediments of greywacke, orkase, and conglomerate, intruded by dikes and stocks of syenite porphyry. The ores occurred in veins and lenses in, along and around these intrusions. In the central part of the East Malartic property the intrusives might be classed as diorites. Indications were that in the east portion of the claim the intrusives were granodiorite and ores similar to those of Siscoe Gold Mines.

Many gold-bearing float boulders were found in one area. The boulders ran a length of 300 feet east and west parallel to the main break or general shearing and for a distance of 500 feet north and south. Similar boulders were found at various places on a straight line parallel to the general strike of the formation for a distance of two miles to the east.

The boulders were very angular, mineralized syenite porphyry, with many fine quartz stringers carrying free gold up to the size of a pea. The main

Malartic-Siscoe-Shawkey-Lamaque gold-bearing break traverses the entire East Malartic property approximately one quarter mile north of the line of strike of the boulders. They indicated an important ore body nearby as they were very angular and up to 30 tons in size, but they were not in place; they had been deposited there by glacial action.

In 1934 East Malartic Mine Ltd. was incorporated to operate the property.

Note: Many of the prospectors in those days were educated, very well educated geologists. Lee Coffin and Frank Buckland were two who were part of the East Malartic. Herewith a story:

A geologist and another guy were driving down a country road. The other guy looked out to a field covered with large rocks. He asked the geologist, "Where did those rocks come from?" The geologist told him, "The glaciers brought them down." The other guy asked, "Where are the glaciers now?" The geologist said, "They've gone back for more rocks!"

Early Financing of East Malartic (1934-1935)

The first concern with any mine was the exploration and identification of the ore body, if any. The money to pay for this was, of course, needed immediately.

Nesbitt Thomson did not immediately take up the financing of East Malartic. Thayer Lindsley did, however, but J.P. needed further financing from the market, and Porteous & Co. from Montreal were contracted as the brokers.

With the burden of financing the East Malartic company on his mind, J.P. began searching for a promoter to provide the initial capital to explore his claims in Fournière township. He had not been impressed with either Fred Crawford or Doug Wright and was looking for someone else to take on this task. Nesbitt Thomson were just beginning with Perron and were not ready to take East Malartic until a later stage of development. It is not clear how J.P. found Porteous and Co. in Montreal; however, he needed someone to make the market for East Malartic, and Geoff Porteous was worth a try. J.P.

just didn't have time for this problem with the operation of Inspiration and Perron demanding so much of his efforts.

In the incorporation of East Malartic there were 1,000,000 shares allotted for the vendors: 350,000 each to J.P. and Lee Coffin, the stakers; and 300,000 for the brokers, as a bonus, to make the market. Porteous needed half the vendors' shares placed in pool so the vendors could not flood the market until the mine reached a later stage of development. He attempted to keep all expenses to a minimum to provide as much cash for the company development as possible. J.P. bought 50,000 shares @ .13 per share, personally, and this money went to the trust company for development of the property. Porteous put out shares on the following basis: 250,000 shares @ .10 and 250,000 @ .15 per share. His commission was .02 per share and the brokerage firms received .03 per share. All cash generated by these sales went to the treasury of the company for development.

Immediately upon contracting Porteous to handle the financing, J.P. began contacting his old mining friends in hopes they would invest in his company. Thayer Lindsley was the first to be invited to participate and had already accepted through Ventures and himself, personally. Parkinson, the secretary of Ventures, wrote J.P. re the agreement that Thayer and J.P. had agreed upon:

> Your letter to Mr. Neelands and the undertaking regarding future financing contained in the confirmation of a firm sale of 200,000 shares of East Malartic Mines, Ltd. stock to Ventures, dated Nov. 9, 1933, have been received...
>
> My understanding of our agreement in this regard is simply that in consideration of our purchasing 200,000 shares of East Malartic Mines Ltd. stock at .16 per share, we are to have the first chance to provide for further financing of East Malartic's future requirements. This would, of course, give us the prior right to accept or reject any other firm deal that you may offer to or be offered by others.
>
> I don't think any formal agreement beyond this letter should be necessary and if this is also your understanding, would suggest that you have the enclosed copy of this letter signed by your partners and yourself and returned to us. We will take care of the draft as presented.

E.V. Neelands was Lindsley's general manager and right hand man and handled all details for Lindsley. Any property Ventures invested in at this stage of development required close attention to their investment and this was Neelands' job:

Re: sampling. If you prefer to sample along the lines you suggest, I think it will be satisfactory to us. I note also that Crowe can be present, which I think will be a good plan so that we shall know just what you are doing. I assume also that there will be no objection to our running a survey line so as to ascertain the exact position of your showing with reference to Ventures' Canadian Malartic. I have asked Jack Millenback [manager Canadian Malartic] to arrange this with you. I trust you are going to have some good luck with your exploration.

Any time a large successful mining company would make an investment, particularly one managed by Thayer Lindsley, the employees who worked for these companies were free to invest their own funds on their own initiative. J.G. Rix of Ventures was one of many who contacted J.P.:

I do not know what the arrangement is as to the sale of stock in your company, but I notice our company has taken over a considerable block and therefore am writing you direct as I would like to gamble a small number of shares, if it could be arranged.

Let me know if about 2,500 shares would be available and how I would go about it to make purchase.

Publicity and press reports on the mine development were always critical so that potential investors might participate. J.P. wrote Porteous regarding the *Quebec Mining News*:

Re: Glidden and the *Quebec Mining News*.

Glidden came in here one morning with the plea that he must have help for his paper or else fold up, so something had to be done immediately. Previous to this I had given him nearly $200 personally. When we started East Malartic you will remember that something was promised to him for helping us and I think you will agree that he has done even more than his share. You will also remember that part of the shares allotted to you were for advertising purposes so that I think you should allow him the 200 shares requested out of this allotment. I am not going to pay him either company funds or treasury shares. Something must be done about this quickly so I hope to hear from you by return mail.

I note you have had to buy East Malartic shares. This is to be expected in any operation [broker Porteous had to buy them to protect his market]. I hope that you succeed in disposing of them quickly.

John Partanen has apparently made another important find on East Malartic, near Goldfields line in the eastern section of the property. Our telephone is so bad that I cannot give you full details now but will do so in a few days.

The first hole has been completed with some interesting sections. No assays are yet at hand; however, it looks as if we will have some very interesting information.

J.P. Norrie

In early January of 1935 the first assay samples collected on East Malartic were received and Lindsley informed J.P. they were not willing to extend their option on the present terms. This setback did not slow J.P. down at all and he negotiated a revised six-month working option where Ventures would provide for a reasonable amount of work to be done. Ventures also asked for more vendors' shares if possible, which would become a bit of a problem later. They also asked for weekly progress reports on the work done on East Malartic. By the end of February the first drill hole showed a mineralized section which assayed .07 oz of gold per ton.

By the end of January, 1935, the following shareholders owned 428,900 shares of East Malartic providing $51,835 for property development:

G. Porteous	10,000
W.G. Warden	12,500
F.H. Jeffery	500
Mr. & Mrs. Sarjeant	200
A. Waltz	18,000
J. Errington	10,000
T. Lindsley	10,000
Mrs. Mailley	1,600
Mrs. Stoker	1,000
J.P. Norrie	100,000
J.P. Norrie (Inspiration)	8,000
Canadian Malartic (Ventures)	100,000
Sudbury Basin (Ventures)	50,000
Ventures	50,000
Hector Authier	10,000
F. O'Hearn Co.	1,000
E.D. Flatt	5,000
A.J. Davis	20,000
Mrs. Panneton	1,000
A.J. Bolton	5,000
Miss Preston	100
Johnson & Ward	15,000
Total	428,900

Inspiration had been doing the exploration on East Malartic all through the winter of 1935 under very difficult weather conditions. Many of the ordinary details concerning the deals verbally agreed upon were difficult to complete just because of the severe weather and the isolation of the northern bush. Thayer Lindsley wanted to meet J.P. to inspect the property but only if the snow was not too deep and the bush plane could fly from Toronto to Amos. J.P. also had difficulty getting Lee Coffin to sign legal papers for Ventures because he was deep in the bush prospecting even in mid-winter.

By the end of May assays from extensive trenching and six diamond drill holes defined an ore body which would support an exploratory shaft, and this would require more money, to sink the shaft. Ventures now wanted more cheap stock but J.P. advised them what their extended option would cost:

When you became interested in East Malartic last fall we agreed to offer you further shares when additional funds were needed. Previous to your participation approximately 200,000 shares were sold to provide funds for incorporation of the company and 248,000 shares were under option to interests represented by Andrew Waltz at .35 per share. These shares were not taken up.

It is now our intention to sell them at .23 per share to provide funds to sink a prospect shaft, as diamond drilling has located the biotite-diorite ore zone in six holes, three of which have shown visible gold in the core. The best hole is the most westerly one where much free gold can be seen.

At a point 300 feet to the west of the original discovery and approximately on strike, the zone has been again located under light overburden in two trenches 100 feet apart. We have here opened up an average width of 18 feet of vein material now being blasted and sampled. The only two samples so far taken assayed .26 oz and .34 oz gold per ton over 3.5 feet each.

The following sections have been secured to date:

Drill Hole	
# 222.5 ft	.098 oz/ton
# 415.5 ft	.047
# 513.0 ft	.070
# 615.0 ft	visible gold
# 716.0 ft	visible gold
# 811.0 ft	visible gold

We have developed a length of 500 feet by trenching and drilling an average width of 14 feet. Average grade approximately $4 per ton of ore. I believe the coarsest gold will at least double the grade indicated by drilling and trenching.

The new show to the west is most promising and indicates that we have a great length of ore-bearing material.

Please advise me if you are interested in taking 249,000 additional shares at .23 per share as in order to interest responsible people we will probably have to make some arrangement with them on the balance of treasury shares.

I believe in the area east of Canadian Malartic. Some large low grade ore bodies will probably be developed. Sladen is getting excellent results from the drilling on his ground.

Andrew Waltz, who was a prominent consulting engineer in northwestern Quebec, sent in a very promising report on East by June, 1935, and Ventures informed J.P. they could not accept the terms, at that time, which J.P. had outlined to them. J.P. wanted Nesbitt Thomson to take up the options in any case and wrote Mel Thomson:

...in my last letter to you I mentioned some facts re: East Malartic. Waltz has already given you his opinion of the property before any ore was outlined underground. Now we have definitely located a valuable ore body for 600 feet with an indicated length of 4,200 feet.

Milling results on Canadian Malartic are remarkable and the stock must soon start to skyrocket. They are now increasing the mill to 300 tons per day and it will shortly be again increased to 600 tons. The performance of Canadian Malartic will soon draw the attention of the entire Canadian public which will help East Malartic financing.

During the fall I expect to be very busy on Perron and Beaufor as we have the new mill to attend to so I want to get the East Malartic finances cleaned up soon. We are ready for a shaft and a pilot mill there now and I am anxious to get this going.

I believe in East Malartic. We have a bigger mine than Canadian Malartic with the same grade. Mr. Waltz can give you the whole picture as he is very familiar with the whole district.

I am sure we have far more to show on East Malartic than we had in the start on Perron and my advice to you was surely right.

I would like to go into the matter of finances with you soon and would be glad if you would wire me upon receipt of this letter if I can discuss a deal with you on Tuesday, June 25 in Montreal.

We are prepared to give you bonus stock and make the proposition worth your while in every way. I feel I am asking you to finance a property that only comes along once in a lifetime and I am willing to gamble on my opinion, in fact I have done so already.

J.P. Norrie

Since Lee Coffin and Porteous possessed vendors' shares, as well as J.P., their agreement on any deal with Nesbitt Thomson was absolutely necessary. J.P. wrote to Thomson:

Since returning here I have discussed with my partner, Lee Coffin, the matter of giving you 300,000 East Malartic vendors' shares and he is agreeable.

Developments on the property continue excellent. We have made another surface discovery which extends the surface show another 400 feet to the east. All recent drill holes look good.

Finally, by the end of July, 1935, the Nesbitt Thomson group were ready to become partners with J.P. again:

In conversation between yourself, Messrs Nesbitt, P. Thomson and M. Thomson, you represented that East Malartic Mines has been duly incorporated under the laws of the province of Que. with authorized capitalization of 3,000,000 shares of $1 par value, of which 1,000,000 shares have been issued for the properties, 435,000 shares have been issued to the public for cash and the balance is in the treasury of the company.

You have further represented that you and your fellow directors of East Malartic Mines Ltd. (no personal liability) are desirous of securing our assistance in further financing of the company, and upon receipt by us of satisfactory legal and engineering options as to the titles of the company etc. you will use your best endeavours to secure an agreement between us and the company under which the company will sell to us 465,000 shares of its unissued shares for the consideration of $69,700, being at the rate of .15 a share, which amount will be payable in nine equal monthly installments.

Further, that with respect to the balance of the authorized stock the agreement will provide that in consideration of the foregoing purchase, we will have the option to purchase as follows:

250,000 shares at .25, $62,500 payable in four monthly installments.

250,000 shares at .40, $100,000 payable in six monthly installments.

250,000 shares at .60, $150,000 payable in six monthly installments.

250,000 shares at .80, $200,000 payable in six monthly installments.

The agreement will further provide that the moneys to be realized from the sale of the unissued capital stock of the company will be deposited with a recognized trust company and will only be released for expenditures, subject to our approval, in connection with the development of the property.

In the event of our entering into an agreement upon the foregoing lines you agree to deposit or procure the deposit of 700,000 shares and at least 300,000 shares of those in the hands of the public with a recognized trust co. to be held in escrow until the termination of any purchase or option agreements that we may enter into with the company, and that further under such deposit agreement it will be provided that the shares so deposited will be voted in favor of the election of directors all of whom will be nominated by us.

P.S. It is understood that in the event of our entering into such an agreement and for services to be rendered that you will deliver or procure the delivery to us of 400,000 shares of the issued capital stock of East Malartic Mines Ltd. (no personal liability).

As the East Malartic property developed, Nesbitt Thomson became more and more interested as J.P. had hoped. The bonus of 400,000 vendors' shares seemed to do the trick even though the property was obviously a very good bet. But as the deal came to a close, Geoff Porteous committed a blunder, which certainly angered J.P. and caused Nesbitt Thomson to delay signing the option. J.P. let Porteous know what he had done in no uncertain terms:

I have just learned that the delay in negotiations with Nesbitt Thomson is due to your circulating a story in Montreal that they have put $250,000 into East Malartic and that you are trying to sell shares at .30 per share advising clients that on the strength of their financing East Malartic shares will soon be at .60 per share.

I have substantial proof of the above and if I am successful in getting the situation out of the mess you have landed in, then I will produce the proof. After all the trouble we have had I surely cannot understand your action in throwing a monkey wrench into the gears but that is just what you have done.

I am trying hard to get things straightened out again and get another interview with them which should have been tomorrow but which has been postponed for the meantime at least.

We will have to turn over all shares to Nesbitt Thomson if they go ahead (i.e. shares in the statement I gave them) so please do not sell any further shares until we find that negotiations with them are definitely off.

After working for some weeks on this promised success I felt much disappointed at the way you have handled things.

In any case take no further action in any way until you hear further from me.

Later J.P. felt guilty about giving Porteous hell, mostly because the deal was back on track:

I expect to be in Montreal again next Monday when something definite from Nesbitt Thomson should be forthcoming. I regret I lost my temper in our conversation last Thursday evening. At such times we sometimes say things we do not mean and I guess this occasion was no exception to the rule. In any case I apologize if I said anything to hurt your feelings.

After J.P. cooled down a bit he forwarded to Porteous the conditions of agreement that Nesbitt Thomson required to complete the East Malartic financing:

In order to complete financing arrangement with Nesbitt Thomson for East Malartic Mines it will be necessary to fulfill the following conditions:

1) That Nesbitt Thomson are free to appoint all directors of the company.

2) That they shall have full control over the expenditures of all funds.

3) That the vendors i.e. Lee Coffin, Geoffrey Porteous and J.P. Norrie donate to them 300,000 vendors shares; J.P. Norrie and Lee Coffin 75,000 shares each and Geoffrey Porteous 150,000 shares.

4) That the vendors are agreeable to giving 400,000 treasury shares free to Nesbitt Thomson and Co.

5) Nesbitt Thomson agree to provide the East Malartic mine with $600,000 in consideration of the balance of treasury shares which is approx. 1,365,000 shares.

We hereby agree to the conditions authorized above.

Whether or not Nesbitt Thomson took up the options at each stage depended on the size and grade of the gold ore body.

The option agreement was finally signed in August, 1935, and all vendors' shares belonging to J.P., Coffin and Porteous were in pool to prevent flooding the market while Nesbitt Thomson made the market. The vendors' shares belonging to Ventures were, however, not pooled and J.P. wrote Lindsley asking that he pool his interests as well.

Ventures wanted more vendors' shares, however, which J.P. was not willing now to give to anyone as they would soon be worth a lot on the market:

Mr. Neelands informed me over the phone that you wished to place a figure at which the stock should be sold, and named .25 per share.

I do not know how this will be handled. However I believe that when a public issue is made, that the first three blocks of stock will be grouped and probably sold at .35 to .40 per share. Thus I trust you will raise your figure to at least this and that you will agree to pool for one year.

Ventures were upset about several issues and did not cooperate with the pooling of their shares. J.P. had to contact others who had vendors' shares asking them to pool theirs for a period of time. One person was his close friend Walter Worden:

Re: pooling of East Malartic shares: I had a promise from Lindsley that he would pool his shares. This was more or less contingent upon his being able to purchase from me some of my vendors' shares, which I was not willing to do.

Anyhow he is peeved because he could not get a cut in the Sladen Malartic deal and because they referred to Canadian Malartic in their prospectus. He now wants to put strings on his pool that are impossible… you will notice the agreement mentions 31 months… this had to be as this is the period of the option with Nesbitt Thomson; however, if you want to have your shares released at any time I will see that it is done.

Even though J.P. was very busy with the engineering of East Malartic, Perron and Inspiration, he still had time for other interests such as the selection of good directors for the company:

...I believe it would be well to retain as directors, Hector Authier, local M.P.P., who has wide mining interests in this section, and has much influence with the Quebec government. I would also like to have Walter Worden, retired civil engineer of Toronto retained. Worden is a good man and along with him suggest the following for your consideration:

A.J. Davis, C.O. Stee, general-manager of Siscoe, W.H. Englebright.

Stee is a good man of high standing in this area. Englebright is mining adviser for O'Hearn Co. Toronto. He is a mining engineer of excellent financial standing. Would you consider Gus Rainville? He has gained considerable wealth and prominence through Pamour and seems pretty close to Murdock, Timmins, etc.

I know practically all the leading mining men in Toronto. However, I believe the ones I have mentioned would be the most acceptable. Please advise me what procedure is necessary in appointment of new directors. I expect it would be better for all present directors to resign and have you name new ones. Coffin would sooner have Worden act in his place.

Early Operation of East Malartic (1935-1939)

The mining communities in this area of Quebec were not very large and gossip travelled fast even to Montreal. Mel Thomson was upset by one of these rumours. J.P. had to explain:

The information you received re East Malartic is absolutely untrue. If you will remember when we were in the Mount Royal Hotel together recently I received a telegram from a man, Buckland by name, who I advised you at the time, was making trouble on the property. This is the senior man you mention, and he is no more senior than any other.

This man is a fairly good geologist but knows nothing of mine operations. I used to send him down to the property to check up on the logging and sampling of the diamond drill cores and surface exposures, and he was so impressed with the property that practically all of his worldly fortune was invested in the East Malartic stock, and for this reason his wife chiefly and secondly himself thought they owned the place and could do everything according to their own wishes.

This chap is rather retiring and is married to an aggressive educated American girl who is the real boss of the family and really tried to be the boss of our job.

We have had quite a turnover in employees at East Malartic and now I find it due to this woman. She extended her efforts to Aub Davis and when this did not work, she went after the man on the property who has been there continuously since work started, and who has done all real work there. Because I would not stand for him being fired, Buckland came out with the ultimatum that he would quit, so in order to save the entire organization I told him to go.

The reports you have heard come from a vindictive liar of a woman who has caused me much trouble, and believe me I am glad both have gone. On the property I have an excellent operator and have as his assistant the man who has been in charge there to date.

You will probably hear more untrue reports sponsored by a dirty woman, who has nothing else to do but make trouble for others and believe me she has lots of time to do her work.

A woman around a mine where they have little to occupy them are the biggest troublemakers on Earth.

The mine progressed quickly with its development during 1936 and 1937 without any major problems. By September of 1937 Nesbitt Thomson wanted to take down more treasury shares to market to their clients as the mine was now a producer and ready to pay a dividend. They wanted to pay .90 a share to the treasury which would be used for further development but J.P. was upset with them as he felt the shares were worth $1 and they had already made much money on the original options. He wrote Andy Davis, director of East and close friend of Mel Thomson:

...I had a long talk with Messrs. Thomson and Nesbitt on Monday. They think the East Malartic shares should be taken down by them at .90 per share which I think is very unfair to present shareholders. In the original financing they were given 400,000 free shares and on the shares sold to date they have made much money. As the situation now stands we have probably the second best mine in Quebec. I believe, and this is confirmed by both Futterer and Reid, that the new shares can be sold to net the treasury $1 per share. Thus I think Nesbitt Thomson should pay this price and by so doing they can make more money than they have made so far as with any half decent market these shares would sell for $2.

During this period Nesbitt Thomson continually monitored all aspects of the operations of the mine, making sure they were protecting their $600,000 investment. J.P. wrote to Thomson:

The cost of the East Malartic crosscutting is high due to the fact that the haulage crosscut and drift on the 335 foot level is 8 feet by 8 feet and has to be close timbered to support the back [roof] in all greenstone areas [greenstone is unstable]. This slows up the work and greatly adds to the cost.

At this time practically all pumping has been charged to this work as shaft water is hauled direct to surface in the sinking buckets. A heavy flow of water is now coming in. In fact our pumping equipment is taxed to capacity. A pumpman must be kept continually on the job. i.e. the higher cost of the work is due to:

1. Oversize of the haulageways.
2. Timbering.
3. Excessive water.

Some of the water will dry up when the overlying low ground is drained. The balance will serve as a mill supply and will probably save us money in that a long pipeline to Piche Lake with accompanying costs will be unnecessary.

Re: ordering equipment, etc., all authorized in your letter has been ordered.

It is absolutely necessary that everything be delivered before April 15. The ice on the Thomson River is now unsafe for heavy transportation. In fact, I am afraid when crossing there now with a light car, so we will have to haul from Amos. The Amos road was absolutely unpassable for over one month last spring and will be the same this year.

Each time Nesbitt Thomson took more treasury shares they were concerned with the pooling of all vendors' shares. There was no problem with J.P. He didn't want to sell his anyway so pooling his was no problem. Lee Coffin was not as cooperative with Nesbitt Thomson. He always listened to J.P., however, because it was due to him that he held so many shares in the first place. J.P. wrote to Thomson:

I have just been in conversation with Mr. Lee Coffin re: further pooling East Malartic stock. He originally agreed to extend the pool on his stock for two months less 30,000 shares, on 20,000 shares of which I can secure the extended pool. Now if he is to extend the pool for a period of four months he wants 60,000 shares free on May 31 with a guarantee that the balance of his stock will be delivered to him on Sept. 31, 1938.

Mr. Coffin has dictated this letter to this point. I have asked Coffin to go to Montreal and discuss this matter with you which he refuses to do. I am sorry that I cannot make a better arrangement, however I believe now it is impossible. Please let me have your comments as I would like to clear this matter up with the other vendors and this cannot be done until some satisfactory arrangement is made with Coffin.

The mill was nearing completion in the summer of 1938 and Nesbitt Thomson were concerned that only the best man possible be hired to operate the mill. Thomson wrote:

When in the north last trip I intended to speak to you regarding a man for the East Malartic mill. I was wondering if you had given this your consideration...

My opinion is that you should get a very able chap the best available... as you are going into very big production, and it certainly requires a man who has had a great deal of experience in the milling game.

We have two very excellent men lined up for the East Malartic mill. Both of these men have wide education and experience, around 35 years of age, and have the okay of Mr. Futterer. It will not be necessary to make the choice for another month.

Always looking to hold the costs down Mel Thomson was concerned with the cost of housing at the mine campsite:

I have a report from Mr. Joe R. Norrie [J.P.'s brother]... giving a brief summary of expenditure during the month of June. I notice the house no. 9 has a cost of $5,300. I was just wondering if this was a dwelling or some other house. If a dwelling, my understanding was that the workmen's houses would cost in the neighbourhood of $1,800 to $2,000, and the managers' houses around $3,000. Would like to hear from you in this connection.

J.P.'s reply:

Re: expenditures on house no. 9. According to the original estimate this house, occupied by Joe, was to cost $4,000. Joe informs me that Mr. Futterer recommended certain changes to make the place suitable for frequent assembly of the staff and thus promote better understanding between employees. I think the charges are okay. However, they should have been brought to the attention of the directors. In future this will be done.

Thomson always advised J.P. when money was getting tight:

Referring to our conversation at the directors' meeting when you were here on Friday last, I feel that it would be very wise for you to keep a closer check on the expenditures at East Malartic, as from what I can learn the forecast is at present time that we will probably go in the hole $300,000. I wish you would check this very closely and let me have your ideas.

In March of 1940 an old partner of the early days of East Malartic, Geoff Porteous, came back to bother J.P. again, but not for long:

In 1936 when concluding arrangements looking to the financing of East Malartic by Nesbitt Thomson, which involved substantial contributions of pooled shares, particularly from myself, I drew to your attention that, at your request, Frank Hewett had been advised through you that he had been allotted 10,000 shares out of the shares in pool, and which had been pooled by you, Coffin and myself. I felt it should be made clear that I had no responsibility to Hewett.

When the final settlement was reached, and before it was okayed by you and me, you gave your assurance that you would take care of Hewett, and thereupon the settlement was agreed. Hewett is now asking for these shares, and in view of your assurance, I would like to know what you have done about it?

Porteous had profited a great deal from East Malartic and had not produced when he had the chance. J.P. had no time for his promises:

> ...during the course of the financial arrangements between Nesbitt Thomson and East Malartic in 1936 or at any other time, I gave no assurance that your personal obligation to Mr. Hewett would be taken care of. In fact the subject was never even discussed between us in connection with Nesbitt Thomson on the East Malartic deal.
>
> I believe all correspondence between us on the subject points to the fact that all responsibility to Mr. Hewett rests entirely upon you and does not implicate me in any way.

The happiest occasion (and the best excuse there is for a party) for any gold mine is the pouring of the first gold brick; everyone wants to see a gold brick poured. Everyone was invited. The secretary of East Malartic discussed details with J.P.:

> For the occasion of the official pouring of the first brick at East Malartic, which will take place some time in January, we are planning to issue an illustrated brochure or booklet describing the mill, reviewing its history, using photos and other statistical information.
>
> At the suggestion of Mr. M. Thomson I have today written Mr. Joe R. Norrie, asking him to supply us with whatever information he can, and it has occurred to us that possibly you can supply us with an article or information which we can incorporate in the booklet.
>
> We hope to make it an impressive booklet which will be widely circulated, and naturally are anxious to get full and complete readable information.
>
> Your kind cooperation in this matter will be appreciated. Just a suggestion, is it possible Mr. McLeod of the Val d'Or-Lamaque News will assist in this matter, but, of course, you know best.

Chapter 16

Frustration Leads J.P. to Resign

by George Norrie

Later East Malartic Operation, 1939-1943

The operation of the mine from 1939 to 1943 became more and more concerned with the soft wall rock and mining systems. The gold-bearing ore at East Malartic was in a hard host rock. When the ore was removed, the soft wall rock, which did not contain any gold, would cave in and become mixed with the gold ore, causing increased milling costs. This was called dilution of the ore. This was a particularly difficult problem at East Malartic and very controversial. The confusion eventually led J.P. and his brother Joe both to leave the management of the mine. The mine operation was hampered, as with all mines, with war restrictions and poor labour supply. Then in 1942 the Howey mining people gained control of the mine and forced J.P. out.

As the East Malartic developed their ore bodies, it became obvious that these soft-rock walls were some of the worst known to exist in Canada. Rumours about mine problems circulated in the small communities quickly, and eventually reached trade papers and mining magazines and brokers such as Colling and Colling of Toronto. J.P. wrote to Watson of Colling & Colling:

I quote from a recent letter to one of your clients. "East Malartic Mines, we understand, will experience higher costs than anticipated through the fact that the ore is inclined to be of a soft nature. It shales off easily and, therefore, it will be necessary to timber some of the stopes and drifts. This will raise cost of operation per ton. Frequently before a mine goes into first production, certain difficulties arise which naturally tend to have a reflection on the price of the shares."

If you are acquainted with the official reports on this mine, and in your position you should be, you will remember that the estimated heads were reduced to $6 per ton to absorb dilution of 20%. Naturally in our cost estimates this was considered.

Since the mine has been in operation three months, we now estimate that dilution will be less than 10% with consequent higher grade. Within one year we expect an operating cost of approximately $2 per ton. As a matter of fact the mine is responding far in excess of our expectations in every direction.

I believe this is just rumour credited to East Malartic and other mines aimed to bring commission to the broker rather than give the public beneficial and accurate news.

Some of the remarks J.P. made about the brokerage business were not received too well by Watson:

We note that you do not contradict the information as we gave it, but simply expand on the principle of dilution with which we are quite familiar.

It is to be regretted that you found it necessary to add your final somewhat caustic paragraph in which you make reference to brokers' commissions etc. We think your statement was unbecoming of a man in your position.

By March of 1939 the method of mining the East Malartic gold ore had become an even more contentious issue. The mine consultants Futterer and Reid were considered among the finest mine consultants in Canada and they decided to recommend a change in the mining system:

In our recent visit to the property Mr. Reid and I discussed a number of operating policies with Messrs Joe Norrie, MacLean and Griffin. Following are notes covering our discussions:

East Malartic mine operation , 1940

First of all, concerning the mining system, it must be borne in mind the same is probably one [of] if not the most highly technical system in Canada. The staff should be given time to familiarize itself with the same. While no unforeseen difficulties have as yet presented themselves, it would be surprising if some did not arise. Until such time as the underground force has gained a much greater experience in the present underground methods it would be unwise to open up any more stopes than are absolutely necessary to maintain normal mill demand. So long as the mill schedule is in excess of what the mine can efficiently produce, the mine will be forced into uneconomical practices, the operating costs will suffer, and the extremely valuable experience gained in present stoping practice must (so far as its application to further operations are concerned) be indefinitely postponed. It would be the height of folly to assume that we have been able to develop a mining system based on diamond drill cores that in practice would need no revision, particularly an unorthodox one of the present type. We believe that the essence of success, so far as the mine is concerned is a thoughtful consolidation of the experience gained to date and applied in an orderly manner to future operation. This experience cannot be applied if the mill demands are continually in excess of what the mine can meet. In fact, both mine and mill must suffer if this condition prevails.

Under the circumstances no criticism of the policy so far is in order. However, the time has arrived when a definite milling schedule, based on the mine's ability to produce, should be set and followed. If in the meanwhile unanticipated trouble should arise in any of the stopes, a drop in production from any one stope should not cause a radical drop in the tonnage milled. In a few words, a steady healthy growth with a minimum of rush tactics is the logical procedure.

Ed Futterer of Futterer and Reid

Six months later it seemed as though Joe Norrie, who was mine manager and J.P.'s brother, was being used by Futterer and Reid as the scapegoat for the mine problems, although they never confronted J.P. directly. J.P. expressed his views to mine president, Andy Davis:

First of all I want to thank you for the frank manner in which you have discussed with me the present East Malartic situation and every other that has ever come up in our business experience together. I believe I have reciprocated in this spirit always and think it is the only happy way to do business.

Since I came back I have discussed with Joe some of the points of which we were talking, however I did not convey to him full import of our talks as this would only have a demoralizing effect on him at a time when full efficiency is necessary in order to protect his position and ours. I believe Joe has done and is doing a good job at East Malartic not alone at the mine but in the building of the community in general.

He has made some bad friends in the town due to his objections to grafting contracts etc., and being successful in this, some of the would be grafters would delight in seeing him out.

Joe thinks the start of the trouble was with Castawick, Futterer's white-haired boy, and that he cannot so far lay much blame on McLean. He says McLean has too little brains to do much unobserved undercover work. It seems that trouble started at a hockey match between Sladen and East at which Castawick selected the referee. After several disgustingly bad decisions of the referee, Joe took a hand and did have a fight with the referee. Castawick then came in. Although the latter was only a battle of words he has had it in for Joe ever since.

Must the manager of East be a sucker for the Sladen mine captain to hold his job? When Futterer comes up here he practically lives with Castawick and I am convinced that he is the seat of the trouble. Joe is also a very strong Mason, which may hurt him.

Apparently no fault can be found with the costs to date and that is the real meat of the situation. Mr. Thomson mentioned the buying. If this were bad it would be reflected in the costs. I am sure they are tough buyers at East. I have spent my entire productive life in building up the Malartic area from black spruce swamp, with no rock outcrops to work on, to one of the finest mining camps in Canada. Surely some benefit should come from it. One of these should be that Joe at least be given a square deal at East Malartic.

When we sold our claims to East Malartic we received no cash in the deal, and in lieu of this, management was to remain in my hands. It is not my intention whether legal or not, to ever enforce such a clause at the expense of East Malartic and all concerned. However, I do insist on a square deal and in this I am sure I have your full support.

I certainly do not believe that Howey have control of East Malartic. [Howey Gold Mines had started to increase their share holdings in East Malartic over several years and were starting to muscle in on operations.] They may have solicited the support of some of the large shareholders outside their own group and with them may be able to talk pretty loud, however, the average man intends to play square and I feel they will in this case.

Up to now it was impossible to secure good costs at East Malartic, due to the turmoil caused by continued construction which will not be finished for some months yet. If a new man came in there now he would reap the benefit of the years of hard work of all of us. His costs would look well on paper at the expense of a ruined career for Joe and myself which we surely do not deserve.

Money is not everything in this world. I intend to retire myself from East Malartic someday, however, I do not intend to be deliberately pushed out for no good reason. I hope to leave with a good record and a good name.

As far as Joe is concerned I think he should be given one year from Jan 1, 1940 to prove that he can make the costs set out in Futterer's estimates to you of Aug. 24, 1939. These costs are subject to some upward revision which can be agreed upon at a round table talk, and would also be subject to increased wages and cost of supplies on account of war.

If at the end of one year he cannot get these costs then I am in favour of his retirement and any further changes in the organization on which the directors may decide. He will at least have been given a square deal and a chance to defend his career which would be entirely broken up if he were let out now.

I am sure Joe is well qualified for this job in both brains and general all around ability and that he will prove it if given a chance. He has done so in the last year. I might say that during the last week or so rumour has been going the rounds up here that Joe is to go. This could only come from Futterer or one of his cohorts. It creates a very humiliating situation.

J.P. Norrie

P.S. I was at East Malartic with Mr. Futterer yesterday. He apparently has no complaints about anything.

Considering all the controversy regarding the mine operation it seems that Joe and the mine managers were given a vote of confidence with these salary raises:

This letter will be authority to increase Mr. Joe Norrie's salary, as manager of East Malartic Mines to $1,000 per month, and would you be good enough to date this back to Apr. 1, 1940. That is, his increase will start on Apr. 1, 1940.

I did not get the suggestion in time to take it up at the directors' meeting on Thursday, but have discussed it with the majority of the directors since, and they are agreeable, and it is not necessary to make a notation of it in the minutes of the meeting.

Joe has recommended and we approve of raises in salaries for Carl Kamm, chief engineer, and W. Griffin, mill superintendent. Their present rates are $325 and $325 plus house, respectively. Recommended raise to $350 and $350 plus house. If this meets with your approval will you please confirm.

Early in 1941 the mine consultants decided to recommend another change in mining methods. The mine had been using a method called cut and fill stoping and now they were switching to a method called shrinkage stoping.

J.P. replied:

The trouble at East Malartic is in the mining and could have been averted if Joe's recommendations to sink the shaft six months earlier had been followed. We were duty bound to try out shrinkage stoping and determine if the bad walls could possibly be handled by that method. We found they could not, so we are forced to change over to cut and fill which will be $1/ton more expensive, but mill heads should be over $1 higher.

For the next three or four months heads will be lower until the new stopes come in. We expect earnings in 1941 to be a little better than 1940 and to increase from then on. The ore and values are surely in the mine and I feel the mining problem has been solved.

Two examples of customized winter transport, 1940

Visitors to the mine was a common occurrence. Since the wall rock was a serious problem at East Malartic, many managers of mines also on the greenstone band would visit East Malartic to see how they were dealing with the problem. Since Hollinger were taking an option on

National Malartic, they sent their men to check out this situation. It was expected that these visitors would use the information with discretion, but a rumour started and quickly made its way around, that one of them, a Mr. Gustafson, said East Malartic dilution was as high as 30%, which caused Mr. Gustafson to write:

Several rumours had reached me to the effect that I was held in low regard in East Malartic circles because of promiscuous statements I was alleged to have made concerning your property. It was, however, only the other day in Malartic when I saw Dr. Goranson and yesterday in Toronto when I saw Mr. Futterer that I found out the full extent of the crimes I am accused of. From them I learned that a rumour - not yet tracked to its source - has it

1) that I or I and my associates following a visit to the East Malartic mine last winter made public certain unfavourable opinions regarding the mine.

2) that I made the definite statement that dilution at East Malartic was 30%.

I can understand your being annoyed if you believed the rumour. The accusation is, however, unjust, and I would like to put the record straight. The facts are as follows.

1) Neither Mr. James Dougherty, the Hollinger underground super who accompanied me, nor I, wrote any report on East Malartic. The sole object of our visit was to appraise National Malartic which worried me as an operating problem because of its low grade, narrow widths and slabby talcose walls. We visited East Malartic to find out how those walls were apt to behave in stopes.

2) We reported verbally to our principals, Mr. J.R. Timmins and Mr. John Knox. We also discussed the matter with several engineers connected with East Malartic, National Malartic or closely related organizations. We did not, however, make public our opinions.

3) It is astonishing news to me that I ever said East Malartic dilution was 30%. That figure is wholly contrary to the data obtained from you and

Dr. Goranson which I have in my notes and contrary to my own impressions of the mine. My recollection of the condition of the wall rock at East Malartic as I reported it was that whereas dilution to date was not great - estimated at around 12% by the mine staff - there was unmistakable signs that dilution would become excessive in the very near future if the mining methods were not modified, and there was even danger of losing a level. I further reported that a wholly satisfactory low cost stoping method might be difficult to achieve. (Since you have since modified your mining method, I suspect that I was essentially right in what I said.) The whole point of describing these wall rock conditions in any case was that when translated into terms of the narrow National ore bodies they spelled trouble if the ore bodies were to be stoped by ordinary cheap methods.

I enjoyed and appreciated your hospitality at the mine, and I don't like being suspected of abusing that hospitality. I also have some regard for factual evidence, whether it relates to dilution or anything else, and I don't relish being regarded as a damned fool who ignores such evidence. I am also extremely sorry that you and your associates have been upset over this matter.

Gustafson

The quality of labour and management at East Malartic was a constant problem during early 1942. The war was not coming to an end and many top employees left to enlist, leaving the mine in very inexperienced hands:

The following key men have left with the exception of Mr. Kamm who is to leave shortly.

Engineers:

Swanson T.	Mine captain
Kamm C.	Chief engineer
Anderson Don	Engineer
Kerr T	Engineer

McNally Pat	Engineer
Brooks Lorne	Engineer
Brassard Howard	Engineer
Thomson Norman	Engineer
Chisholm Bill	Engineer

Accountants:

Graber Carl	Chief accountant
Raybold Noel	Assistant warehouseman

Electricians:

Anderson Fred	Assistant chief electrician

Mechanical staff:

Stamp D.	Shop foreman

Laboratory staff:

Whithard S.	Chief assayer

Note: All but three of the above are in or about to enter the armed service.

The controversy and concern about East Malartic continued. Andy Davis could not understand why these problems could not be remedied and as president of the company he needed explanations. He wrote to J.P.:

I have just received in the last day or so the last production sheets for East Malartic Mine Ltd., for the month of May, and I find that our production for the month of April was 37,198 tons and a value of $222,148.86, and I find that taking our production for May up to the 30th of May, which would give the same number of days as April, our tonnage milled was 37,449 tons and the value $221,778.99. Now of course with the extra day that we had in May over April it puts our production slightly over April, namely $229,021.23, and the total tonnage produced in May was 38,748 tons, or average of 1,250 tons per day.

Frankly speaking, I am worried about the situation at East Malartic. For over a year now we have had to be making excuses, first of one kind and then of another, and every time the matter has been discussed in my hearing, we have been assured that in the course of a couple of months or in the course of a few weeks, or something of this kind we would be through the troublesome part and we would be away in good shape again, but month after month goes on and our production does not improve. True, our grade is up over what it was several months ago but now our production goes down, and the net result is that we are not producing any more in dollar value than we were.

As you know, at the end of this month will be our second quarter and production figures will have to be given and I am very much worried that when the public is acquainted with where our production is that the stock is going to get a decided slap in the face.

I freely admit that I am not a practical mining man and therefore am at a decided disadvantage, because I do not know from my own knowledge about different matters, but can only take what I am told, but the point to me is that we have gone along now as far as we should go and it seems to me there must be some drastic action taken of some kind in order to get us up where we belong.

I appreciate the fact that there is labour trouble, and bad labour turnover, that the replacements of labour are poor, but on the other hand from what I can learn of other mines throughout the country they are managing in some way or other to hold their production up and not have it down as we have.

The bottleneck seems to be in getting the tonnage hoisted to the mill, and therefore it seems to me that those in charge and responsible should put forth redoubled efforts and if necessary spend considerably more time underground until they get things straightened out so that our production is up where it should be, and it seems to me that if those who are in charge have done everything they can do and cannot improve the situation then we should have a round table conference as to what is to be done.

I am trying to put this matter as fairly as I can but I have got to the point now where I am decidedly worried about the situation, and am putting it up to you as general-manager of East Malartic Mines for your reaction to the situation and what you think should be done to improve it, and I will appreciate receiving an answer from you just as quickly as possible.

A.J. Davis

Clarence Stee was J.P.'s friend and now a mine consultant. He knew all the history of these Malartics and how difficult the road had been for J.P. He couldn't believe, now that all the most difficult work had been done to develop East Malartic, that it was possible J.P. could lose control of the mine:

The one outstanding impression that I got from the meeting Friday was that the East Malartic is now being run in accordance with the wishes of the Howey. Regardless of the opposition which you so ably presented they got just what they asked for, made no concessions, and departed.

The fact that you discovered the mine, organized the company, nursed it through its early trials, got Nesbitt Thomson and Andy Davis to assist in the financing and finally, after a large mine was proven you invited the Howey interests and their consulting engineers to come share in the fruits and that the latter's interests represent a little more than 10 per cent of the stock, was apparently given no consideration at all.

This is just the beginning. You will find that new places will be demanded on the board. The grip will be tightened. Yourself and Davis, and Thomson interests will fade out of the picture and the "invited guests will take possession of the house". Personally my stake is not great. It will not matter much. But I should regret very much to see such a thing happen. I believe there is such a thing as justice and fairness in the world of business. If action is taken before it is too late something can be done.

Why not examine the list of stockholders very carefully and add up the friends and possible friends against the enemies and see where you stand. Your name alone would carry tremendous weight. Your long line of successes will not be put aside lightly. The Nesbitt Thomson and Davis organization will measure up to anything they can offer. If a battle is necessary, I am sure you would have the bigger guns. You have time on your side. What they can do in the next few months will not help them. I am quite sure of that. What Ed said on Friday when he would promise nothing bears this out. This is my reaction. I thought I would pass it on to you as my view of the road that should be taken.

C. Stee

Futterer & Reid, the regular mine consultants, were optimistic following Stee's report. Mr. Reid suggested the difficult wartime labour conditions would improve eventually but that the mine policy needed to be revised, as suggested by Stee:

I have during the past ten or twelve months, been apprehensive of the immediate future of the gold mining industry, and of East Malartic in particular. This is because of our lack of priorities as to men and materials. As to the future of the industry I have no fears and believe that East Malartic, when conditions become normal, will then have a long life and a profitable one.

The immediate future will require much thought and careful planning in order to arrive at the proper solution of East Malartic problems now in evidence because of adverse conditions steadily developing due to the shortage and inefficiency of labour, and also to the loss of a number of capable foremen and engineers.

I am personally of the opinion that the solution of the problem will not be solved by a change in mining method, nor in the operating executive. It goes deeper than that. I do believe that a radical change will have to be made in mining policy. Much thought should be given to this in order to have the mine in a healthy condition when peace is restored and efficient labour and capable direction is available.

It is possible that developments in the East ore body will permit the milling of say 500 tons a day of a much higher grade of ore than is now being milled. If this were possible much could then be done to regain our former position as a real money maker.

If a change in mining policy is decided upon I will be only too pleased to assist in the proper solution, and I am sure that Mr. Futterer will be very glad to give his advice in this connection.

Reid

The Howey Gold Mines had secured increasing shareholdings in East Malartic for the past several years and were now clearly trying to take control of the East Malartic board of directors. J.P. wrote his closest friend on the board, Andy Davis, explaining his opinion regarding the Howey group:

I was sorry not to have a chance to talk with you following our last directors' meeting in Montreal and have verified first hand just what the Howey people have in mind. In any case I have always felt that a definite plan was set up by them shortly after they came into the East Malartic picture and what is to happen at the Toronto meeting is just a culmination of that plan. It would have come anyway probably sooner had the results at East Malartic been better.

I think the plan was to trump up excuses to get Joe out first. Then Worden, Stee and myself and if possible yourself and Mr. Thomson.

As I have said this plan has been in my mind for a long time so that, although disappointed, I am not surprised. I have not heard whether you are to carry on as president or if the executive office is to be moved to Toronto. However, I expect moves favourable to Howey in both instances. During the last several days I have gone over carefully all of the correspondence with Futterer and Reid since the beginning of the operation and must say I am struck with Futterer's many favourable references to Joe. Particularly in his last letter to me of April 10, 1940 when he recommended that Joe's salary be raised several times during 1941 and early 1942 climaxing with his favourable remarks at the last shareholders' meeting which you will recall.

From this I take it that Futterer was not aware of the Howey scheme and that each time he praised Joe he was throwing a wrench in the gears. In other words the pressure was coming from behind the scene and that generally Futterer was throwing unwilling balls prepared by others and which was part of the general scheme.

I must not forget either the conversation I had with Futterer and Reid in the Sladen guest house in Aug. 1941 when they came up to dispose of Joe. When the story was unravelled they were convinced that Joe was doing okay and they decided not to act at the severe displeasure of those who sent them there. I have not recalled all of this in Joe's defence but to show that it was part of the general plan set down a long time ago. However, with all credit to Joe, I will say that the operation at East Malartic has deteriorated since Joe left and continues to do so. I almost forgot to mention the day that Futterer came up to take over from Joe. He made two attempts to enter the office and each time returned to Mr. McClusky. They finally entered together and McClusky remained in the office during the entire conversation. I think right here we can put our finger on the one who has been responsible for most of our troubles.

The reason given that friction was the cause of the present upheaval of course in my opinion is the only one available to complete the plan because there has been no serious friction during recent months except for a few objections to the continued use of a mining system which I am sure most of us agree is no good, and which as directors it certainly was our duty to criticize. I might mention here that any trouble we have ever had has always been started and carried through by the Howey people. This too could have been avoided had the Howey people agreed to an independent report on the mine which I have always advocated. I think it is certainly due now after the report of Mr. Stee. The proposal of Mr. Birks, of Howey, that he was willing to buy or sell looks fair on the surface. However, as far as I was concerned he offered to sell me 150,000 shares (in a mine which he and his associates had nearly run out) which I would have to buy with my own funds. Or he would buy 50,000 of my shares with company money. I would say he had me over a barrel.

The question now comes to who is to remain on the board. If Mr. Thomson desires to serve, that is his business, however, I do not think he should, neither do I think you should, as I feel that I will be let down in either case, more particularly if you stay on. We have been pretty close in mining matters since we sat around the table in Nesbitt Thomson's office before Perron was really born. I have enjoyed my association with all, you, in particular. I hesitate to mention this, however, I believe it is necessary in order to bring my case more forcibly home to you, in that I have played the game with you fairly at all times. A large part of your holdings in East Malartic, Malartic Goldfields and Inspiration were secured through me at very low prices, and in many other operations such as Dubisson, National Malartic, Seventh Malartic etc. on which I spent much time to secure. I feel that the confidence I had in these operations and the reports I made on them kept you in until the shares became valuable.

In fact I will venture the statement that neither Nesbitt Thomson or yourself have ever made so much through any other individual. It may be well to call your attention to the fact every operation that the Howey crowd ever touched ended in failure.

They went in on the Howey when the dirty spade work was done, and shortly after their hands were in control, it progressed into a salvage operation. Little wonder they made good costs, when there was no development work to do. The other two operations I refer to were El Dorado and Gunnar Gold.

Figures say that Howey operation will bring East Malartic to a like fate. I definitely believe it and will venture the opinion that the life of the mine will be short. I have a firm belief that we will have a chance in the not too distant future to get back into East Malartic and make a mine of it with great credit to ourselves.

Since its inception I have given much of my time to Inspiration and my end of the business has been a distinct success. Starting from scratch we have made money year after year. More money than any other company in the business without one cent of compensation to myself except one .02 dividend.

During my years of association with you I have covered more ground than any other three engineers in the business. Few mineral finds have been made in the east that we did not have a finger in. With so many things to think about and do, it is a natural consequence that some mistakes have been made. I understood that some of these have now been singled out and that I am accused of being erratic. Which is the best course to follow? To confine one's energies to one operation and do the job with great caution and be friends with everybody, or attack the industry as a whole as I have done with great profit to us all. My erraticness has been profitable.

I understand Mr. Thomson picked out the Malartic timber limit as an instance of this, at least I take it from certain remarks which he made. His remarks were based on the opinions of certain Quebec officials who were displeased because they had not been greased for the timber we have been getting. Last spring we were advised that due to the limited amount of timber left in the country i.e. 100,000,000 board feet to satisfy the requirements of all the mines, that we could have 750,000 board feet for the year while our needs were 3,500,000 board feet at each of Malartic Goldfields and East Malartic. I started last May to secure the necessary timber for our mine and with the help of the local member succeeded in doing it without costing us one penny except my expenses to Quebec. During the summer it became quite apparent that we should buy a limit of our own. This was recommended by the local timber agent. I also discussed the matter with Futterer and Reid and Birks, who were strongly in favour of it, before I recommended it at the meeting. I believe the figures I recently sent you support my recommendation.

I think that most of the mistakes with which I am credited are exaggerated during a time like this and can be satisfactorily cleared up if people's figures are secured.

I understood too that I have been criticized for employing too many of my relatives. Unless we can help our own why should we remain in this country at all? Anyway with their assistance some very nice mines have been developed. I think it will be a long time before Joe will be replaced at East Malartic. Bob Halet at Goldfields, did a good job first at Perron and later brought in one of the finest ore bodies in the north. His cost record is good. We cannot be associated with too many men like him.

I have also heard that the firm of Nesbitt Thomson have remained cool toward me for holding out for $1 per share as against their offer of .80 per share when the last finances were secured for East Malartic. Surely after the development of such a large and valuable ore body the shares were worth the money and as a director I felt it my duty to stand for the higher price. As a matter of fact $1.50 per share would have been more like it.

In our telephone communication today you mentioned that your staying on the East Malartic board would help any future financing by Nesbitt Thomson. In the first place mines have to be found before they are financed and in the second place I am sure they would not hesitate to finance anything recommended by me, as everything I have ever recommend to them came through big.

Well! I have tried to set down here some of the highlights of our association and in so doing try and influence you not to remain on the East Malartic board. If you do, it does not mean that our happy association will necessarily have to terminate; however, I will feel that I have been let down and when I request this of you I do it knowing what I would do were the conditions reversed.

I have a firm belief that my knowledge of this country will lead to more mines and if we stick together we can all make plenty in the bright mining days that lie ahead. Do not hang onto a straw and let the load go by.

J.P. Norrie

P.S.: I made up my mind long ago not to fight with the Howey people as I had long anticipated a change coming and when it did I wanted to be in their good graces and remain on the board. Of course this plan was disrupted when the mine conditions became so bad that I could stand it no longer. I am now being penalized for taking a stand in that I wanted a change in the mining system in effort to bring about better mining conditions.

Finally in January of 1943 J.P. decided his time was not worth the effort on the East Malartic board and tendered his resignation to Andy Davis:

The first mining system at East Malartic Mines Ltd., started in 1938; shrinkage stoping with backfill placed on a log mat on the surface, worked with a certain degree of success to the 670 foot level. It provided a large tonnage of ore at low cost but was responsible for wide ground movement and subsidence which imperiled the shaft and surface plant.

In an official statement by Messrs Futterer and Reid, consulting engineers dated Dec. 11, 1940, it was stated that the mining system in use to the 670 foot level would not be suitable at greater depths and it would be changed to a cut and fill system for most of the stoping below this level. The system, except in a couple of stopes, was never installed. In a letter from Mr. Futterer dated Jun. 5, 1941, I was advised that a new system was to be used, as a matter of fact a combination of two systems whereby approximately 40% of the ore is mined by shrinkage stoping and the balance by an expensive square set method.

I have outlined the general history of the situation to show that the mining system at East Malartic has always been experimental while we were supposed to be paying for experience.

Sometime previous to August, 1942, it became increasingly apparent that this system was a failure in that the dilution is too great and therefore the heads too low in that portion of the ore being mined by shrinkage stoping. That 60% of the ore will never be removed because the cost of removing it will be higher than the recovery and the continued use of the system will make necessary the sinking of a new shaft and the removal of the surface plant and mill due to subsidence, and if this comes to pass new finances will be necessary as present diminished profits, even with limited development work, will not supply the funds.

Since August 1942 I have strongly advocated the employment of an engineer with wide experience in the East Malartic type of ore body where in general the ore is harder than the walls. Such a man I am sure would quickly change the mining system and surround himself with men convinced that this type of ore body can be mined safely and at a good profit.

As my endeavours in this respect have not received favourable consideration I feel it my duty to resign as vice-president and director of the company. I therefore tender my resignation from these offices.

J.P. Norrie

Lee Coffin wrote J.P. when he heard that J.P. had resigned from East Malartic:

Many thanks for your letter re East Malartic. I was sorry indeed to learn that you are out of the picture, after all your efforts during the early days to make the show go. Perhaps I alone know that had it not been for Jim Norrie, there would have been no East Malartic.

As far as J.P. was concerned the Howey group would not be able to manage the East Malartic and he and Lee Coffin began to dispose of their large stock holdings:

Lee, I will be glad to act for you in the sale of East Malartic shares in any capacity you desire. The market seems to be still strong, in fact higher than I got for mine. So you should be able to get them out without difficulty.

I note you are sending 169,000 shares of East Malartic to the local branch of the Canadian Bank of Commerce and that you want 69,000 shares sold.

Upon receipt of your letter I got in touch with Streit in Toronto re selling. The bid last night was $1.23 which I think is good under the circumstances so I advised him to start selling today at this price. I will check the sales slips when they come and will have the bank make deliveries and take payment for same and deposit to the credit of your account. I will send you copies of all sales slips.

I was advised today that Howey are becoming alarmed re their East Malartic investment and that they intend to liquidate part of their holdings. I expect they will help support the market until they have completed selling.

I am very glad to be able to help you in this way, Lee. You can depend I will do my best on the job.

J.P. Norrie

J.P. would never again become part of East Malartic and its operational problems. But then, he now had all his time available for exploration and development of new properties, his first love in any case.

East Malartic Mines, Limited
(No Personal Liability)
(Incorporated under the Quebec Mining Companies' Act)

PROFIT AND LOSS ACCOUNT

Bullion Production—less Shipping and Mint Charges			$2,818.714.56
Miscellaneous Income—including Rentals, etc.			12,089.77
Bank Interest and Income from Investments			4,785.49
Profit on sale of Investments			968.75
			2,836,558.57
Development and Exploration		$707,170.47	
Excavation and Backfill		121,022.95	
Mining		455,114.66	
Milling		257,734.33	
ADMINISTRATION AND GENERAL EXPENSES:—			
Mine	$129,016.82		
Head Office	29,673.27		
Insurance	16,032.52		
Taxes—other than Income	21,808.77		
War Time Donations	11,500.00		
		208,031.38	
			1,749,073.79
			1,087,484.78
Underwriting Expense—written off		8,333.28	
Preproduction Expenses—written off		124,024.92	
Shaft Sinking—written off		24,965.64	
Depreciation on Plant and Equipment		191,480.48	
Provision for Provincial Taxes and Federal Excess Profits Tax (estimated)		139,000.00	
			487,804.32
NET PROFIT FOR THE YEAR			$599,680.46

Chapter 17

The Prospector's Role in the Search for Gold

by George Norrie

One fact is certain, there would be no gold in circulation unless someone decided to search for it. The prospector is the searcher, the one who provides the first and most essential piece of the story. J.P. did not prospect all the gold properties he managed but he certainly knew who deserved the most credit and respect, and he always returned to that part of the business. When gold was finally produced from the mine, he moved on, but he made sure of one other thing: the prospector always benefited financially in any property that he (J.P.) promoted. His heart was always where the prospector travelled.

Herewith a story:

When Rex Beach appeared as guest of honour at the Writer's Club in Seattle, he declined to make a speech but agreed to answer any question put to him by the club members.

"Tell me, Mr. Beach," said one young lady, "to what one thing do you attribute your success?"

The famous author considered the question briefly and then replied, "I can best answer that by telling the story of a Swede prospector in Alaska. He was owner of several rich mines, and all his friends and acquaintances wondered how he had managed to become so successful. One night one of their number asked him.

"Ay never tolt anybody before," he replied, "but ay will tell you. Ay yust kept diggin' holes."

J.P. wrote this account of the history of early mining days in the Harricanaw district, which was found among his papers.

The mines of the Harricanaw area represent wide differences in host rocks, structures, grade, wall rock, etc. I doubt if any other group of mines in Canada covers such a wide range of conditions, each having a bearing on the location of new mines.

The Harricanaw district is that served by the upper waters of the Harricanaw River and composed of the mining camps of Cadillac, Malartic, Val d'Or and Pascalis. Early transportation to these camps was from Amos, either by the Harricanaw and its tributaries in summer, or by tedious overland trails in winter. Many of us here tonight will remember arduous passages by both routes.

Prospecting interest in the Harricanaw area began during the construction of the Canadian National Transcontinental Railway through the northern part of the district in 1905. Had this railway been constructed 40 miles further south it may and probably would have unearthed a golden mile in its course, and advanced mining 30 years, and would have been contemporaneous with the development of the large Ontario camps. As it happened, Quebec derived great benefit from experienced prospectors and developers from Ontario and also received much development money from Ontario mine production.

The story of the actual development of this region was characterized by courage, skill and foresight of a high order. The oldest prospectors in the Harricanaw area were James Sullivan and Hector Authier who found gold in July, 1911, on the south shore of Demontigny Lake on what is now the Sullivan property.

The early history of Sullivan was fraught with many difficulties. However, under the guiding hand of Pierre Beauchemin it eventually became a substantial producer and to him goes most of the credit for its success.

One of the early and most colorful prospectors and developers was Stanley Siscoe who found gold on Siscoe Island in 1915 on a small showing later called the "c" vein. Siscoe indications then were not very impressive. The Siscoe mining syndicate, after many pitfalls, developed into Siscoe Gold Mines Ltd., at one time the only real gold producer in the province. Siscoe was the one property that really flourished under the old price of gold, chiefly under the capable management of C.O. Stee. Siscoe has paid in excess of $7 million in dividends.

The Stabell was one of the interesting developments of the district. Found in Dec., 1921, by Joseph F. Stabell it attained quite substantial production through the efforts of Jack Hammell.

The Shawkey property in the southeast corner of Demontigny Lake was found by Jack Masso in 1911 and through the efforts of Bob Shawkey reached the production stage.

Lamaque was staked in 1923 by R.C. Clarke. It developed through the Clarke-Reid syndicate under George Kruse and the Reid-Authier Gold Mines Ltd. until in 1932 the property was optioned by Major MacMillan for Teck-Hughes. Since then progress has been rapid. Under the management of Herb Reid production and ore reserves have grown to substantial figures.

After the Teck-Hughes option, Reid-Authier staked the property to the north where the original Sigma discoveries were made by Heber Bambrick. The property was optioned for Dome Mines Ltd. by J.G. McCrea under whose capable direction it came into production in 1937 and has been developed into one of the leaders in the field.

The district's contribution in base metals is quite substantial from the properties of Golden Manitou and Lacorne Molybdenite. Both have had an intermittent peace-time history but are doing an excellent wartime job, the former under the management of Andrew Robertson and the latter under Ted Fell.

Prospecting and development in the Pascalis area was probably retarded by the fallacious idea that gold could not be found in the granodiorite. More courageous prospectors such as St. Barb Sladen, Pierre Beauchemin, George Bussiers did penetrate these parts as the Perron and Beaufor mines were actually staked by Sladen in 1928. Some of his posts can be seen today very close to one of Perron's most productive veins. Such are the fortunes of the prospector and those that follow him [Sladen did not hold his interest in these claims, but did with Sladen Malartic]. Interest in the area really developed in 1931 when Jack Mathews, Pierre Beauchemin and George Bussiers staked the Perron, Beaufor and Bussiers [Cournor] mines. The Bussiers reached production in 1932, followed by the Perron, which became the leading producer in the district under the management of K.A. Davis, and later the Beaufor which has also been a substantial producer. The Pascalis property, sponsored by Ventures Ltd., was staked shortly after the others. It has considerable ore in sight and should someday reach production.

The Golden Sidewalk of the Harricanaw district was found early in 1924 by Austin Dumond while in the employ of M.J. O'Brien interests. At the same time James Thompson staked the Thompson Cadillac, and Peter Graham the Graham Bousquet. Much work was done on all these properties with the O'Brien emerging as the most spectacular producer of the area, as rich shoots of almost pure gold were found to extend to depth. Profits were made from earliest operations by shipping high grade from the property until a mill was built in 1932. Under the direction of Robert Bryden and later H.E. Sparks, the story of the property has been one of progress and increasing production in gold and arsenic, the latter playing a most important wartime role. Other straight gold properties in the area are the Central Cadillac, Wood Pandora and America Mines and Minerals; all of these, now closed, have produced substantial amounts of gold and will probably do so again when wartime restrictions have been removed.

When Amos was the centre of activity in the early days, samples of gold were exhibited around town by Indians and Joseph Ladoux, which was definite proof that gold did exist somewhere to the south. The Indian samples were coarse gold in quartz while the Ladoux gold was in a heavy, black, forbidding rock which really did assay, yet could not have come from a mine because the gold was not in quartz. The Ladoux samples, however, did come from the Number Two ore body of Malartic Goldfields.

Excited by the rich Indian samples, Wilbur Gouldie set out in a determined effort to find their source. While travelling through the bush one bright morning in August, 1923, he came to a freshly squared post. From the post he noticed a well-cut trail, and looking along it he could see gold glistening in the morning sun. The source of the Indian samples had been found.

In the spring of 1923 three budding prospectors, J.C. Carroll, Hal Kennedy and St. Barb Sladen left Ottawa in the wake of a geological friend, Dr. H.C. Cook, from whom they hoped to learn much in the old mining areas of Ontario. Bored by the inquisitions of these youths, Dr. Cook advised them to go, among other places, to the wilds of Fournière township, Quebec, and follow up a rumour that gold had been found there by John Mack in formation he knew to be favourable. The advice of Dr. Cook proved to be timely, for these young fellows made the first real discoveries, which started the now Malartic camp on its way. In July of the same year they made a gold find which was enlarged in 1924 in what is now the Number One shaft area of Canadian Malartic.

The stakings of Carroll, Kennedy and Sladen were incorporated into Canadian and Sladen Malartic Mines. They were ably developed by Messrs Andrew Waltz, Jack Millenback, E.V. Neelands, Robert Basserman and others, with the help of St. Barb Sladen.

Now entering the picture is my good friend Lee Coffin, one the ablest and most prominent old-time prospectors of Quebec. His connection with East

Malartic, Sladen Malartic and National Malartic is legion and will live long in the annals of Quebec prospecting. Lee Coffin came to Senneterre from Boston (Harvard graduate) in 1920. He too was searching for the source of the Indian gold and in 1923 made a discovery to the south of Sladen. With several Indians, he cut what has since been called the Coffin Trail from the Piche River seven miles west to his gold discovery. This trail is roughly the south boundary of the East Malartic property.

During the early Malartic days float boulders were found east and south of the easterly projection of the then known ore zones. The significance of these boulders was recognized; however, the whole area was held by many stakers. In the depths of the Depression in 1932 when all ground had run open, the area was systematically staked by Messrs Coffin and Norrie with the boulders as a base. Careful geological study by Dr. Frank Buckland and others, and prospecting by John Partanen soon found the source of the float in the East Malartic Number One shaft area on the north contact of the greenstone band. Later, drill holes spotted by Sladen and Norrie found the Number Three ore body, which really made the mine. It came into production in 1938 under the direction of J.R. Norrie.

Like the Indians, Ladoux would not reveal the source of his black rock samples. He was a trapper as well as a prospector, so along his trap lines was the logical place to look for the source of his samples. It was eventually found leading in a southeasterly direction from the rapids on the Piche River, and on it some samples of float. It was estimated that the float did not travel over a mile, so with this in mind the Malartic Goldfields property was staked in 1933 with the trap line roughly the south boundary. Prospecting during 1934 by John Partanen revealed no rock outcrops. After spending $175,000 on approximately 60,000 feet of diamond drilling, much low grade ore was found in Oct., 1938, along the south contact of the greenstone band, and the higher grade ore body on the north contact. The property came into production in 1939, under the management of R.A. Halet. The source of the Ladoux float was found in the Number Two ore body in 1942, approximately 4,000 feet to the northwest of its occurrence.

Geologically the western mines have been developed in or subsidiary to the Cadillac-Malartic break which is a band of igneous rocks known as the Blake River Volcanics, of varying widths up to 4,000 feet enveloped in Temiskaming sediments, bounded on the south by the Kewagama and on the north by the Cadillac series.

The Blake River Volcanics, known locally as greenstone, have been intruded by diorite and feldspar-porphyry dykes, which are ore carriers, as are (in places) the sediments and greenstone. The eastern mines in the area are found in or subsidiary to a granodiorite batholith which extends from Siscoe to Perron.

Gold production in Quebec in 1925 before the development of the Abitibi mines was valued at $37,909. Production from the Harricanaw district alone in 1942 was $21,877,051 from 2,981,344 tons, an average of $7.34 per ton. Grade varied from $4.02 per ton at Canadian Malartic to $14.38 per ton at O'Brien. These figures mean that a good-sized gold mine is depleted yearly.

Roads were started from the towns of Amos and Senneterre and were finally connected up in 1937 to serve the entire mining community in all directions. In Dec., 1938, the branch line of the Canadian National Railways was completed from Senneterre to Noranda. This greatly reduced the mining transportation problems.

Probably the most outstanding development of the area was the construction of the northern Quebec transmission line through the entire length of the district in 1934. Without this cheap supply of electric power many of the mines would be unable to operate.

Financing, always a serious problem in any mine development, has been well and satisfactorily done. It was probably made easier in the Harricanaw by the success achieved in the older Ontario mining camps. Those most prominent in financing properties in the district were Nesbitt Thomson, Ventures, Anglo-Huronian, Dome, Teck-Hughes, Draper Dobie and other concerns too numerous to mention.

The entire area is now served with excellent means of transportation and travel in the way of good hotels, buses and railway service. Stephen Leacock says "It has developed to the point where Indians freeze in the summer time and bell boys deliver ice in the winter."

Some Early Prospectors

J.P. also recounted how Wilbur Gouldie played an important role in the discovery of the Malartic area mines but never reaped his proper rewards from his long years in the muskeg and bush.

When the East Malartic was at its earliest stages in January of 1935, J.P. was certain that he had a gold mine and one of the first people he persuaded to invest in his prospect was Wilbur Gouldie.

Jan. 3, 1935

Dear Jim:

Enclosed please find two cheques totalling $651.05 in payment of 5,000 shares East Malartic stock.

I suppose your drill is in there now and I hope you have luck in hitting your ore body, and also hope you hit other ones.

The market is picking up and a good market is looked for in February and March.

We are getting very cold and stormy weather here, there isn't much improvement in the old legs but I guess they will be ready for spring.

Wishing you all the success possible in 1935.

Yours sincerely,

Wilbur D. Gouldie
Gouldie House, Dwight, Ontario
Lake of Bays

Joe Errington was a great prospector and friend of J.P. who paid this tribute to Errington after his death in 1942:

In passing it is fitting that I should pay my respects to the life of Joseph Errington. He was in my opinion one of the greatest men that this country has ever produced, and with his passing the prospector and all of us have lost a worthy friend. He was to the prospector and young engineer the inspiration that led their trails unflinchingly towards the dawn of a brighter day. Men of Errington's qualifications are needed now; his interests were broad. He faced the insurmountable tasks that were placed on his shoulders, most of them concerned with mining; he, however, did not at any time forget the human element that in most cases is interwoven in all projects. Through his capabilities as a mining executive he made many staunch friends; through his human qualities he will remain forever a fond memory in the life of those who knew him. The results of the work of Joseph Errington and many others are obvious. Our mining industry has prospered. Our Canadian mills are now grinding an aggregate of 20,000,000 tons of ore annually. Our ore reserves at the present time are sufficient to carry us for many more years at the present rate. Alarming, however, is the fact that our prospecting industry is not keeping pace with our fast diminishing ore reserves.

J.P. Norrie

Alfred Reeves was a prospector from Toronto and one of the first prospectors J.P. contracted to go into the bush for Inspiration. J.P. had plans to send several parties into the Chibougamau area during the spring and summer of 1935. Frank Buckland and Harry Norrie would be one party (for Harry's account of that trip, see page 475) and another was John Partanen with his partner. Each party would prospect a certain area. There was lots of virgin bush for all of them and it was obviously one of the geologically interesting areas to J.P. These letters talk about the financial deal that each prospector wanted. They also tell just why many prospectors wanted to explore for J.P. He would promote and develop their finds no matter how small, where most large mining companies were only interested in the large find. Even though J.P. had not developed most of his mines at this time, the mining communities already knew who would treat the prospectors fairly:

Mar. 7, 1935

Dear Norrie:

Yes. I am ready to go in and while I would prefer to go on a 50/50 and no wages, I cannot do so. I have spent three years in Tiblemont developing and no sale yet & taxes eat up a lot. So I'll need some wages.

But you can be assured we will do our best. My partner is one of the best.

I am glad to tie up with you because I know if I get anything you can have it developed and we keep going further afield.

Now if you say so I'll come up at once and we can arrange.

I have a group in Travernier. I would be glad to take three or four men in too and shoot into it right away. A splendid break. Only four miles from R.R. If you will put up for the work and taxes, six claims for ¾ interest, I'll come along and do it before I go on to Rose Lake, Chibougamau area.

If I am to go in for you I would like to be in before break up as one can travel a lot on the crust looking formation up.

This Travernier break I think better than Murwood & Tiblemont break. Assays from blank up to $15 of gold per ton. A splendid range. We have it for ¾ mile. Two claims wide.

We can arrange wages and interest, okay Norrie.

Believe I'll get a sale in silver. Nipissing have taken over my old ground.

When I get a good thing for you in Quebec I want you to send me to where I got float. Gold $7 per ton, lead 6%. Silver over 700 ozs. I'll be glad to hear from you.

Alfred Reeves

This letter shows the real interest of the prospector, which was the percentage interest that he would get. They all really wanted the greatest maximum interest in anything they found. The prospectors usually couldn't afford what they wanted:

Mar. 21, 1935

Dear Norrie:

Last night's paper here speaks well of area you spoke of, Rose Lake and easterly.

Re our plan. If you can see your way to give me say $50 month plus expenses, basis of 55/45 int., or $100 month and 60/40.

I would go on the 50/50 but I have certain obligations to meet.

We would like to take our own canoe, even if we took it in by dog team to Rose Lake.

Yours truly,

Alfred Reeves

These letters give a good idea of the life and difficulties of a prospector working in the isolation of Chibougamau:

July 1, 1935

Gentlemen:

Herewith some samples for assay. We get no colours. But they should be tested. The quartz is the best we have seen and we go in again to see if it improves.

We have this rusty bell for over 20 miles.

North of here looks interesting but we must get into it from Gull Lake or Olga Lake. It is too rough to pack in from here.

Are not on the lake you had planned for us. We are many miles more north. We are perhaps 20 or 25 miles from Rose Lake.

We are now east of main camp and will scout more there for a few days.

We find a range of pink ironite, south of it several miles of grey granite. Plenty of sand planes. Considerable altered greenstone. Not much quartz. The quartz we send you looks clean, healthy. Best we have yet. The first pyrrhotite we have seen. No copper yet.

We have been north. Hard to travel. We hope when we get moved to get into easier from north.

We have been west perhaps 12 to 15 miles. South we find same conditions. So now we are doing east.

We hope to see plane soon. We try to be at main camp every Saturday for clean up and bake.

We find no fishing tackle. We need by next plane about three slabs Swift premium bacon. Not this stuff here. So much waste.

We need fly oil. Preferably oil of citronella & olive oil. This paste is little use and we are nearly out.

When plane comes a nice treat is six or eight loaves bread.

We have seen several planes this week. One going west later on it returned east right over us.

We hope to get moved soon. We will look over the east for a few days yet. We have an out camp there now about six miles from here and we hope to travel a good deal this week as well as look over the quartz and pyrrhotite area.

Glad to get some oil, bacon and mail and to get moved.

Yours truly,

Alfred Reeves

These letters show clearly how important the bush plane was to these guys. There were several companies flying in and out. It didn't make any difference who moved them as long as they got moved and made a find for that season's work:

July 17, 1935

Dear Norrie & Coffin: Rose Lake, Que.
Prospectors Airways Camp

We walked across country and got in here last night. A hard trip. About 25 miles.

Radio man took a message for you. He said it would get through to you via Rouyn and I might get an answer.

We will stay here say two days and then back to camp. We did expect message or letter here from you.

We have samples at camp for testing. We want to be moved. Either to Gull Lake or Waswanipi. Somewhere where we can move in canoe. We are at our main camp every Sunday. So your plane would know when we would be there.

We move out in our small tent and then another stretch with tarp so we can cover a yard area.

New find on Wedding River so they tell us here. When you send in plane, we need some fish tackle. About three slabs Swift premium bacon. Some fly oil.

If Mr. Coffin home, better send my boots (high ones) boots are at Coffin's house in pack sack I think.

Hope to have plane soon to move us. We tried to go down Izerhoff Creek but we found it no good for a canoe. We want to move to get somewhere to make a find. The season is going along.

Alfred Reeves

P.S. If you can't find my boots, get me new pair. Dayfoot Elks. Get few nails in size 7. Something to stand. Two tins Dubbin.

It was difficult for J.P. to know where they were at all times but he had already instructed Alfred Reeves to use his own judgement. If an unexpected tragedy happened, such as a storm or attack by bears, no one would be there to help:

Aug. 1, 1935

Olga Lake, Que.

Dear Mr. Norrie:

We arrived in camp last night. Found your letters. Also bacon, fly dope, fish tackle. Thank you.

Prospectors Airways seem strange people. Sending mail out they sent Cyril Knight party mail back to Toronto. We did not seem welcome. I asked if we might stay. Said we could pay our way. Oh. Yes and gave us blankets. I said we might stay a couple of days and see the showings.

Second day he asked McKay how long we were going to stay. The weather was terrific. Rain rain. However, we caught Ressigoer later 2nd day and flew to our camp. The first camp. Now called Inspiration Lake. We had no mail so we did not know what to do.

Mr. Coffin had suggested north side of this anorthosite bell so I thought we would see more country this way than any other. So we are here.

We had a tough trip to Rose Lake across country. No bedding. We slept out and I tell you it was cold. We were out all night the night of the moon eclipse. Frost.

You know we were not on the little lake you thought we were. We were way north on a lake four miles long. Hope you have samples Ressigoer would express to you. From that are about 25 lbs. I sent them so you could see the nature. We could not get colors. We found that bell for over 20 miles.

We just got in last night when the worst storm I have ever seen on a lake broke down for 10 minutes. Thought we would lose canoe and tent by falling trees. A green, big spruce a foot through behind us broke off up 15 feet. A bunch of birches saved our tent. We are okay and thankful.

Yesterday we found a bell of schist, as samples I'm sending you. The best looking thing we have seen. We had to come out for supplies and today is too wet to return so we are fixing up and drying out.

It does look very interesting. We have it over 100 feet wide. We spent no time looking for mineral in it. We kept opening it to see how much there was. So we hurriedly broke a few pieces and hurried on. We are out about five hours from camp. We must be left alone a while to see if we can unearth anything. Our main camp, the Skyways pilot knows and he will bring mail. If we get colors or a quartz vein of right size we will stake I think.

Air pilot brought us sack bread. That is all expense except two days at Rose Lake. Plus $2 for the Quebec govt. surveyor putting us across Rose Lake. I feel like leaving some stuff at Waswanipi H. Bay Co. Post. You know there is more stuff than we need to fly around. So should our stuff here prove worthless we hope to go via Waswanipi Post prospecting on our way.

Met Knight's man. He says a great lot of men south but all wishing they were elsewhere so we are perhaps as good here as anywhere. At all events I feel we have a piece of interesting ground that may show something.

I'm glad you left me free. I feel like so much more depends on me. Of course if you get word of a new find Dominion Skyways pilot gets our camp every week if you say so.

Fly dope sent in not much use. Hot days one sweat and it will get in our eyes and it says don't let it in your eyes. So if you will send one penny royal, one pine tar and four castor oil (mix in one bottle) we will be glad to have. Thanks for the fish tackle.

Glad the bears did not get our camp. They were close. Excuse hurried writing but I have a lot to do to get ready and away again. If you can get assays and send us results of this lot I would be glad because it means so much.

Dominion Skyways Senneterre will drop mail very readily. But it means a day coming and going. Doing my best and surely will be glad to bring in a find.

Alfred Reeves

P.S. I will ask pilot to bring bread on his trips. It saves so much time. He can get it in Senneterre. We like it unwrapped it don't mould so readily.

Pilot Ressigoer seems very very good. No kick. Accommodating. He flew our canoe with no trouble and no kick. We will be away early in morning if weather okay.

Aug. a good month for me. Aug. 1928 my silver find; Aug. 1929 copper; 1933 gold; Aug. 1935?? Well we'll try.

Don't see anyone; two parties since leaving Amos. Just as well perhaps. Don't get any news. We don't tell pilots our business or where we are. They don't ask. Just as well. If we can get kick in this area you'll like it I'm sure. Lake and river transportation.

Sept. 2, 1935
Opawicka Lake

Gentlemen:

Arrived here okay. Picked up Bell River camp and Olga Lake camp.

I'm sending some stuff out with plane to store in bakers at Senneterre.

We told Airways to look for us. If we are not out by Oct. 15, Skyways pilot gone to Prospectors Airways. We will keep moving and later go out via Eagle River & Megiskane. We will be along the Opemiska Road cutting camps on our way out.

Yours in hurry,

Alfred Reeves

P.S. No table to write on.

Wilfred and Paul Croteau were prospectors who would have a long association with J.P. They would work for Inspiration, and their partnership would culminate with the Croinor many years later.

April 9, 1937
Senneterre, Que.

Mr. Wilfred Croteau

Dear Wilfred:

I have your letter re summer prospecting. I am outlining a proposition which I think is better than last year. This proposition covers two men. If Paul [Croteau] cannot go with you I thought you may be able to get Bob Browning.

The proposition is:

You supply all equipment and we will supply grub, transportation, expenses etc. Everything except equipment. We will pay $600 for the season i.e. $300 to each man. We will also pay a bonus of $1,000 i.e. $500 for each man for any discovery which stands the examination of our engineers. Interest to prospectors 40% to us 60%.

If this arrangement does not suit you please outline one that does.

J.P. Norrie

J.P. would sometimes form syndicates to finance prospector friends; in this case, Scotty McLeod, as far away as the British Columbia and Yukon fields.

Mar. 31, 1942

Dear Jim:

I spoke to Karl about your suggestion in connection with sending two men to scout the area in the vicinity of the Ȧlaska-British Columbia highway and he was quite agreeable to participate with you and several other parties in financing a syndicate to send two men out there, but he left it entirely up to me to decide whether I wanted to form a syndicate for that purpose or not.

If I were to form a syndicate to scout the road area it would mean that I would have to quit working for Leitch Gold Mines and it would require about $10,000 to finance ten men in that area for a season, and in the fall I imagine I would be on the unemployed list, which I could not very well afford to be.

I told a friend of mine by the name of William Dennis about your suggestion and he approached some people he is connected with about it, and some of them seemed to favour the idea but they are in exploration work themselves and may want Dennis to go out there for them, however, he is to see them tomorrow and will let us know definitely.

I am going out to the southern part of Yukon Territory for Leitch this spring and if you anticipate sending a party into that area I would be very glad to co-operate with them in every way.

Mining Corporation and Coniagas are sending two men into northern B.C. probably in the road vicinity and I think there will be quite a few parties going out there this spring.

I expect to go to Ottawa tomorrow and will be back here in a day or so.

I will drop you a line later and give you all the dope on what is going on.

W.A. (Scotty) MacLeod

Decline in Prospecting in Quebec

J.P. wrote to the legislature of Quebec about the decline in prospecting in the province:

Mar. 18, 1940

Mr. Adolphe Routhier
Dept. of the Attorney General
Quebec, P.Q.

As you probably know, for the last several years my interests have been more allied with the financial end of mining rather than the prospectors and exploration end, however, I am close enough to the latter to know that the prospector has been practically eliminated from the province of Quebec along with the mining prospects brought in by him.

We all recognize the fact that the prospector is the lifeline of our industry, without the prospector the industry cannot survive.

I firmly believe the plight of the prospector in Quebec is much worse than in Ontario where they have a powerful prospectors' association of upwards of 1,000 members while in this province we have no association of any account due to the very limited number of competent men in the industry.

I believe also that we must have promoters and financiers who must have sufficient encouragement to proceed with the financing of poor surface showings many of which have turned into important mines.

The small promoter and financier is probably the most important of the lot as they will tackle problems which larger ones will not touch. As an instance I will cite the most recent development, namely the National Malartic in which Noranda, Consolidated Smelters and Quebec Gold Mining Corporation all had an interest. Yet they were afraid to put money into it as the gamble was too great. I succeeded in interesting a small brokerage firm, E.H. Pooler, of Toronto in putting up funds. This money resulted in the development of a large low-grade mine which I feel sure will prove to be a great asset to the province.

I do not believe the larger financial houses will find much change in the present security laws. They realize the prospector is now tied up and thus is in a poor bargaining position, whereby they can make arrangements favourable to themselves, to the great disadvantage of the prospector in whose welfare they are not particularly interested, as when a good mining prospect is now available for them they can take on bond or industrial financing to fill the gap.

We need in Quebec not only the development of many competent prospectors but many competent promoter and financing houses who confine their activities to mine development alone. We have neither at this time. We all know the Ontario mining industry was built up without security laws, that the booms created by the development of such camps as Cobalt, Timmins and Kirkland Lake and even Noranda spread not alone over Quebec and Ontario but Canada and the world. Much good was derived from these bursts of activity, particularly in the province in which they were located. Even if the odd Rubec and other sore spots did mar the broad picture, everybody made money, everybody was enthusiastic.

We had no boom connected with the development of the Malartic camp which I am sure must be considered a major one.

The enthusiasm which accompanied the development of a major mining camp generally lasted for a period of years and resulted in the location of another substantial camp, for instance Cobalt led to Timmins, Timmins to Kirkland Lake, Kirkland Lake to Noranda. Let us have more enthusiasm and I feel sure another major mining camp will be found within 50 miles of the town of Amos. Let us not cramp the wide activities such a camp will develop and which will probably lead to many others.

In any case the activities of the fraudulent promotion are generally headed off before anyone is seriously hurt.

Our mining operations today are headed by an honest group of men who have a general broad interest in the development of the mining industry. These men provide a real handicap to fraudulent developments. Rubec conditions were well broadcast by them before the shares reached abnormal prices and thus the life of the promotion was short lived.

Now to answer your questions:

1. I believe restrictions relating to the escrow of vendors' stock should be entirely removed. Escrow of shares should be left to the buyer and seller.

2. I believe the profits of promoters, financial agents, etc. could be limited to say 40%, however, I believe this should not be too closely governed to the point of discouraging this phase of the industry. The law which recently was applied to the activities of Johnson and Ward is a serious deterrent to fraudulent promoters and may be enough to govern this phase of the industry.

3. I believe all promoters and financial agents dealing with the development of mining properties should be registered with your department and be compelled to furnish you with a substantial bond guarantee of their good faith and actions.

If I can be of help to you at any time please do not hesitate to call on me as I am interested in seeing a great renewal particularly in prospecting and development.

Our best source of development money is Ontario mining dividend cheques; certainly something should be done to attract them to Quebec.

J.P. Norrie

The Prospectors and Developers Association

Toronto Stock Exchange Review
January, 1955

Often far afield and out of touch with current affairs affecting their interests, these hardy lone-wolf finders of Canada's mineral wealth have for 23 years maintained their own organization, known as the Prospectors and Developers Association.

Originally formed by 13 successful prospectors, the association now has a membership of over 1,100 drawn from every mining area in the Dominion.

During its growth period, the association has fostered sound improvements in security laws, mining legislation and in taxation on behalf of the prospecting fraternity, and at the same time has been directly responsible in attaining many new roads to mushrooming mining districts.

The association, under the leadership of President Viola MacMillan, is holding its 23rd annual meeting and convention at the Royal York Hotel, Toronto. Pertinent details relative to the industry will be presented to members by a score of speakers at that time.

Historic speech by Norrie remembered

On October 18, 1945, the *Val d'Or Star* reprinted the text of a speech J.P. made to the Prospectors and Developers Association, for that speech, said the *Star*, was as fine a tribute to Jim Norrie as anything which could be printed:

Mr. Chairman, Mrs. MacMillan, ladies and gentlemen,

I consider it an honour to be called upon to address you this evening and help in my small way to further the interests of the prospector. Firstly I would like to congratulate and thank both Mr. and Mrs. MacMillan for their efforts and organizing skill in building up the Ontario Prospectors and Developers Association into a forceful and efficient organization and thus elevating it to dominion status. Surely they must have had a better conception than most of us of the necessity of such a step as now when the job has been completed we can all see the necessary field for such an association. I think unlimited praise is due both of them for extending their activities to this new mining field and for their coming in person to carry them. Here again I must congratulate them on their foresight because I believe they are now seated in one of Canada's really great mining centres which in a very few years will demonstrate its great productive capacity in gold and base metals.

When I was asked to speak to you this evening I chose the prospector, not alone because it seemed appropriate for the occasion, but the added reason that I think the prospector needs encouragement from every branch of the mining industry, because he is our lifeblood, and unless he is given a better chance to help replenish our rapidly depleting ore reserves, then our industry is absolutely headed for decline and in a matter of a few years will cease to hold its place as a dominant factor in our economic life. Real opportunities for industrial growth will come to those countries that are best able to supply the raw materials of industry, ...in peacetime as in wartime. No raw materials are more important than the products of our mines.

The result of the prospector's efforts in the past has provided us with our livelihood. Firstly the promoter, the financier and the broker, with the property on which he raises the money for development; and the operator, the geologist, the engineer, the accountant, and the miner, with the task of bringing the property into production. Therefore, if for no other than our own selfish reasons we are firstly interested in the prospector's welfare.

In my opinion the prospector is struggling under handicaps, many of them needless, which must be removed before the prospecting industry can produce efficiently the new mines which must be found to replace the ore now being mined. The solution of his problem will be tough and I am afraid prolonged . However, necessity will force a solution even if it comes late.

It may be interesting here to mention that most of the ore bodies in Rio Tinto in Spain were covered with many feet of overburden and rock capping. These ore bodies were discovered by ancient Roman prospectors, and modern science has failed to find a single ore body in the great area formerly worked by the Romans. It does not seem to be understood how this great prospecting job was done; however, it has been noticed that in all the ancient paintings of two of the earliest Christian saints in Spain, St. Abdon and St. Seven, they hold in their hands twigs which look very much like willow wands of the water witch. Can it be that this doodlebug method was used by the ancient Roman prospectors with concrete results? I hardly believe it; however, the Romans seem to agree with the principle: the one who finds an ore body is a saint, and the one who fails, a rank impostor.

My remarks do not cover all the ills of the prospector nor am I capable of pointing out all of the practical factors which may benefit him; however,

I believe a discussion of the prospector's problems can best be done under the following heads: history in Canada, factors contributing to his rise and decline, and factors and suggestions which may help solve his problems.

History of prospecting in Canada dates back to the earliest explorers and French Jesuits who traversed the country chiefly north and south along the great waterways such as the Mackenzie, the Ottawa, the Nottaway, etc., the Arctic and Hudson Bay. In old reports of their travels several references to minerals may be noted. Some of these have since been developed into producing mines. I believe the first real prospecting in Canada was in Nova Scotia around 1860 where the art of locating the source of float boulders was highly developed. This was followed by British Columbia and the Yukon rush 1898 so that when the mineral discoveries in Ontario were found we had a fair number of Canadians well equipped to carry on. These with migrants from California and other parts of the United States and Canada formed the basis of our original prospecting stock. The finding of silver and cobalt at Cobalt in 1903 gave great impetus and incentive to the prospecting industry and prospectors rapidly fanned out in every direction to find, over time gaps of approximately five years, the now famed camps in Ontario and Quebec.

I might say here that I believe new finds will be made in even shorter periods in the future than in the past and some of the surface indications that have come to light within the last five years will probably develop into Kerr Addisons and Malartics which, although only recently developed, were actually discovered by prospectors many years ago. In other words I am not worried about the paucity of mineral discoveries in the immediate past as in most cases it takes many years of exploration and development work to fully realize the value of the prospector's original find.

The finding of silver at Cobalt under ideal conditions and without restrictions caused repercussions which extended over the Earth, and as each new camp was found, …[interest] increased in intensity and volume until capitalists and adventurers from all over the world were attracted to the Canadian north.

I believe restrictions have seriously reduced this free advertising of our country so that in order to attract capital today promoters have to spend great sums of money. Security laws even further hinder him.

Factors contributing to his rise and decline:

The first prospectors of the north in the areas best known to us had a virgin territory to work in and throughout the early years most of the outcrops in the area south of the Transcontinental Railway were hurriedly gone over and from these most of the producing mines were developed. The finding of each one of these mines in a period without restrictions created a great incentive to the prospector to carry on and find other great mines. The mining camps were wide open. Security and mining interests were freely traded. Money was plentiful. Life really had spice that has been unknown in the finding and development of our more recent camps.

This succession of events laid the foundation for a progressive and prosperous prospecting industry which came to a climax shortly after the discovery of Noranda, or when government restrictions to protect the supposed unwary began to develop. The news of this prospecting prosperity and great mine development was freely advertised throughout the world and attracted easy mining money from every direction. All this helped to build up the economic life of Canada. I will venture to say that one third of the city of Toronto and a good part of any Ontario city owes its growth to the mines of the north. Quebec has yet to receive full benefit from its mines; however, that will come. Most of our gold producing mines have of their own volition gone pure and today have forgotten their romantic past, which, in reality, was in a big way responsible for their being, and thus little objection comes from them to purifying laws which so seriously affect the prospector of today.

In this purifying process laws were enacted to protect the mining investment of the layman. These laws proved to be a great handicap to the prospector and in my opinion were the greatest cause of his downfall. I do not believe that laws will prevent a sucker from parting with his money and for every dollar "saved" from our industry, the economic life of Canada suffers thousands.

We well know that many of our mines have been found, not by the large exploration companies, but by prospectors financed by a few individuals or friends in the village or town from which they came. John Smith, for instance, when he was ready to start out in the spring, had no trouble to raise a grubstake. He went to the local doctor, butcher, baker, merchant, etc. and from each secured from $200 to $500 to build up his grubstake. It was easily secured. He cannot do the same today because these men on account of income and other taxes just cannot afford the money. In other words, laws have deprived the prospector of his means of support.

When a prospector did make a discovery he had little trouble disposing of it even if it were not a silver sidewalk or golden mile, and thus repaying those who grubstaked him, because there were no security laws to tie up most of his interest in a non-transferable pool.

Today we will suppose that John Smith does succeed in raising a small grubstake and with it he is lucky enough to find some ore indications. Probably with what limited work he is able to do on it the prospect will not look too attractive so he will probably not get a cash payment, or in any event not a very large one, so that he will be forced to accept payment in pooled shares in a company in the operation of which he has no voice. The fact that security laws forced the pooling of vendors' shares places the prospector at a distinct disadvantage when making a sale of his shares because he is absolutely at the mercy of the purchaser. For the pooled shares that he receives there is no market so where is John Smith to get his next year's grubstake? Certainly he cannot carry on like this for many seasons. The fact is he has not been able to, and this with other obstacles has put him out of business.

Factors and suggestions which may help solve his problem:

I believe the field for the prospector is still great in the Canadian Shield. True, most of the outcrops south of the Transcontinental Railway have been worked on; however, 90% of the Shield lies north of the railway, where very little real prospecting has yet been done and where from my experience outcrops exceed the approximate 10% figure to the south. I believe many mines will still be found by diligent prospecting in the old mining areas, by projecting contacts, following up float and other semi-scientific means. Before this can be done the profession of prospecting must be made attractive to the point where at least a living can be made. I believe present difficulties have forced many prospectors to be simply claim holders. Laws are not necessary to prevent the holding of unworked claims if the incentive to work them is created. It is necessary not only to revive the prospecting industry to its original state; it must be made more progressive. More prospectors must join the ranks to satisfy the increasing demands of the future.

To get a clearer picture of what I have in mind I will point to the prospecting days of Cobalt, Porcupine, Kirkland Lake and Noranda when most in the prospecting industry were happy. Everyone made a living and many made fortunes. Several things must be done to bring back those days.

The Dominion and provincial governments have done a good job in geologizing and mapping the area generally opened up by the prospector, and their maps and reports are very valuable; however, the prospector must precede the geologist, and the prospectors must be prepared as in the past to pave the way for all comers. Mining companies, all of whom derive much from the prospector's efforts, must be prepared to provide off season work for as many as possible and this I think most mines do on an ever-increasing scale. Transportation difficulties will be easier after the war with the great influx of aeroplanes then available.

I am definitely not in favour of government subsidy for prospectors, either individually or collectively. Nor do I favour any government undertaking the development of our mineral resources. I believe a large part of the prospector's troubles can be solved by permitting a return to the old time conditions with which most of us are familiar. This can best be done not by budgeting a portion of taxes already collected but from money left in the hands of the taxpayer. With this in mind I gave the idea to Mr. and Mrs. MacMillan some two years ago that subscriptions to genuine prospectors should be income tax free. With their usual aggressiveness they took the idea to Ottawa and were successful in having 40% of $500 subscriptions to recognized prospecting syndicates up to a total of $5,000, exempt from income tax as applied to strategic minerals. I notice in the last budget that the total subscriptions can now be unlimited. This, however, is of little help to the prospector as he deals generally in small figures.

To carry this idea further I suggest that amounts of $500 in each syndicate up to a total of $5,000 be income tax free as applied to all valuable minerals and that in amounts above $5,000 only 40% be exempt. I am sure that this concession will send the prospectors back to the bush and they will stay there in ever-increasing numbers.

The $10,000 syndicate with some alterations should prove of considerable benefit to the prospector. I believe that companies should be permitted to subscribe to the syndicate and that the issue of formal share certificates would perhaps create more interest in it.

The great mining industry of Ontario was built up without restrictions imposed by security laws, but in an effort to benefit the investing public much harm has been done the prospector. My opinion is that the pooling of vendors' interests should not be under provincial control but should be left to the buyer and seller. As the security laws now stand, the vendor or prospector is at the mercy of the purchaser when making a deal on his claims. A more free exchange of vendors' interests would certainly be of great benefit to the prospector and his backers. I believe the present provincial laws should be amended to embody as much as possible these suggestions.

Of course all taxes which directly affect the mining companies also hurt the prospector by removing the incentive to invest funds in his activities. These include corporation income taxes, excess profits tax, royalty taxes, municipal and school taxes, etc. It is hoped that these taxes will at least be reduced to pre-war figures at war's end.

Theoretically a mine, being a depleting asset, should not be taxed except for local school and municipal taxes until the shareholder or investor has regained his capital, and then it should be taxed the same as any other industry with a lower depletion allowance. Out of the 100,000 mining claims staked in Canada and the millions of dollars spent in them, approximately 125 producing mines have been developed. Out of these only about 65 will ever repay the invested capital. So that under present conditions 60 mines will pay taxes to the government while the shareholders will be left holding the bag. As the present tax situation absorbs in excess of 50% of the profits of a mine, a correction of this situation would prove an enormous boost to the prospector.

Did you ever ask yourself the question why an investor should buy a mining stock? The answer used to be, in view of the great number of failures of mining enterprises, that most investors were after a nick in a rich prize such as Hollinger, McIntyre, Dome, Noranda, etc. The present tax situation has removed the prize.

In the establishment of soldiers in civilian life after the war I believe the Dominion government should set up a school for prospectors in each province in Canada. A thorough course of at least one year in mineralogy, geology etc. should be given in conjunction with extensive bush training. The bush training should be given by experienced prospectors each with a party of from eight to ten students. It should last for at least one full prospecting season.

I believe the extradition treaty recently discussed by the *Northern Miner* and other Canadian periodicals, if consummated, will do considerable harm to the prospecting industry. Every power at our command should be used in an effort to have it deferred.

I was very much pleased to note the activity of the prospecting association and its president in opposition to the establishment, in our mining industry, of foreign unions. I believe great praise is due them for their foresight in being able to appraise the general harm that would befall the industry should these unions gain a foothold, as I think they would destroy more mining companies than prospectors could find. I am sure the problem of mines can best be solved by the local industrial relations committee under government control. With this solution we are all in favor.

I am sure that if my suggestions are adopted the prospector will be happy and when the prospector is happy we all are.

[J.P. Norrie]

Chapter 18

From Bush to Bullion

by George Norrie

Malartic Goldfields

The story of Malartic Goldfields is another rare example of "bush to bullion". When J.P. was mine manager of Malartic Gold Mines he was totally confident that the mine would produce gold at a profit even though the price of gold was only $20 an ounce. Both East Malartic and Malartic Goldfields claims were financed and staked under his direction, while working at these mines. Working for Ventures at Malartic Gold Mines obviously inspired J.P. to venture out himself and begin his own mining and development interests. Now that the price of gold was $32 an ounce he knew for sure he could make a profit mining gold, and more importantly he knew exactly where to find it.

Some of the Malartic Goldfields claims had been held by Clarence Stee, Mine Manager of Siscoe Gold Mines. Stee was J.P.'s good friend and through him J.P. became interested in this area. John Partanen was watchman at Malartic Gold Mines and J.P. financed him to stake a large tract of land just east of the East Malartic claims, which would become Malartic Goldfields.

Whenever J.P. made a deal with a prospector to stake claims, he made up a standard contract to protect both himself and the prospector:

...whereby the parties hereto have agreed between them as follows:

1) The party of the second part [Partanen] is to prospect in the townships of Fournière and adjacent townships in the province of Que. during a period of two months in the year 1932, commencing as early as possible.

2) The party of the second part is to supply one man to assist him in the work, as well as all food, equipment, supplies and licences required.

3) Any and all discoveries of minerals considered of sufficient importance are to be staked by the party of the second part.

4) The parties hereto are to have equal interest of fifty per cent each in any or all find or finds made as the result of this said prospecting, and no disposal is to be made of any such find or finds without the concurrence of both parties hereto.

5) The party of the first part is to contribute two hundred dollars per month in cash or its equivalent for this said prospecting during the period of two months above mentioned.

J.P. Norrie & John Partanen

J.P. held these claims even though he lost out on Malartic Gold Mines. During his venture to the Arctic he also had maintained these claims at considerable expense to himself. In 1932, after he returned from the Arctic, many of the claims in this area came open because of the Depression. At that time J.P. staked even more claims adjacent to both his interests at East Malartic and Malartic Goldfields. He now held exactly the ground he wanted from a geological point of view.

This was an even greater financial burden on him as his only means of support in 1932 was his salary at Bussiers. His vendor's shares in Bussiers were not of any great value or any value at all.

In August of 1933 J.P. was just at the point of making his deal for the Tiblemont Island property and needed money badly. He wrote D.G. Wright of Doherty Roadhouse explaining why he needed to sell his Bussiers shares:

I have another nice showing over near Malartic which needs $5,000 for immediate development. To get quick money I would like to sell 15,000 Bussiers at the best price you can get. Please see what you can do for me in this and I will give you a cut in the prospect in question. It is known as the Partanen Malartic show and was staked by him when he was watchman at Malartic and paid by me.

From August, 1934, to June, 1935, most of J.P.'s efforts were concentrated on managing the Perron and incorporating Inspiration and East Malartic. This was certainly enough of a burden for any man. The Perron was just developing and Nesbitt Thomson had not committed their finances to either Inspiration or East Malartic although they were just about to do that. Nevertheless J.P. also continued work on Malartic Goldfields and advised D. Wright on the capitalization of the Malartic Goldfields company:

...the Malartic Goldfields is capitalized at 4,000,000 shares $1 par value with 1,250,000 issued for properties and 2,750,000 shares in the treasury. No shares have been yet sold and all vendors' shares are pooled for two years and are deposited with a trust co. in Montreal.

We wish to option the treasury stock in blocks to provide money for work. Write your own ticket on the option if within reason and you will be given a good lot on the deal.

Stee of Siscoe held some of this ground for many years and is a director of the company. He, too, recognizes the importance of this break.

I hope that you can do something for me on this as we are anxious to get to work immediately. You will have the full support of us all.

While Nesbitt Thomson were committing to East Malartic, they were not prepared to take on another large project such as Malartic Goldfields. At least not until J.P. or someone else demonstrated that the property was a sure bet. So J.P. turned again to Geoffrey Porteous for initial financing dollars to develop the property further. This arrangement didn't last any longer than the partnership on East Malartic and in the summer of 1936, when Nesbitt Thomson were sure of Malartic Goldfields, Porteous was out and Nesbitt Thomson, with all their financial influence, were in again:

...I tendered my resignation as president of Malartic Goldfields, Ltd. to further the best interests of the company, and to comply with suggestions in several of your previous letters, and also by you

Prospecting at Malartic Goldfields in winter, J.P. is on the right.

here in Montreal, that it would be advisable that all business of the company henceforth should be carried on at Amos, or the company's property. It did not follow that my resignation meant that I would not cooperate and promote to the best of my ability the welfare of the company, and this I will continue to do in any way possible upon request. The books and all papers belonging to the company will be forwarded to you in the near future. In yours of Jan. 1 you expressly asked me to deal with Mrs. Forbes, and I have done so adequately.

I note that you are anxious to "settle this business in as peaceful a way as possible". I would appreciate your reviewing again the situation, and as noted above I will be glad in every way to cooperate with you.

The property was diamond drilled during 1936 and 1937 to the point where the ore bodies were indicated and in June, 1938, Andrew Waltz made a positive geological report in preparation for the shaft sinking. Waltz had made his reputation as mine consulting engineer at Canadian Malartic. He had brought that mine into production for Ventures from 1932 to 1936. Although he received much credit for his efforts, it was J.P., when he was mine manager of the old Malartic Gold Mine, that had "made a mine out of

Malartic Gold Mine". These were Oscar Hershey's words, the consultant for the old Treadwell Yukon, and he acknowledged that J.P. had really done the most difficult development work.

In early 1940 the Goldfields mill was increasing its production:

The mill is now treating 600 tons per day. The present capacity is approximately 750 tons. With some additional equipment it will treat 1,000 tons.

Due to late delivery of the new hoist our development program has been seriously held up so that only 600 tons will be milled for some time. This means approximately 725 tons hoisted.

We think it best to adopt cut and fill mining throughout the mine. Several such stopes are now being prepared. This will raise operating costs to at least $4 per ton. Heads from cut and fill stopes should be better than $8. Operating profits should be $3 per ton hoisted or $3.50 per ton milled.

Until we get a definite mining system established it is difficult to forecast accurate cost and profit figures.

J.P. Norrie

The method of mining selected for Malartic Goldfields was cut and fill. This required a great amount of fill to replace the ore hoisted from the mine. The closest supply was on the Dubuisson property and since Dubuisson could not easily drill their property due to heavy overburden, it would serve both companies to use Dubuisson backfill:

Would you kindly make application to the Quebec Department of Lands and Forests for permission to remove backfill from lots 8, 9 and 10.

You will recall that in the development of the property of Dubuisson Goldfields we were unable to penetrate the overburden in this area with diamond drills and as this is the best prospective part of the Dubuisson property, the removal of the overburden will greatly facilitate diamond drilling, surface work, shaft sinking and other development work. Thus the removal of the overburden will serve its purpose to both companies. For Dubuisson it will make development work possible and for Malartic it will supply the necessary backfill for the safe removal of all the ore in the mine.

Bob Halet, a doctor of geology, had been recommended to J.P. by Dr. Jim Gill, a well-known professor of geology at McGill University. J.P. made Bob Halet manager of Malartic Goldfields, and, incidentally, Halet married J.P.'s daughter Pauline.

The mine had only been in production for about one year and already the wartime restrictions and quality of labour were becoming a serious problem when Halet wrote to J.P. about this problem:

Staff bonuses: as this company is beginning to earn substantial profits and will sooner or later begin to pay dividends, it will be natural for staff to expect salary increases.

I would prefer to pay a bonus based on profits rather than an arbitrary increase in salary. I believe it gives the men a greater stake in the enterprise and hence a greater incentive to increase the efficiency of the organization.

The labour turnover has become quite a serious problem, especially among underground employees. At the present moment, out of 180 men working underground, only 29 have worked here longer than a year, and 38 more between six months and a year. Almost half of them have been here less than three months... during the month of April 66 men were hired and 46 men left.

The situation will probably get worse before it gets better. In the competition for labour with other mines in the Malartic district, we suffer from the fact that our men have to be transported from Malartic to the mine every day. This is a considerable nuisance to the men and costs them about $8 per month.

It would be a partial remedy if the company were to absorb some or all of the transportation cost. I do not believe we can try to attract experienced men to this mine, there are none available. However, we can try to keep the men we have, on whose training we have spent much time and money.

J.P.'s letter to his friend Elmer Bray in New York expresses just what a struggle it was to develop this mine from the muskeg:

I take considerable pride in the Malartic Goldfields development as it has been a lone wolf struggle from the start. In the early days I faced, not only the tough job of developing a mine in muskeg, but the adverse criticism of many of the mining fraternity which made financing all the more difficult. In certain cases I did try to enlist help by promising shares in payment when same were released from pool or in other words when the property developed into a mine. No such help was ever forthcoming.

Rosario Roux was mine recorder located in Amos. Rosario had been a very good friend of J.P.'s and he wanted to make sure his office was given full credit by the government of Quebec:

When paying your claim rentals please send the cheque to the Bureau of Mines at Amos rather than directly to Quebec City. The local department do us many favours and by paying rentals here it helps out their showing for the year.

J.P.'s management and engineering report to January of 1942 gives the early history of Malartic Goldfields and the latest development of the mine:

The property of Malartic Goldfields comprises 65 claims with an area of about 2,700 acres in Dubuisson and Fournière townships, Quebec. It occupies the valley of the Piche River, a small sluggish stream which flows from Piche Lake to Lake Lamoyne. It is locally known as the Thomson River. It, as a matter of fact, fairly accurately defines the general structural geology of the area. Several years before the property was staked the structure was indicated by aeroplane maps of the area.

Previous to 1940 when the CNR completed the branch line through the area the town of Amos was the mining headquarters for the reason that a good waterway by way of the Harricanaw River and a series of lakes provided the only means of transportation. Familiar faces in the area in the early days were those of St. Barb Sladen, Lee Coffin, Hal Kennedy, Wilbur Gouldie and many others. All of the men named except Gouldie (now deceased) have done well from their properties and today hold important interests in the various mines of the area.

Early work on the old Malartic Mine, now Canadian Malartic Mine, suggests that important shear zones and perhaps mines would be found far to the east. The prospect of this was further confirmed by the presence of a long line of angular float boulders found on the surface for many miles to the east. The position of these boulders, which we now know came from the surface exposures of the leading mines of the area, was approximately 1/2 mile to the south. Geological work done in the area by Drs. Gunning, Ambrose, Buckland and many others, established the fact that a 2,000 foot band of greenstone traversed the country for many miles in a southeasterly direction and is now known as the Malartic greenstone band. This band roughly parallels the line of the float boulder occurrence to the south and the Piche River to the north. All the producing mines of the area are in or related to this greenstone band.

The factors responsible for the staking of the Malartic Goldfields were:

1: The Piche River structure suggested by aerial maps of the area.

2: The presence of float boulders to the south.

3: The surmised presence of greenstone enveloped in sediments as found in the area to the west. This we knew to be favorable structure even though the presence of the band with fairly regular contacts was not yet recognized.

With these main factors in mind we were able to plant the claims of Malartic Goldfields fairly on the later-found ore bodies so they are amply protected in all directions without any rock outcrops in the immediate vicinity.

Float boulder containing free gold, which led to the discovery of East Malartic and Malartic Goldfields

A few remarks here regarding float boulders on the Malartic Goldfields property may be interesting. Away back around 1920 a French Canadian trapper by the name of Jos. Ladoux found a heavy black gold-bearing rock on his trap line which extended southeasterly from the rapids on the Piche River. Ladoux did not reveal the location of the boulders but did exhibit them and made their high gold content known. Before staking the property we spent several days prospecting in the area. We found the trap line and after diligent search found some of the boulders. These boulders we now know came from the most recently located ore body on the property.

The main part of the property was staked in 1934 by W.B. Cram, Allan Murray, John Partanen and others. When the Malartic Goldfields company was incorporated in Jan., 1935, these claims along with some previously staked by C.O. Slee were included, 99 in all. With money provided by ourselves and friends the property was thoroughly prospected in 1935 and the winter and spring of 1936. During this period the outlining of the greenstone band was set out by dip needle survey. The band was found to have an average width of 2,000 feet and to extend east and west across the property for a distance of 12,000 feet.

On July 16, 1936, an agreement was entered into with Nesbitt Thomson and Company for financing the property to production.

In the fall of 1942 it was clear that Malartic Goldfields had to curtail development operations and production until the end of the war was in sight. J.P. wrote:

I was at Malartic Goldfields yesterday and went over carefully all expenditures that possibly can be curtailed. The Number 2 shaft should go at least 900 feet below the 450 foot level so that mining operations on the level will not be hampered when sinking starts again. In the light of present conditions we have decided to discontinue sinking today at approx. 25 feet below the level so that this work should be all cleared up within two or three days.

We have approx. $5,000 yet to pay on mill additions, and future expenditures on the mill should be light. It should be completely finished on Sept. 1. Two small motors only remain to be delivered.

Little head work remains to be done on the Number 2 ore body. A crosscut of 25 feet will have to be driven from the shaft to the ore body on the 300 foot level. The 450 foot crosscut can wait better mining conditions.

As you know, much timber must be used at Malartic Goldfields. Last year our contractor started cutting late in the fall. Due to shortage of and inefficient labour, coupled with an early breakup, 14,000 logs were left in the bush. With all cutting conditions even worse this year it will be necessary to start cutting now so that men can be used when they are available. This will mean that the contractor will need financial help of approx. $3,000 per month at 5% interest up to Jan. when his income will begin to come in.

As we secure our timber at least $3 per thousand below market, I believe we should finance him, which you authorized this morning. The contractor's name is Ed Jolin, a Frenchman - a very good man.

When the master mechanic Mackenzie at East Malartic found Joe was through he immediately resigned as he never did get along well with McLean. The latter then promptly fired him. We have decided, with your permission, to employ him at Malartic Goldfields as he is a particularly good man.

J.P. Norrie

Timber and timber limits needed to supply the mine were a continual problem at Malartic Goldfields as well as at all the mines along the soft greenstone band:

In 1940 and 1941 when timber requirements at East Malartic and Malartic Goldfields assumed abnormal proportions, timber cutting permits were difficult to secure from the Quebec Dept. of Lands and Forests and much of my time was spent with local officials and those in Quebec in securing the necessary permits.

In the spring of 1942, Mr. L.P. Massicotte of Amos approached me re: a timber limit owned by Carl Erickson and himself covering the southeast quarter of Malartic township.

I talked with Mr. Massicotte several times re: the limit until a few days before the directors meetings in May, 1942, when Massicotte informed me that he had been approached by different parties re: the purchase of the limit for pulp cutting. At the time there was a great demand for pulp wood by the paper companies as the previous year's drive had been held up due to low water, and the limit in question, being located near the railroad, was well situated for cheap removal of the wood.

The price originally asked for the limit was $80,000 which after much trading was reduced to $55,000. Then I asked for a 10% commission of $5,500 which reduced the price to our companies to $49,500. Massicotte informed me that he must have an answer from us the following day.

Mr. Martel, local head of Dept. of Lands and Forests, informed me that only 100 million board feet of timber remained in the area adjacent to our mines and as East Malartic and Malartic Goldfields were using annually approx. seven million board feet, that this limit should certainly be purchased by us as a reserve for the future. He also mentioned that we could not expect to continue to get such large cutting permits at the expense of other timber users in the area.

J.P. Norrie

In preparation of the 1942 annual report J.P. discussed the inexperienced quality of labour at Goldfields:

I am enclosing Malartic Goldfields progress report. As usual I have given no cost figures. These you can add if you think advisable. These figures will fluctuate from period to period as long as the war lasts, due to the fact that labour, particularly at Malartic Goldfields is seasonal.

During the winter we had plenty of men, such as they are, and all possible development work with consequent higher costs, has been done. We have 70% French on the payroll, many of them farmers [farmers were not experienced miners]. Shortly many of these will leave which will reduce the development, perhaps eliminate it. We will be able to carry on during the summer until these men return in the fall.

Due to the fluctuation I do not think cost figures should be given, however, if you think so it is okay with me.

Late in 1943 the end of the war was in sight and J.P. wanted Thomson to apply to the metals controller for permission to sink the No. 2 shaft.

Will you please apply to the metals controller for permission to sink the suspended work on No. 2 shaft at Malartic Goldfields Ltd. from 500 feet to 1,500 feet.

The No. 2 ore body has been well developed on three levels so as to place it in condition for quick production after the war. It will be necessary to sink the shaft now so that the deeper levels can be opened up when men are available. Timber for the work has been on the property since sinking was discontinued some time ago.

Men for the job can be switched from other development work so that additional men will not have to be hired.

Malartic Gold Fields Limited

(No Personal Liability)

PROFIT AND LOSS ACCOUNT
For the Year Ended 31st December, 1941

Bullion Production—less Shipping and Mint Charges			$1,841,668.94
Miscellaneous Income—including Rentals, etc.			7,552.92
Interest on Investments			1,488.49
Profit on sale of Investments			175.00
			$1,850,885.35
Development and Shaft Sinking		$258,529.40	
Mining		544,590.32	
Milling		175,029.65	
ADMINISTRATION AND GENERAL EXPENSES:			
Mine	$101,313.86		
Head Office	19,806.92		
Insurance — Fire, Liability and Unemployment	16,081.59		
Taxes—other than Income	2,661.43		
		139,863.80	
			1,118,013.17
			732,872.18
Preproduction Expenses—written off		69,705.60	
Depreciation on Plant and Equipment		93,081.91	
Provision for Provincial Tax and Federal Excess Profits Tax		113,164.95	
			275,952.46
NET PROFIT FOR THE YEAR			$456,919.72

Chapter 19

True Stories from the Stopes

by Harry Norrie

Malartic Goldfields 1938

J.P. had drilled all down the west side of Goldfields and halfway across the south side with disappointing results. Then he and Bob Halet, the mine geologist, on a trip to the property, decided to lay out a series of holes on the diagonal. The fourth hole intersected what became known as the "F" ore body and Goldfields was born and became a mine.

When this ore body and some smaller adjacent ones were ready for production, the money to build the mill was obtained from the Bank of Toronto, the first time in Canada that a bank had advanced money for such a purpose.

Most of my work in the Malartic area was at Goldfields: early on as a mucker, shovelling broken rock into mine cars, later graduating to the position of rodman for Jerry Thomas on the transit (theodolite). One day we were surveying No. 47 stope. Jerry set the transit up at the foot of the access ladder. I was in position in the middle of the stope, standing on broken muck, with the rod in position for Jerry to take a reading. Suddenly the muck gave way beneath my feet. Some guy at the draw point directly below me in the drift had opened a gate to let the broken rock flow into a mine car. Surprises like that have a tendency to inject bursts of energy into your legs, giving you the agility of a demented squirrel. That day I passed Jerry on the ladder.

With the incorporation of East Malartic and Malartic Goldfields, diamond drilling programs were started at once with up to four drills working on each property at the same time; and rather than contract the work to outside companies, Inspiration was formed to do the drilling and Miners Inc. to handle advanced development.

Since J.P. felt that the new science of geophysical surveys had not advanced to the point of reliability he still depended on surface geological surveys and diamond drilling.

At Goldfields his method was to cross-section the potential structure by overlapping drill holes but the first attempt at this was not successful. There were a number of mineralized intrusives along the southern part of the property, but here again drilling met with failure.

With the large amount of sampling that had to be done it was practical to set up a permanent assay lab on the property. One summer J.P. sent Roy McKay and me to work in that part of the Goldfields' lab where samples were prepared for assay. Roy ran the crushers, splitters and grinders while it was my job to screen the samples and further reduce the reject from the screening. In this we used what was called a bucking table; that is, the reject was placed on a heavy metal plate and ground down with a heavy half-moon-shaped tool until everything passed through the screen. It was always a thrill when flecks of gold were seen flattened out on the metal table or clinging to the bucking tool. On the completion of this work the samples were passed through a window into the lab where they were fire assayed for gold and silver content. At some mines any visible gold found in a sample is thrown out, supposedly to give better overall assays, but I feel it is more likely an effort to make heroes of management when the ore body mines out above expectation.

The manager at Goldfields at the time was a Mr. Downey. One afternoon the manager at East Malartic came to Mr. Downey and directed him to have Roy and me sent to East for work on the following Monday morning, and that he would see that we would be put at work that we deserved. In other words, back to the bull gang. I puzzled a long time about how I might refute this order without being accused of running to J.P. for help. The answer was slow coming.

I had already planned to go to Amos on Saturday anyway and to be back at work on Monday morning. I moped around the house with the problem on my mind until in the evening I noticed J.P. and Marg on the balcony talking quietly. Then an evil thought and the circumvention of the work order came to mind. I joined J.P. and Marg on the balcony and casually said, "I've got a new job." J.P. was surprised and said, "Where?" I told him that Roy McKay and I had been directed to go to East for work on Monday morning. J.P.'s simple reply was, "Stay where you are." We did and heard no more about it. Blood cooled here and heated up there. I think J.P. had come to the conclusion that work in an assay lab would be more beneficial to us if we were to study mining engineering, rather than spend our lives rising to management of a bull gang.

It was around this time when I had, for some reason, to take a trip to Montreal. I was just leaving the property when I noticed a couple of guys lighting a fire to burn slash - that is, they were piling up old tree branches and scrub growth that had been left after a firebreak had been levelled around the property to protect the mine buildings in the event a forest fire headed their way.

A stiff wind was blowing and it struck me that these guys were pretty dimwitted to start a fire so close to the bush, and as I watched, that is what happened. At that time in the north it was not possible to put out a bush fire and little effort was made to do so; they burned on until they reached a lake or a wide river or were extinguished by periods of rainy weather.

My trip took me first to Val d'Or where I would board the train for Montreal. I had some time to wait and before I left, a dark heavy cloud of smoke from the Goldfields fire was rolling over the town.

It may be hard to believe but shortly after leaving Senneterre our train came to another bush fire burning along the tracks. The train came to a stop to allow the crew to determine if it was safe to continue. It was decided that since the track ties were still intact it would be safe to go on. Regardless of the condition of the tracks, when the train entered the burning area a wave of heat could be felt as it came through the windows causing a moment of panic for me and other passengers.

The train arrived in Montreal at about eight in the morning. I went directly to a restaurant in the station to have breakfast but found all the tables occupied. Two friendly looking guys were seated at a table set for four. I asked if they would mind if I joined them. There was no objection. As I sat down one of the gentlemen, who had been reading the paper, turned to the other fellow and said, "It says here that Val d'Or burned down last night." Another moment of panic but I soon found out that Val d'Or was left standing but that Pascalis had burned.

When I arrived back at Val d'Or I heard that during the Goldfields fire one of the employees, whose house was on the side of the road in the direction the fire was heading, had rushed home, taken all of his furniture and other possessions out of the house and piled them on the side of the road so that if his house burned he would still have furniture and stuff. A spark from the fire landed on his possessions, but his house was saved.

In any gold mining camp there are always a few guys around who are well dressed, well fed and out of work. They are professional gamblers and high graders. Two gamblers in Val d'Or made a bet about who could rake in one hour the most money from a table without the other players beefing. One guy raked $92 the other $89 and it was a low stakes game.

There are always ways to steal gold. At a camp in Colorado the local minister would visit frequently, particularly the mill where the gold was recovered. He felt it necessary to heap his beneficence on these poor sinners. Especially he would question the workmen about their jobs; about how the gold was collected. After being shown he would run his fingers through the concentrate, then cleanse them just over his vest pocket. When it was found he had a flat metal tobacco can in the pocket, he was not encouraged to return.

Inevitably East Malartic mine was robbed. After a day of pouring gold bricks, two worth about $50,000 each and a small one worth about $15,000, the bricks were locked away in a vault. The next morning they were gone. But the thief left pretty good evidence who he was and where he could be found. With a white chalk he had made straight lines around the lock, and with a torch had cut along the lines, a familiar practice in a machine shop. The two large bricks were recovered but not the small one. The mechanic had sold it for quick cash to a guy in Val d'Or who just as quickly sawed it up like a loaf of bread and sold the slices to ready buyers lined up at his gate.

A couple of guys in Val d'Or stole a bunch of zinc concentrate from one of the local mines [in the final recovery stage in a gold mill, the gold is picked up with zinc]. The thieves were told that if they heated the stuff the zinc would be driven off leaving the gold. They made a kind of a bomb to put the concentrate in, then rented the use of the furnace in the local post office and fired it up to a high heat. The bomb exploded and destroyed the furnace. They had forgotten to leave an escape hole for the zinc fumes. When the soot in the furnace was assayed for evidence, it was found to run very high in gold content.

I was working underground in a stope above the fourth level at Malartic Goldfields. My partner was a guy with the unlikely name of Bud Sad. We came to our work station and found two miners debating the properties of dynamite. One boastful fellow held a stick of dynamite in his hand. He said, "I can take a piece of this and light it with a match." Long streams of sweat immediately ran down my back revealing an equally wide band of yellow. The boastful friend broke off a piece of dynamite, set it on a rock and put a match to it; and it did, it burned with a pale blue flame. If he had struck it with a hammer at the same time he applied the match, it probably would have split the rock, and me.

Bud Sad was an American from Fargo, North Dakota, related to C.O. Stee. When the Second World War broke out he decided to go home and join the army. A large group of his friends threw a party for him the night before his departure. They escorted him to the train station the following morning. The train had to be stopped a mile down the tracks to throw his friends off.

Bud was sent to the South Pacific. On Saipan Island he was shot several times by a Japanese sniper. He recovered and came back home to Fargo. Then back to Malartic, but like I said, mining is boring so he quit and I haven't seen him since.

When a miner states that he has drilled and blasted a round he means that he has advanced a drift (tunnel) for a certain number of feet. With stope work, a round represents the number of holes drilled and blasted leaving a shift's worth of broken ore. At the end of the day, conversation with the shift boss might go as follows: "Did you finish your round?" "Ya, I finished the round." "Did you blast?" "Ya, I blasted." Whereupon you were free to finish the conversation in your head. "Dumb."

In the Malartic-Val d'Or area, what is called the break is a wide region of fault zone, running in an east-west direction for several miles, where the rock has been sheared and metamorphosed, leaving soft weak material called greenstone that must be supported when an ore body is being removed.

The ore is granite-type rock called diorite that long ago intruded the greenstone in large and small masses. The interval between the greenstone and diorite is very sharp and easily followed. When mining the diorite the policy was not to break into the greenstone. If it was left undisturbed it was strong enough to remain unsupported for a height of about 20 feet before splitting off.

In a stope or drift the first few minutes of a shift is spent scaling down loose rock that was left hanging after the previous blast. This is done with a metal rod flattened on one end to pry into cracks and free the loose rock.

When Bud Sad and I left the dynamite burners we went on into the stope to set up the drill and start our shift but on our way we had to go around a couple of large chunks of rock that would have to be re-blasted before being drawn from the stope. To get around the large chunks it was necessary to squeeze up close to the wall, but that led under a length of wall that had been blasted into, leaving a dangerous overhang. I questioned Bud about passing that way. His less than reassuring reply was, "Well, it hasn't come down so far."

The leader of the opposite shift was a fellow called Davidson, that I had gotten to know in one pub or another. A few days later I left to return to college only to receive a sad letter from Bud. Davidson was dead. He'd walked once too often under that overhang. A young fellow from Cape Breton, he had hired on with an experienced miner. It was his job to work scaling down and with setting up the drill machine. During the remainder of the shift he changed the bits on the drill steels while the miner was running the machine. On his last shift he and the machine man tried for a long time to pry down a large chunk of loose rock but could not dislodge it. Finally, the machine man told him, "Leave it. Just don't walk under it, walk around." He followed this advice all during the shift, but when the holes were drilled and loaded with dynamite, leaving the machine man to light the fuses, he took a chance and walked under the loose rock just as it came down. As I said it was his last shift.

On surface we all know what bootleg means. Underground a bootleg is part of a drill hole that hasn't fired, leaving dynamite in the hole. If a still partially drunk or unobservant miner drills into a bootleg, he's going to get himself a hell of a headache.

Early one evening, greatly in need of the uplift that can only be found below the neck of a bottle, I wandered into the Morocco Club downstairs in the Chateau Malartic Hotel. There I found a couple of friends already enjoying an uplift, Max Bailey and a guy called Turner. We joked for a while until they had to catch a bus to go on shift at Malartic Goldfields, seven miles east of Malartic town. My uplift took some time to level off and sustain. At about eleven o'clock my two friends returned in great distress.

A mine shaft is rectangular in shape; its size depends on what it is to be used for. An exploration shaft is usually quite small, generally two or three compartments. When the mine goes into production a much larger shaft is needed, a production shaft.

The Goldfields production shaft was to be sunk to 1,200 feet and was now at about 500 feet. As a shaft is sunk, a timber framework of 8 in. by 8 in. timbers is put in place to form compartments. When the rock is broken at the bottom of the shaft a large metal bucket up to four feet in diameter is used to remove it. The bucket is raised and lowered by a sturdy, braided-wire cable. The bucket is kept from swinging by a wooden framework called a crosshead that rides up and down above the bucket and is held in place by guides attached to the 8x8 timbers. The guides are heavily greased to reduce friction when the crosshead is in motion. The cable passes freely through the crosshead. Also hanging in a corner of the compartment is a cord to send signals to the hoistman on surface.

Bailey and Turner were shaft men in the Goldfields production shaft. In getting to their work at the bottom of the shaft the miners are supposed to ride inside the bucket, but they rarely do unless a foreman is watching. They usually stand on the rim of the bucket and hang onto the cable. On this occasion Bailey and Turner and two Swedish miners were standing on the rim. Part way down the shaft the crosshead hung-up on an obstruction, but the bucket kept on going and started to swing; it caught up on a wall timber and tipped. When the bucket tipped, the two Swedes fell. Max fell too. Turner, safe on the top side of the bucket, shouted, "Christ, there goes Bailey." But Bailey said "No, I'm all right." He had fallen into the bucket.

Turner, the shift leader, was supposed to hang onto the signal rope letting it slide through his hands as the bucket descended in order to signal the hoistman in case of a mishap. Turner grabbed the signal rope and stopped the bucket, then signalled to go back to surface, picking up the crosshead on the way, leaving two dead men on the bottom of the shaft.

Max quit mining, joined the army and survived several years fighting in Europe during the Second World War. He came home and opened a store in southern Ontario.

In the middle of the century, 1930-1970, shaft sinkers were a select group of men who ranged across the country from job to job. When they went on shift it was usually in a state of high intoxication. Alcohol was part of their diet. Their work was difficult and they fully understood that early on in their shift they would become sober, the alcohol would have become a part of their clothing along with several pounds of sweat. They knew their work and did it to perfection, even though they killed themselves every once in a while. In all probability they will have had many children and never known any of them, or cared.

Weak men need not apply. Only a few grew old. Shaft men come in all sizes, but there is one characteristic they have per copy, tough. Every shaft has traumatic moments. In one case a large chunk of rock broke from the lip of a station above and came roaring down the shaft. The shaft men heard it coming but could do nothing. Except one guy, trying to escape, ran hard into a wall and fell down. He got up and tried again. It still didn't work. But the Devil was asleep and they all escaped with minor injuries.

Deep in the darkness of a mine - the underground - loneliness, like moisture, becomes part of the atmosphere, but hostile, frightening, dangerous. On surface, loneliness has a more gentle nature, softened by sunlight.

One day in the yard at Goldfields I found an old man toying with a bit of work that didn't need to be done. I stopped to talk for a while - generalities at first, but we came to be friends. There was a loneliness in the old man's eyes. He was like a chip of wood floating on an expanse of water, drifting, until water-logged it sinks to the bottom. He was out of place, not a part of bush life. It may be that he had applied to J.P. for work and J.P., with kindness, seeing that a daily wage would be sufficient to sustain his modest needs, sent him to Goldfields. His name is still somewhere in my mind but, with the opaqueness, I don't seem to be able to retrieve it. It might have been Cosco - Mr. Cosco - let it be.

Sometimes I would see him through a window, and shirking work that needed to be done, I would go out to talk with him - really more to listen. Gradually he started telling me stories of his life. As a young man he had been to the American southwest. He was in Silver City, New Mexico two years after Billy the Kid was shot there by Sheriff Pat Garrett. To Mr. Cosco, Billy the Kid was nothing more than a man, a snaggle-toothed, murderous little buster, and a picture still in existence confirms that; at least in appearance. Americans love to glamorize some of their worst murderers: Bonnie and Clyde, John Dillinger, Willie Sutton, Billy the Kid. They make up movies about them - nowhere the facts - just entertainment - like handsome Robert Taylor playing Billy the Kid - Robin Hood! Self protection - nonsense. At the time it was said that Billy the Kid was shot in the back by the sheriff because he was afraid to face him. Billy is buried near Vaughn, New Mexico. On a later date with four American friends, I had reason to drive through Vaughn but we found it not to our advantage or interest to visit the grave site.

I haven't tried to draw a word picture of Mr. Cosco's physical characteristics - it's not important - readers, if any, will draw their own picture; to say he was a fine, intelligent, lonely old man is enough.

In another of Mr. Cosco's stories, still in the southwest, he told how he had been hired to deliver two beautiful horses to a man in a distant town which meant that he had to spend the night in an inn between towns. It chanced that the madam in the local whorehouse saw the horses and wanted to buy them but he had to decline her offer since he was not the owner. Later, in a saloon, he was sidled up to by a guy who whispered that he had better get out of town or he would not survive the night.

Now, sometimes in bed on sleepless nights or mindlessly driving, here or there, my mind gets focused - like spotting a lone crow on a telephone wire will draw your attention - I focus back on Mr. Cosco.

One of J.P.'s faithful old followers was a guy called Tony Mazarik, Polish, starting at Stadacona (maybe even before), to Malartic Gold Mines, Treadwell Yukon and East Malartic. The old fellow finally wore out and should have been shown a little compassion in his old age. Beyond J.P.'s knowledge, at East Malartic he was given an absolutely meaningless job clearing a small path of bush that was of no use to anyone.

Old Tony treated me something like family. I guess he figured after being with J.P. so long he really was family. He had no one else. Many times he gently corrected me about one thing or another - there was ample room and occasion.

I found him one day working on his little plot of land. He had dug up a handful of long stringy yellow roots and was very earnest in telling of their medicinal benefits that he had learned as a young man in Europe. That was the last time I saw Tony. I had known him for a long time.

Malartic 1940 - Gunfire in the Streets

On a cold, dark, damp mid-October evening, I was wandering aimlessly along Malartic's deserted main street. I had just passed the Goldale Café when suddenly I heard the sound of several detonations. My immediate thought was "fire crackers". What fool was celebrating when there's nothing to celebrate? Had I known at that moment that I was in direct line of fire of two police issue .38 calibre handguns, only one of which was in the hand of a policeman, I would have immediately made a 90 degree turn and vanished.

The town of Malartic had only recently been organized into a recognizable community with a population of upwards of 3,000, made up mostly of miners and their families supported by a few butchers, bakers, bankers and green grocers along with the odd parasitic bootlegger.

We now had a mayor with his several attendants, a police chief with a couple of outriders, and a fire department capable of at least sounding an alarm, but equipped with little else.

Malartic's main street was only a short section of the main highway running through the district from Noranda to Senneterre with a few subtle streets running off at various intervals and directions to nowhere in particular.

Malartic town was the urban centre for several producing gold mines: West Malartic, Canadian Malartic, Sladen Malartic, East Malartic, Malartic Goldfields, stretched out, east-west, a distance of about ten miles, plus many other properties where mineral exploration had not yet determined their economic future - good or bad. There was only one oversight, Malartic was named after Malartic township where it was supposed to be located. But it was in Fournière township, one township removed. I hate to think this was an error of J.P.'s since he was a boss of the first Malartic mine, long before there was a Malartic town.

With the sound of detonations that night I hurried my steps in their direction expecting to find merriment of some sort. But arriving in front of Benny Clement's store, I saw that crime had broken out. The store fronted on Main Street opposite Charlie's Hotel and across a stubby side street from the police station, both of which came into play as the crime scene expanded.

Benny catered to the clothing needs of both men and women and, to protect modesty, divided the shop into two departments with a stout wall between but with a unique alteration: Benny cut a section out of the wall at the front end of the store where he set up his cash register and where his sales force brought him the cash while he rested comfortably in a sway-backed chair. (In this narrative we are not interested in what went on in the men's section, so let's get to the girls: Benny's girls.)

In the women's department there were three exceptionally beautiful lady clerks, the senior of whom was the most beautiful of them all: Tess Corbeil, about five feet eight inches tall with a perfect number of pounds spread becomingly over a delicate bone structure. Her face alone would have demanded the crowning award for the world's beauty queen. She had long dark hair cascading to below her shoulders. The most striking feature in her facial structure was her eyes: large, bright, usually smiling, magnetic with low smouldering fire: magnetic to draw deserving swains: fire to burn the unworthy, and fire also, perhaps, to punctuate the final reward.

Next was Annetta Blanchette, nicknamed by some awakening wit, Annie Blanket. Annetta had less height than Tess and more weight to lend comfort on a cold winter night. Well, let's not be specific; comfort at any time, day or night or day of the year.

Now we come to the cause of all that night's raising: the last of Benny's girls. She had not been long in the town and after that night would not linger. She was small, delicate, doll-like, and she had a boyfriend, a criminal; well, perhaps not really a criminal, but after that night his destiny was questionable.

With embarrassment I must admit that I no longer remember the name of that precious little creature, Benny's third girl, so let's, with little imagination, call her Bo(u) Tigue. At a point in time, Bo denied her boyfriend further passage and sent back his engagement ring. With this finality his Gallic blood rose several degrees to a point where fire could be seen in his eyes as well as in other regions that will not be identified here. Bo lived in Malartic, but her boyfriend lived and worked in Val d'Or although his father still had residence in Malartic.

At the sound of denial the boyfriend wore out the knees of his trousers seeking readmission but to no avail. With love life terminated, it was reported that he went to a close friend in the Val d'Or police force and requested the use of the officer's side arm, a heavy duty Smith and Wesson .38 calibre fire arm, the appearance of which would have scared Caligula. The gun was allowed out on loan and with the next bus in line the boyfriend took off for Malartic and his ill-thought-out adventure of revenge. It might be cogent here to remember that the Malartic police station was just across one of those stubby little side streets from Benny's store.

The boyfriend arrived in town just after dark, got off the bus and walked purposefully into Benny Clement's men's and women's clothing store. Bo was behind the counter. He walked up to the counter, leaned over it, threw the ring down and said, "Here's your ring back." He pulled the gun out, and saying, "And here's something more for you," he fired two shots directly at her - and missed! An impossibility? But I tell you the truth, that is exactly what happened.

There are now one or two minutes of rapid activity. Bo ran to the back of the store and hid behind a rack of ladies' dresses. Benny jumped forward and cut the boyfriend off from following Bo. Bo, peeking from behind the dresses, saw Benny's intervention, and running behind him bolted out the front door. Now the boyfriend saw her leave and went to follow her, but then turned around and said to Benny, "And here's one for you," and fired a shot directly into Benny's face. But the god of survival was riding on Benny's shoulder. The bullet entered the extreme side of his mouth and exited just below his ear touching neither bone nor tooth. Patented death - that healed in a few days.

The boyfriend ran after his girl, and just before she could reach a little hotel adjacent to Benny's, on the east side, he emptied his gun at her. One bullet burned the flesh on one shoulder and another found a fleshy mark just at the upper part of her ribs where the breastwork swells out and creates such a sensation for vulgar imaginations. No vital organs were touched, no ribs shattered and only two cute little dimples were left to excite the successful suitor.

And where were the police while all this was going on? Oh! They were there too, but just a little after the main event. The chief of police was at home, having earlier finished his regular day shift but he had left the protection of the town to one of his outriders, a beefy, not too bright young man. With the sound of gun fire in the street, the officer, being familiar with the sound, having downed the odd squirrel here and there, ran out the door firing his weapon as he went. His first shot passed through a window and into a wall in a room on the second floor of Charlie's Hotel occupied by a miner trying to get a little sleep before going on graveyard shift at one of the local mines. The officer's second shot went wild as did the third, fourth, fifth and sixth. If the policeman's bullets killed anything it could have been a nighthawk or night owl but most likely nothing.

Still there was that love-starved young gunman to apprehend. All that dark night he could not be found. But that should not have been unexpected since the thick bush, in that latitude, is just outside everyone's back door. And that's just where he was all night long. But with the dawn, wet and cold, he left the unfriendly virgin forest and went to his father's house where after a period of reasonable persuasion he went with his father to the police station.

The case went to court. Jacques Miquelon, a lawyer who later represented the crown in our district, acted as prosecutor. Jacques was one very good reason why I liked and admired the Quebec French people so much. He was clever, energetic, full of humour and could never pass an uninhabited piano without thumping out a bunch of familiar tunes to combine with the voices of any number of standing locals, French or otherwise. Jacques spoke English as well as, if not better than, his native French.

As the trial went on, there was endless comment in the community with the weaker minded pretending deep concern and outrage that Jacques would even think of seeking a long prison term for the overheated boyfriend, and perhaps it was even the young lady's fault for causing his torment and momentary mental imbalance. But Jacques was very serious when he insisted that he was going to put that guy so deep in the hole his only light would be the intermittent glow from an off course firefly. The judge, perhaps remembering the heat of his own early passion, took pity on the young man and estimated that three years would be about enough time to bank the fire.

So it was. Any one of those stray bullets, unintended as it may have been, could have caused me grievous harm. Winston Churchill once said, "The most exhilarating thing in the world is to be shot at - and missed." In my situation, not knowing I was in the line of fire, I had no reason to feel exhilaration, but ever since then, on hearing a loud detonation from any source, I immediately fall to the ground and crawl on my belly like a reptile.

Chapter 20

The Greenstones

by George Norrie

National Malartic

We now believe that gold ore can be found in any favourable structures within a radius of say five miles of the East Malartic ore body. As you probably know this condition exists in Porcupine with Hollinger as centre. This theory applies generally to all major gold camps in Canada.

J.P. Norrie

The National Malartic mine lay within this radius and sooner or later J.P. would do his part to see whether his theory applied to this property.

By 1937 J.P.'s efforts had developed first the Perron and then East Malartic to the point where gold production was imminent. Malartic Goldfields was also going to be another major producer.

Inspiration was the extension of J.P.'s physical limitations. Through Inspiration he became involved in many more prospects at the earliest possible stage of their progress. National Malartic was one of these.

With production at hand for his present mines J.P. had time on his hands. He had only one solution for that. Find more mines. Although he had demonstrated his ability to actually direct the underground success of any mine he supervised, he found this drudgery, boring, compared to the thrill of the first discovery and proving the existence of an ore body.

The National Malartic property was staked by J.P.'s friend, prospector Lee Coffin. It seems J.P. always felt that he owed a debt, or at least a responsibility, to his friends in any mining capacity, particularly the prospectors. This was one reason he pushed the property at that time.

Gus Rainville was the president of Quebec Gold Mining, the company that took control of the Bussiers, then closed it down. Remember Ron Macaulay, the vice president of Quebec Gold Mining, the guy who drove J.P. out of Bussiers? Six months later the mine closed down because of incompetent management. Rainville had not stopped J.P. from leaving Bussiers, but now it was obvious that J.P.'s successes had changed his mind.

Quebec Gold was now in control of the National Malartic claims. Lee Coffin was not always easy to deal with, partly because he was usually deep in the bush. Even when he appeared in town it usually took the involvement of J.P. to persuade Coffin to trust or agree to any deal concerning his properties.

Gus Rainville needed J.P. if only to deal with Coffin. He wrote:

We are just about to finish up on the National Malartic (Demara group) and expect the agreements to be signed during the course of this week. Coffin has at last consented to "play ball" with us. As soon as the agreements have been signed, we will proceed with the drilling and I see no reason whatsoever why the Inspiration should not get the contract. You might give me your price say for 5,000 feet to start with. I expected to be up at the Cournor before now, but have been laid up.

As with most mining developments, the primary mining company would split their bet, particularly at the beginning, by bringing in partners in a minority position. This was the case with National Malartic:

Received your letter, and Noranda and Smelters are of the opinion that the claims should be retained and the taxes paid, and that they would be willing to pay their share. As usual your friend Coffin did not reply to my letter.

In reply to your request for a list of the shareholders of the National Malartic there are 2,231,678 shares issued to the following:

5 shares to the directors

1,035,423 shares to Quebec Gold Mining Corp'n

697,000 to Lee Coffin
3,000 to Waldo Howland (a Coffin nominee)
100,000 to Benjamin E. Martin
100,000 to Noranda Mines Ltd.
100,000 to Consolidated Mining and Smelting
170,000 to Inspiration Mining and Development
26,250 to H.J. Stewart (as part of his contract in the Coffin deal)

Rainville

The National Malartic consisted of two groups of claims as shown along the greenstone bands. The town of Malartic was incorporating and was to include part of the south claims.

Congratulations on the pouring of the first official gold brick at East Malartic. I wish it had been possible to be present to honour you and cheer you for the great success you have made of this property.

By the way, we have been notified by Mr. A.O. Dufresne that the six claims of National Malartic wedged in between East and Sladen are to be included in the block that is to form the municipality of Malartic. I presume that we can have no serious objection to this. The only drawback is that we may have to pay taxes on any buildings we may erect.

Rainville

As with Perron, Beaufor and Pascalis, veins and ore bodies follow the laws of nature and are not inhibited by property lines. The best ore often heads over to the neighbour or can be mined economically only together with the neighbour. This was the case at National Malartic. The neighbour this time was Sladen Malartic.

Sladen agreed to develop the National mine, at least until the ore body was established:

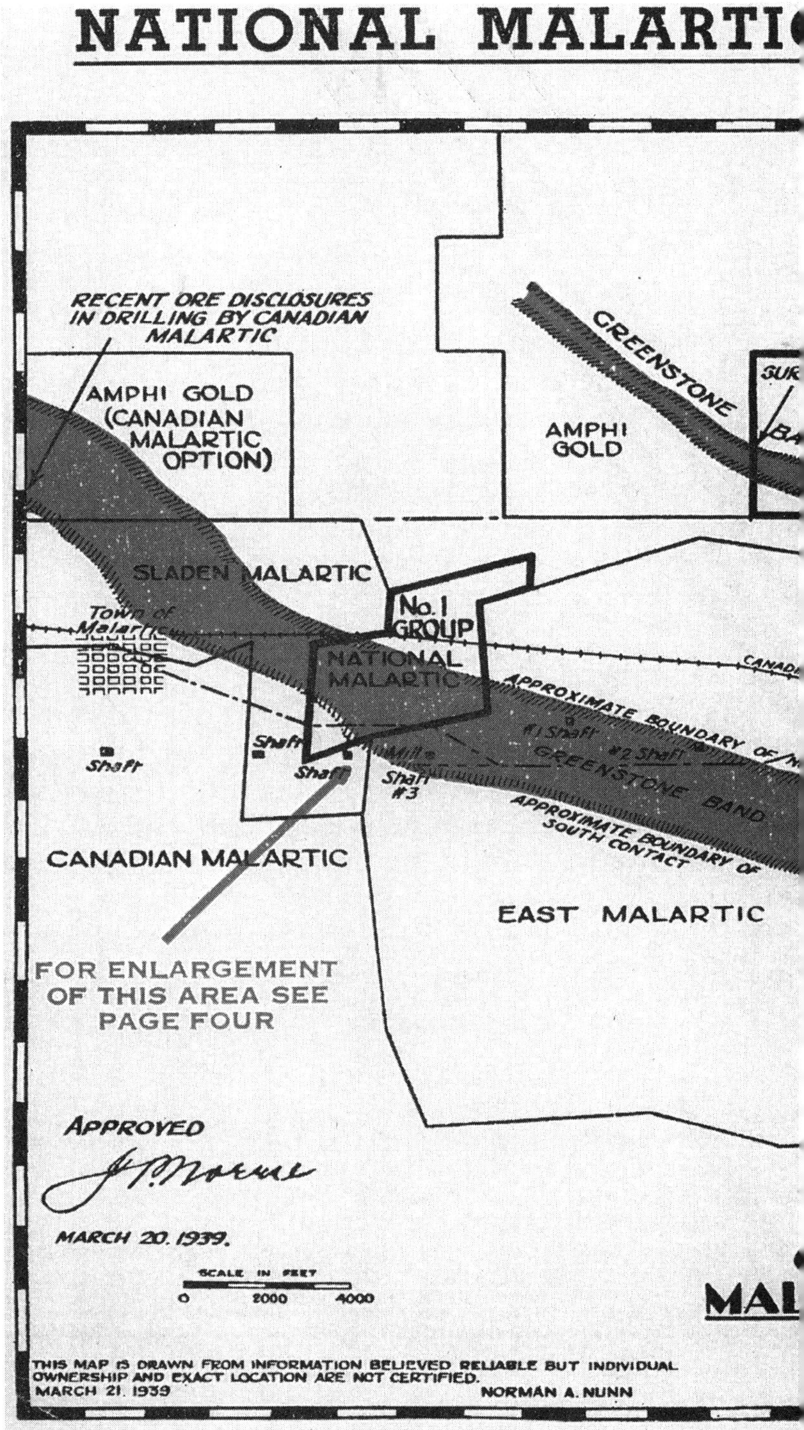
NATIONAL MALARTI
RECENT ORE DISCLOSURES IN DRILLING BY CANADIAN MALARTIC
AMPHI GOLD (CANADIAN MALARTIC OPTION)
AMPHI GOLD
GREENSTONE
SLADEN MALARTIC
Town of Malartic
No. 1 GROUP
NATIONAL MALARTIC
APPROXIMATE BOUNDARY OF
#1 Shaft
#2 Shaft
GREENSTONE BAND
Mill
Shaft
Shaft #3
APPROXIMATE BOUNDARY OF SOUTH CONTACT
CANADIAN MALARTIC
EAST MALARTIC
FOR ENLARGEMENT OF THIS AREA SEE PAGE FOUR
APPROVED
MARCH 20. 1939.
SCALE IN FEET
0 2000 4000
MAL
THIS MAP IS DRAWN FROM INFORMATION BELIEVED RELIABLE BUT INDIVIDUAL OWNERSHIP AND EXACT LOCATION ARE NOT CERTIFIED.
MARCH 21. 1939
NORMAN A. NUNN

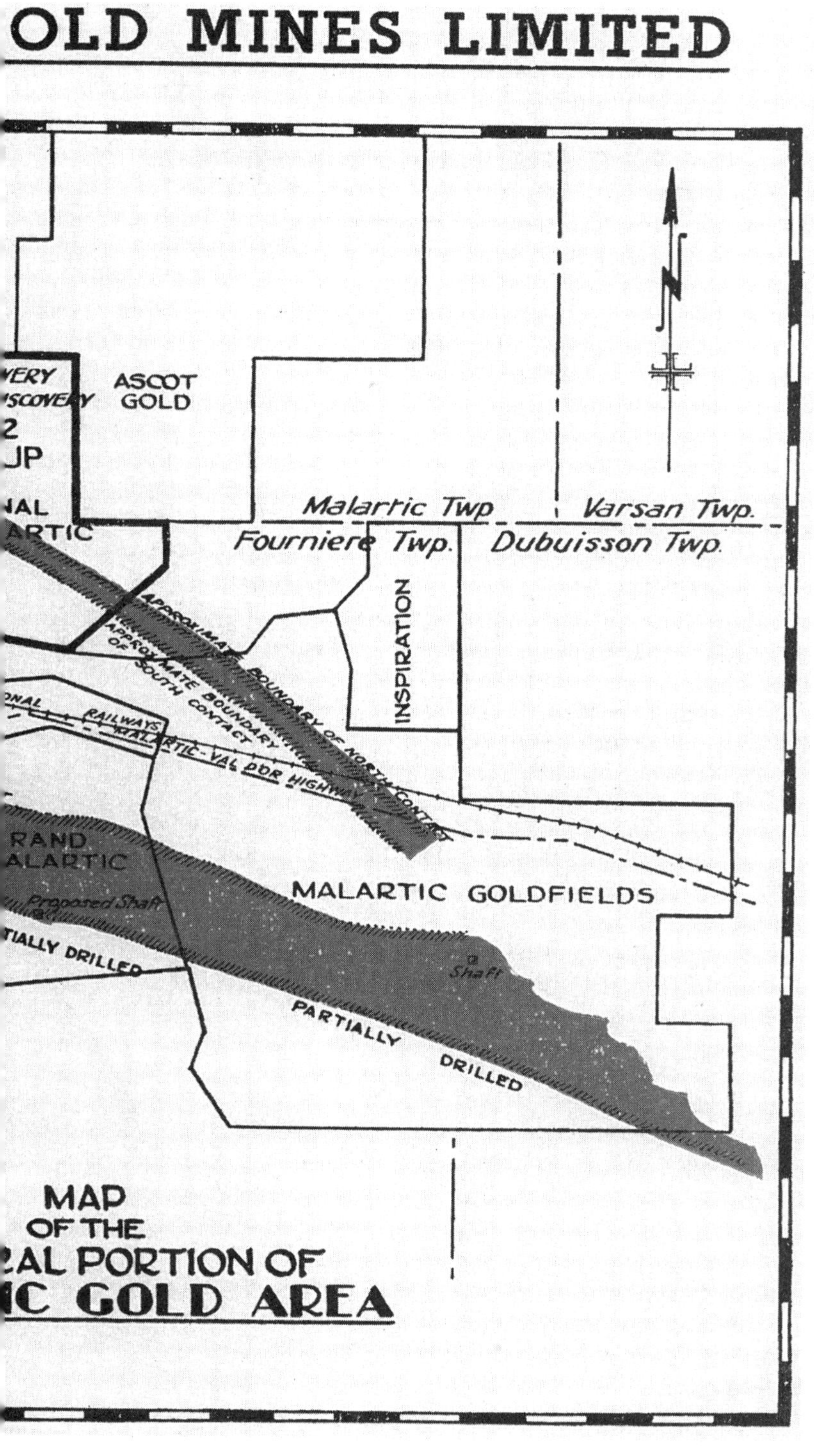

The greenstone band. Many mines were discovered along this strip.

Our general managers Futterer and Reid have authorized us to confirm that Sladen Malartic Mines will carry out drifting and crosscutting for National Malartic Gold Mines on the following basis:

Drifting and crosscutting in waste at cost plus 10%. Drifting and crosscutting in ore at cost; Sladen Malartic to keep all ore from this work.

If for any reason other than acts of God, fire etc. we find that we are unable to continue this work, we will undertake to notify you at least four weeks in advance.

Although Noranda, Smelters and Inspiration were large shareholders with Quebec Gold, the property did not yet warrant these companies taking up their option towards production, at least at that time.

Nesbitt Thomson, either through Inspiration or independently, were not interested in financing National; they wanted a sure bet. J.P. did not stop there, as usual, even though the odds were longer. If he could raise more money for development, the shares of Coffin and his own interests would benefit. In this case he persuaded one of the smaller brokerage firms, Pooler & Company, to market the shares. He gave these reasons why they should take on the property:

Exclusive of the ore indicated in the south and new east zones, it may readily be calculated from the length and width reported that sufficient ore is indicated in the main and branch ore zones, to a depth of 500 feet, to supply a 750 ton mill for five years.

The Malartic area is featured by low mining costs. Sladen Malartic, handling 625 tons daily, has shown operating costs of $2.40 per ton; East Malartic, handling 1,500 tons, $2.05 per ton; Canadian Malartic, milling 670 tons, $2.35 per ton. It can therefore be seen what splendid profit-making possibilities exist in the ore which has been already indicated on the National Malartic property.

By January of 1941 Noranda and Hollinger decided to take an option on the National shares, possibly because of war-time restrictions. Noranda needed more ore to feed its mills and National was available.

This was a surprise to J.P. and to Coffin. This put Coffin and J.P. in exactly the position J.P. wanted when he first put the properties together. The shares were now worth a considerable amount.

Lee Coffin wanted J.P.'s advice:

> I hear that Noranda and Hollinger are taking over National, which I take to mean that Mr. Hall [Oliver Hall, Manager of Noranda] likes it better than Mr. Futterer. I have also heard that the Howey and the Inspiration holdings have been liquidated, only time will tell which crowd made the best play.
>
> With the market on National, this morning, of .275-.30 cents, what should a prospector do? I well recall your advice of the past and I doubt if you have changed your opinion but I would be glad of your present reaction to the deal or your advice, if you care to give advice, on stock market questions. As you know I have most of my National still in my box but should I turn it into cash, what do I do with cash in times like the present? Why not bet on someone making a new gold mine? If gold bricks are no good I'll have to go back to the farm and hoe potatoes anyway; that is, if the farm is still there.
>
> Hope all goes well with you and that you are making the Seventh Malartic.

In all of J.P. Norrie's dealings a handshake, a spoken commitment or a promise was all that was needed to progress with the all important search for gold as quickly as possible. The formalities could come later. This was the case with the Perron and it obviously became a problem later. Although there were many legal agreements that he insisted upon, particularly to protect the interests of the prospector, as far as J.P.'s own financial protection was concerned he didn't waste a lot of time with lawyers.

Now, as the National Malartic was proving to be a mine and Pooler & Co. had provided some financing, and Noranda and Hollinger were interested, Gus Rainville, the president, was slow to fulfill his original promise of bonus shares, and a squabble eventually ensued.

J.P. wrote:

I do not believe you have given full consideration to the facts of the situation when you say that $4,000 is a fair settlement in full for my services.

I will enumerate some of the most important items. In the first place I was successful in closing a deal for you with Lee Coffin. I do not believe any other person in the world could have done so at that time... I spent a lot of time on this at considerable expenses as Coffin was located in the bush and had to be found on several occasions to sign papers etc., for this you promised me an option on a block of National Malartic stock which I never received.

After work on East Malartic got under way and the ore was found in the contact zone I used this argument with Inspiration and Howey to get money to drill the property. True we did not find ore in the contact at National, however, you promised me an option on 25,000 shares of stock at the same price that Inspiration and Howey took it down, which I did not get. You did offer me an option at .07 not at .05 which the interested companies paid.

From close contact with East Malartic developments I concluded that ore might be found north of the contact on National. After getting my idea in shape with several reports, I succeeded in interesting Pooler Co. whose money really pointed the way to a mine and later installed sufficient confidence into Quebec Gold whereby they joined Pooler in the financing. For the latter you verbally promised me shares which I did not get.

Through my influence you have always had full access to the records of East Malartic and Sladen and the ideas of their men all of which was and still is invaluable to National Malartic.

In spite of all the outside reports you have had on the property which varied all the way from a complete turn down to the mine having millions of tons of $7 per ton ore or better, my advice and forecasts on the property have been 100% correct at all times.

I am very sure that you would have to pay any consulting engineer, whether experienced in the area or not, a fee of at least $300 per month. My local experience therefore has been worth more to you. Anyway it has resulted in finding a valuable mine for you.

You have spent a lot of money throughout the north trying to locate a mine. They are as you know very rare. This National situation chiefly through my efforts has just fallen into your lap. In view of this there should be no hesitancy on your part of recompensing me well for what I have done. I honestly believe that from the consulting angle alone, aside from all other services rendered, a fee of $300 per month is more than reasonable. I am naturally anxious to continue my connection with National Malartic; however, it must be on a mutually fair basis. After having considered the points I have outlined which no doubt you have overlooked I feel sure that your own spirit of fairness will bring us together on such a basis.

I believe that a decision to sink on National should soon be arrived at. Of course, the overburden can be penetrated anytime, however, this could be better done not later than March when during the first part of the month the frost is heavy enough to help hold up the walls and during the latter part not severe enough to damage the concrete collar.

Our men have done an excellent job on the Pascalis shaft. It should be complete to 1,550 feet by Mar. 15 when we would like to move them to National.

I am sure they will do equally well on National Malartic. The price we have quoted you on the National sinking I believe is fair. We only expect to get a small part of our investment returned from profits on the job.

J.P. Norrie

Rainville and National Malartic Board of Directors eventually sent J.P. a settlement for his services on the National mine development. Even though it was not what Rainville had promised, he gave Rainville his report on the Noranda, Hollinger offer, for his annual report:

> Your letter has been received with enclosed copy of the offer of Noranda Mines Ltd. and Hollinger Consolidated Gold Mines covering National Malartic Gold Mines financing.
>
> I do not believe you can secure a better deal for National financing at this time. You will be provided with $148,500 to continue with immediate development and later enough money to build a good sized mill if the same is warranted.

The National Malartic was one of many mines that would be affected by the Second World War. Although required for balance of payment reasons, gold was not considered a strategic metal, which meant machinery parts and supplies would be difficult to acquire even for producing mines. The labour supply was also of poor quality and good men were snapped up by the producers.

If National Malartic had been developed on its own merits, it probably would have closed down by 1940. However, Noranda could spend money on development as the production would be required to supply their operating mills.

As the war reached 1942, the War Regulations Board became more difficult to deal with and development of National closed down. The stock value dropped to almost nothing, as well. J.P. wrote a prospective investor, A. Colwell, in Saint John, New Brunswick, with his advice on National, May, 1942:

> Another gold prospect in which we have had a hand is National Malartic. This also is closing down and if you can secure a few shares say around .03 or .04 per share, money should be made on them when the war is over.

After the war in 1948 the Sladen Malartic was running out of ore. As it was adjacent to the National Malartic, which was still not producing, the two mines were merged to form Barnat Mines Ltd. and the mine continued using National Malartic ore after the Sladen was completely depleted.

Partanen Malartic

John Partanen was one of the many prospectors whom J.P. had come to know very well. He was from Finland and a bachelor who had spent his life in the bush in the Canadian north. He staked his Malartic claims when J.P. was at Malartic Gold Mines in 1928. The majority of these claims became Malartic Goldfields while one western section became Partanen Malartic. J.P. wrote to Toronto stockbroker, Doug Wright :

> I have another nice showing over near Malartic which needs $5,000 for immediate development. To get quick money I would like to sell 15,000 Bussiers at the best price you can get. Please see what you can do for me in this and I will give you a cut in the prospect in question. It is known as the Partanen Malartic Goldfields show and was staked by him when he was watchman at Malartic and paid by me.

Although the main block of Partanen claims had been incorporated into Malartic Goldfields in 1935, this eastern block was incorporated into the Partanen Malartic then controlled by an American group.

Nothing important was found by 1938 and Partanen was living in Toronto having had a foot amputated. J.P. again was very concerned about the welfare of this friend and prospector and wrote this letter for Partanen, who was unable to write, asking for release of his vendor's shares:

> I am one of the vendors of Partanen Malartic Gold Mines and would like at least 60,000 of my vendor's shares released.
>
> A recent Ontario law allows me one free share for each four treasury shares sold and as my pooling agreement signed by me has expired I feel entitled to the shares requested.
>
> I am a prospector without funds to continue work in my line and respectfully request that you authorize the release to me of 60,000 vendors' shares.
>
> John Partanen

In the spring of 1939 an American group was considering a diamond drilling program and J.P. asked them for the contract: he also had in mind that if the opportunity arose he could put together a deal. This was always his motive, particularly when his friends were also involved:

> I understand you are shortly going to do some drilling on the Partanen Malartic property. If so, we would very much like the opportunity of doing the work for you. We have done very much drilling in this area and are well prepared to do good work for you.
>
> It might be well to mention that the writer carried this property for many years during the Depression and with Partanen is one of the vendors of the property.
>
> I mention this to lay stress on the fact that no attempt will be made to secure advance information from drill cores; in fact, every effort will be made to guard this for you.

This was another prospect along the greenstone band, making it more interesting, geologically, as with all properties along the contact zones. J.P. advised the brokerage firm of Gordon Jones & Co. who were trying to arrange financing with the Americans:

> I have made very careful study of the Partanen data. Geological conditions on Partanen are identical with those on Malartic Goldfields and National Malartic and very similar to those on East and Rand Malartic.
>
> I do not believe the engineers and geologists so far on the property were familiar with Malartic conditions judging from the work done to date.
>
> I am very enthusiastic as to Partanen's future and will be glad to take charge of future work.

Jones notified J.P. they were doing their best to get the project moving:

I note with interest what you say about the diorite zone in East Malartic and the possibilities on Partanen. I can assure you there is no question about Mr. Lucas and the writer carrying out their part of the arrangement with Partanen just as soon as we are able to get the parties together, and the writer will not miss any opportunity to push this deal along just as quickly as possible. I would like to see the deal completed this summer and be in a position to make plans for the future.

(Gordon Jones & Co.)

Glen Wilton (J.P.'s personal lawyer) advised J.P. who the company vendors were:

According to information supplied to me there are 700,000 vendors' shares of which Mr. Lucas proposed that 20% or 140,000 shares should be made available as a bonus for the financiers in proportion as they take up the options. The holders of the vendors' shares are said to be as follows:

Morris Rothenberger	350,000
John Partanen	200,000
J.M. Harper Jr.	75,000
E.J. Thomson	19,000
Denny Estate	14,000
Dennis Denny	42,000
Total	700,000

The negotiation process involved with the many partners of a mining deal were often slow and discouraging. It was now 1941, well into the war, and at this stage of development it was impossible to develop a property unless essential to feed an existing gold-mining mill. Even then it would be difficult because of ever-tighter war restrictions. The Partanen closed down to await the war's end.

Quemartic Gold Mine

Quemartic was another property just a short distance below the greenstone band. Although not as attractive, it still needed attention.

J.P. and his partners sent John Partanen in to open the property for G.P. Campbell, president of Quemartic, in the winter of 1937:

A campsite has been spotted on the property and a road cleared to the main road. This work is being done by John Partanen, the staker of Partanen Malartic. Construction of the camp is now under way.

I do not know Mr. Morden of whom you speak. However, I presume he is competent to take charge of the property.

I am enclosing a bill for $8.20 covering two camp stoves which the carpenters took along with them for heating purposes.

Campbell

Four years went by without event and then, whereas J.P. usually provoked the action, this time the action found him. Paul Michaud was a life insurance executive and a director. He contacted J.P. in Sept., 1941:

Not until recently did I discover that J.M.R. Corbet, you, and I are trustees regarding the pooling agreement when the original transaction was made re the above.

Corbet has closed his office and is now running a small machine shop, and I am quite confident Quemartic is the least of his worries. Meantime, the company is managed by a board of directors more or less dummies or appointees of Corbet's and have no interest in the company whatever.

I have checked up quite recently through the treasurer whom I know very well, and find ten claims are in good standing, and the treasury has $10,000 in govt. bonds, and $2,000 in the current account.

The last issue of the *Northern Miner* which, no doubt, you have read, shows that of a number of companies along the favourable belt, the Quemartic is the only one, to my knowledge, that has not tried to develop a mine. I do not know that it has any chance, mind you, but if anybody knows that, it would be you. I would naturally be keenly interested in having your opinions in this matter, because I feel that you and I have enough shares to clean up the present board, and appoint our own slate.

I hear you are in Toronto from time to time. Although I have spoken to you on the telephone, I have never had the pleasure of meeting you. If you prefer, I would deem it a privilege if you would let me know the next time you are in Toronto; we could talk this over.

Michaud

Of course J.P. agreed with Michaud's optimism:

I am enclosing a small sketch showing the location of the Quemartic property with reference to a fold in the main Malartic formation. The Sladen and Canadian Malartic properties are on a subsidiary structure, leaving the main greenstone contact as shown. There might be a similar break leaving the greenstone just west of the fold, which I have shown in dotted pencil lines.

A promising gold showing was found about one mile south of the Quemartic property without geological reason so far, so that anywhere an ore body could be found in the immediate area.

I believe 5,000 feet of diamond drilling could be done on the property with a fair chance for success. Stock could be taken down from the present treasury with the understanding that a new company would later be formed giving prospective financiers an option on the new treasury.

The Quemartic, from what we know now, has less chance of developing ore than those on the greenstone band. Still, I think the chances are worth while.

This property was obviously not as interesting to J.P. as most properties. No friends were involved, it wasn't on the greenstone and of course, the war. The only positive was the distance from East Malartic which was the centre of that theoretical circle.

Often investors or shareholders would write J.P. for his advice. People he didn't know at all. Mrs. Fortis of Hamilton, Ont. was one.

J.P. replied:

The report I made on Quemartic mines was previous to the formation of the company and before work was done on the claims.

I am not acquainted with the results of the work. The claims of the company are well located in the heart of the Malartic district. I believe that a systematic drilling campaign might uncover something of value.

Quemartic went unnoticed for obvious reasons until the end of the war.

Lacoma Gold Mines

The Lacoma Gold Mine was a gold property that was owned by other interests. J.P. had watched the progress because of its location along the Malartic break. Work had now ceased and J.P. thought Inspiration could gain control of the company and develop the property. He told Harry Carmichael, his friend and director of Inspiration, his ideas in 1938:

I believe Lacoma is a bet, however, little underground work has been done and that poorly. Continuity of good ore sections has not been proven. I would say that the success of the property depends on that as they really have patches of good ore.

J.P. had persuaded Pooler & Co. to take on the National Malartic project and it proved to be a small mine. Eventually it became good enough to be optioned by Noranda and Hollinger. Lacoma was a very similar opportunity and Inspiration had acquired a large interest so he offered Pooler his geological opinion in 1939:

Enclosed please find the reports of L.P. Worriner, geologist at Perron and H.E. Silver, geologist of Inspiration and J.J. Harris.

The only man yet of any consequence on the property was a young geologist by the name of J.J. Harris, who pretty well described conditions there but was unable to name what he described.

The property was turned down by Bulmot and Wells, two British Columbia engineers sent in there by Mr. Stobie. These men knew nothing of local conditions.

I kept in touch with Lacoma developments while underground work was being done and always liked the property.

Very recently we found that all Lacoma ore is confined to porphyry intruding a diorite which itself intruded greenstone. A very favourable condition, and working on this idea I am sure much ore can be quickly developed. With approx. 600 feet of underground work, 7,500 tons of $12 per ton ore has been developed by a hit and miss system and by men who did not understand the conditions.

I believe $40,000 should be raised at once.

I am sure the Lacoma is a good bet and with reliable management a producing mine can shortly be made of it.

J.P. Norrie

J.P. then laid out suggested financing for the Lacoma. He had a standard plan of financing for mines of this size:

My opinion is that the approx. 1,200,000 of vendors' and treasury shares available at .03 per share should be taken up, as I feel sure that the present market can be boosted to .06 or .07 without any support other than the reports which can be put out and the news of the property's activities.

When Stobie handled this property last summer, I am sure he put little behind the market. The stock went up to .18 from good progress reports. In this property we have something on which to make reports.

The plant is practically ready to go and I am sure the market would go to .07 on good reports alone.

The deal would call for a reorganization when this money is spent on a basis of 1,000,000 shares to the present holders, 2,000,000 shares be in the treasury on which an option would be given at .15 for the first million and .35 for the second on a sliding scale if desired.

I am sure Lacoma is a good bet and the way I have outlined the procedure is best in the meantime. However, if you do not think so and want to go ahead by increasing the capital of the Que. Co., I am with you 100%; however, please give my ideas further consideration.

Just as financing began to materialize, the war began to take its toll in 1940. J.P. explained this to his friend, Brad Streit:

> Your letter re Lacoma has been received. It was my intention to reorganize Lacoma Company when we ran into the big European push in the spring and everybody became jittery. I believe further work there should be left until next spring when I believe money raising conditions will be easier and the underground work better coordinated with surface conditions.

After working on a deal for Lacoma for three years the financing never succeeded and J.P. recommended what Inspiration should do:

> I do not believe we will ever go further with Lacoma so I can see no reason why Rod Kennedy and yourself should stay on the board unless it is to protect the interests of ourselves and associates.
>
> By the time money can be raised for the Lacoma operation most of the present holdings such as plant etc. will have been dissipated. I believe by that time we could get the property very cheaply perhaps for restaking if we were so inclined.

So another mine was put on hold until after the war.

Note: J.P. often took on what many people thought would be hopeless mines, but as one of the Norrie family's favourite poems illustrates, attitude and hard work sometimes can turn grains of sand into pearls. Virginia still recites this family poem at 85 years of age.

The Oyster

There once was an oyster
whose story I tell
who found that some sand
had got under his shell.
Just one little grain,
but it gave him some pain,
for oysters have feeling,
for all they're so plain.
Now did he berate
the working of fate
did he curse out the government
call for election?
No. He said to himself
as he lay on a shelf
since I cannot remove it
I'll try to improve it.
The years rolled around
as the years always do
till he came to his ultimate destiny
stew!
And that one grain of sand
that had bothered him so
was a beautiful pearl
all richly aglow
the tale has a moral
for isn't it grand
what an oyster can do
with a morsel of sand

Chapter 21

The Granodiorites

by George Norrie

Beaufor Gold Mines

The Beaufor gold mine adjoined the south boundary of the Perron mine. Some of the richest gold veins worked on the Perron mine led directly over onto the Beaufor property. (Some of the Perron veins also led across the mine boundary towards the Pascalis gold mines to the east.)

The Beaufor mine had developed in 1931 to the point where it looked like a very promising gold mine. Clarence Stee, manager of Siscoe and friend of J.P., wired him:

> Siscoe drill on main land have cut 16 veins varying from one foot to five feet in the first hole and five or six veins in no. 2 hole one of which is blue quartz. Should have quite an excitement in the near future. This information is very reliable although the assays are not made yet. I do not know anything about values.
>
> Beaufor Gold Mines received an offer of seven hundred and fifty thousand dollars for half interest in the Beaufor from Noranda.

The Beaufor, however, did not prove to be a producing gold mine under Noranda development. The control of the property then was taken over by Pierre Beauchemin, of Sullivan fame. Thayer Lindsley of Ventures and Jules Timmins also became large shareholders. But by 1935 the property still had not proved a viable gold producer on its own merits.

As J.P. realized that control of the Beaufor was essential to the development of the Perron, he quickly began to contact all the principal

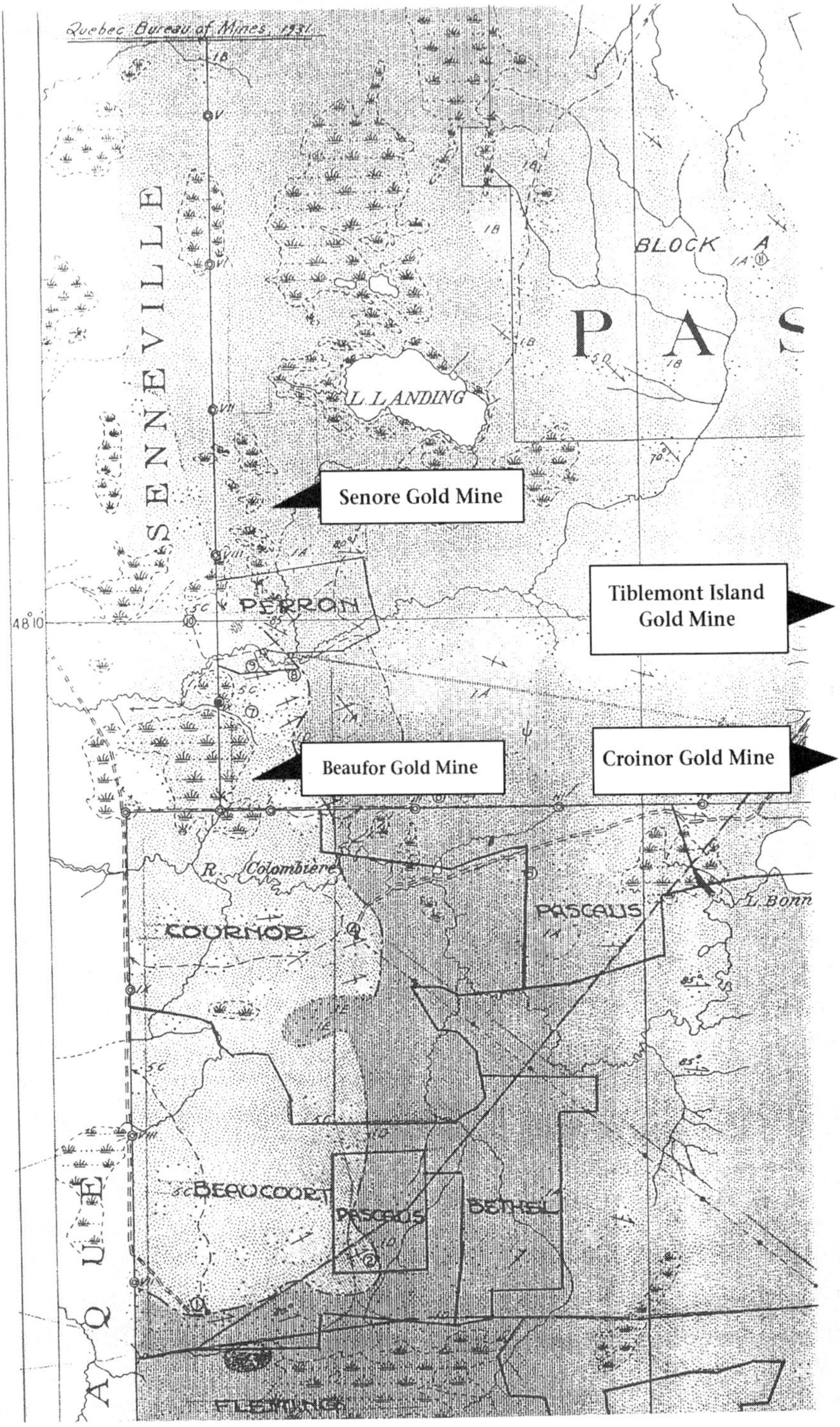

Map showing the location of the Perron mine and associated gold mines, including Cournor, Pascalis, Beaufor, and Senore.

investors and discovered that Pierre Beauchemin, Thayer Lindsley and Timmins were all eager to cooperate with the Thomson group, who would take the lead in the Beaufor and develop both Beaufor and Perron in a manner profitable to all.

J.P. outlined to Mel Thomson how the financing control could be arranged so that the rewards would be sufficient to warrant the risk: the Beaufor principal shareholders were willing to put a portion of their holdings back into the treasury of the company so that Thomson could manage the financing profitably.

The development of the Beaufor was not entirely clear to the Beaufor shareholders, but it was essential for the Perron and would be more profitable for both to be operated as one.

Quick action was essential for the Perron, and J.P. made sure all development information was secured and that there would be no leakage of information to the public. No visitors were allowed underground until the financial dealings were in place.

Almost immediately after the agreements were signed, information leaked out about the mine's future. Overzealous investors talked to the press about how they saw the mine's future development. When this leakage appeared in the press, Nesbitt Thomson were very upset. The new partners did not realize just how important this was to Nesbitt Thomson. The security of information from any investment that was controlled by them was essential. Their own investment and the value of their clients' shares depended on accurate information about their mines. Nesbitt Thomson remedied this situation in no uncertain terms.

A further problem arose from the news article when it was reported that J.P. would be assistant engineer to Andrew Waltz. This upset J.P. He had spent many years preparing for the operation of mines like Perron and Beaufor and demanded all the credit he deserved. This was also remedied by Nesbitt Thomson immediately.

J.P. was in complete charge of the development of the Perron and after one year it proved one of the top junior gold mines in Canada. When the control of the Beaufor was assumed by Nesbitt Thomson, J.P. assumed complete charge of the Beaufor as well, and the new board agreed that Beaufor should have their own engineer under the direction of J.P.

Pierre Beauchemin was the largest shareholder of the old Beaufor and also contributed his advice on the development of the new mine:

I went over the Beaufor with Mr. Beauchemin. He wanted some work done on the south claims.

I have carefully studied the south claims and think there is an excellent chance of finding something on them as the geology is right. I have five men now working on the ground. The man in charge can look after Beaufor interests in general.

Underground Mr. Beauchemin wanted at least two machines put to work. One of these is driving the south crosscut on the second level out to the new shaft.

We have also found a vein in the 323 north crosscut. We have a machine working on this level on Beaufor.

I believe work should be authorized on all levels of Beaufor and we will get at it when the opportunity offers. In any case I believe two machines can be kept continually busy on good ore development.

J.P. Norrie

With J.P. in charge there seemed to be harmony among all concerned, which was not always the case when mines were operated on a shared basis.

Beauchemin wrote to J.P.:

While at the meeting of Beaufor yesterday, I was very pleased to read the letter you addressed to Mr. M.A. Thomson following our recent interview. I am pleased to see that you have already started the surface work according to the resolution of our board.

As a result of our discussion we feel that we should have a young engineer in our employ, working under your supervision, to check the work you are doing on the Beaufor property and the charges we are assuming. We discussed a young man whose name is Mr. Duval. The board felt that Mr. Duval should make a special inventory of all the company's assets, including material equipment, goods and stores etc. and forward same to the company as soon as possible.

As to the underground work, I would like if possible that you operate one more machine on the west side of #4 second level, also, if possible, to crosscut from the east face of vein #4 on the second level going north in the direction of the good intersection of the diamond drill hole #2.

Another good place to start would be to continue the drifting in the northwest face of Mathews vein second level. The board would appreciate any suggestions you may have as to the development on the other different levels already opened.

We wish you all the luck we feel you deserve for your work on Perron and I will be at the mine by the end of next week. We will then discuss any details about the points mentioned in this letter.

Since the days of the Bussiers J.P. had been used to the idea of splitting his mining holdings. He had done this himself with his Bussiers shares and the shares of Long Lac Gold Mines owned by Ventures. He had lost everything he owned by investing all his wealth in Malartic Gold Mines, which closed down. During the Depression all investors in northern Quebec lost on mining ventures and later in the 1930s it was still a common occurrence. After the Beaufor had been operating for one year, Pierre Beauchemin asked J.P. to consider an exchange of shares. J.P. replied:

I have thought over the question of exchanging shares and would be willing to exchange around 30,000 shares of Beaufor on a basis of three Beaufor for two East Malartic. Let me know if you are interested and I will forward the certificates to you.

The Perron and Beaufor developed together harmoniously for the next two years and the profit was very satisfactory for all concerned. In 1937 J.P. gave a detailed engineering report on the mine operations:

A three compartment shaft has been sunk and a level established at 625 feet. A crosscut from this shaft has been driven north to the Perron south boundary a distance of 1,000 feet to explore the north ore zone. In this crosscut favourable formation was encountered in the 500 feet adjacent to the Perron boundary.

In the area between the surface and the 625 foot level two important veins have been developed in the old Beaufor workings. The Mathews vein was followed in an incline shaft to a depth of 150 feet and exposed for 200 feet by drifting. This vein assays approximately $9 per ton over an average width of 3 feet. This work with original surface drilling has pretty well exposed the ground in the north zone between the surface and the 325 foot level.

The area between the 325 foot level and the present 625 foot level has been explored with seven diamond drill holes, total footage 3,197, two of which were drilled from surface on incline and five direct from the Beaufor workings.

On the 625 foot level a flat hole is being drilled to the south zone now at 995 feet. One narrow vein section of good grade has been cut.

Only minor problems ever arose with the Beaufor and Perron management. This was usually concerning who was in charge and was always the decision of the board of directors:

If you will remember at the last Beaufor directors' meeting the question of an engineer for Beaufor was left to me. So I have left R. Halet on the job. He will be back at the mine in a few days. A good geologist is acting in his place. The drilling on Beaufor is very difficult due to boulders, however the first hole is well under way...

Ore indications on south Beaufor claims are within 2,100 feet of the Cournor [the new name for the old Bussiers] shaft and could best be developed by crosscutting from this shaft.

A magnetic survey by Koulomzine has indicated possible ore in the area between the #4 shaft and the south boundary of the property. Diamond drilling of the most important of these indications is now under way.

The Beaufor was bounded on the south by the old Bussiers mine, which J.P. had opened up in 1931. The Bussiers was now in operation again. The new mine was called the Cournor and was owned by the same people, Quebec Gold Mining Co. with Gus Rainville as president. The Beaufor south claims, mentioned by Pierre Beauchemin, had developed new ore bodies but could not be mined economically by the Beaufor on their own. The Cournor shaft was close to the south boundary, however, and J.P. again used his ability to bring partners together. Frank Buckland was now mine manager of Cournor and J.P. reported the progress of the combined venture of the two mines to Rainville:

> I went over the entire Beaufor mine with Frank a few days ago. Frank is doing an excellent job there. His experience in mining small ore bodies is being used to good advantage. Due to the ore being in small lenses all over the place his costs although very good are found to be high.
>
> In the ore body on Perron we have recently found some very spectacular ore. This ore body extends to Beaufor to the extent of about 2,000 tons. The 501 vein, now the 508, is very good, above the 500 foot level about .65 oz gold per ton. Below it is more broken up into smaller lenses with a grade of better than .2 oz gold per ton.
>
> In the drilling just near the north boundary of Beaufor Frank has some good ore sections. My opinion is that the Beaufor will produce many small stopes of good grade some of which have not been uncovered yet. This condition coupled with transportation costs leads to high total costs; as I said before Frank is doing an excellent job and will keep costs as low as possible.

The story of the Beaufor was a story of cooperation. This was obviously due to the confidence all mining and financing men had in J.P.'s ability and knowledge. Alone the Beaufor would never have developed and would have been another casualty of the war which was just around the corner.

Cournor Gold Mine

It is not clear whether Gus Rainville came to J.P. to open up the old Bussiers or not. Whatever happened, the old Bussiers opened up again in 1937 under the new name Cournor.

Quebec Gold Mining Company had certainly come to the realization that J.P. knew what he was doing when it came to all aspects of mining: exploration, diamond drilling, development, sinking shafts, underground development and milling the ore. They had made a costly mistake in allowing him to leave Bussiers in 1934, now they wanted him to direct the newly reorganized Cournor mine.

Frank Buckland was a professional geologist and prospector but J.P. put Frank in charge of the underground operation of the mine. J.P. had confidence that this would work okay with his supervision.

The Bussiers had lain idle for so long that the equipment was either damaged or out of date. Ingersoll Rand discussed the equipment requirements:

Office Memo
To: J.W. De Gruchy, Noranda
From: Wm. Carter, Montreal

The Bussiers Mining Co. are going to re-open under the name of Cournor Mines Ltd. F.C. Buckland is going to be manager. This is being financed by a group of friends of mine, Gus Rainville of the Quebec Gold Mining Co., Jules Timmins and some others. I have been working with Frank Buckland for several days on the revamping of their whole plant. They are going to do away with the steam driven boilers and two steam driven compressors and retain their motor driven compressor. There is quite a changeover in the proposition and I have not all the details worked out yet. Jim Norrie is being retained as consulting engineer and I would not like you to approach Jim until we get our picture all lined up with Buckland.

The purpose of this letter is to tell you that Frank McLellan, mill designer, who designed the Perron mill is going to redesign the Bussiers mill. Among other things, he is going to be interested in a secondary crusher. He is talking about a Newhouse Allis Chalmers type. I am wondering if you could not write him and get his inquiry and see if we cannot quote on a Telsmith Gyrasphere. Let's deal only for the minute with this crusher matter...

After only six months Frank Buckland had developed considerable ore:

Buckland has located ten new ore sections. He has worked out a new zonal theory and now thinks the ore is confined to two main zones similar to Perron in some ways. During the months of April and May at Perron we have been taking out grade widths at expense of grade as an experiment to see how far we could go. In May our heads ran less than $7 per ton with a profit of $20,000. I believe Cournor can equal or better this. It will take a few months to impregnate mill solutions. When this becomes complete I believe a good profit can be made.

J.P. Norrie

Rainville was always aware of J.P.'s influence on Nesbitt Thomson and wanted to sell some stock:

Mr. A.J. Davis did want some stock. I believe he is now pretty well tied up in the market. I will discuss the matter with him.

J.P. Norrie

Rainville also wanted to use J.P.'s name in the Cournor mine publicity, but J.P. had been reprimanded for this by Mel Thomson before and didn't want to upset them again:

Unless it is absolutely necessary I wish you would not publish this as Nesbitt Thomson do not want my name published in connection with outside ventures. Mr. Nesbitt raised hell with me once before about this in connection with another property.

J.P. Norrie

The first gold brick was an important publicity event for the mine. Rainville to J.P.:

…I was thinking of organizing a trip about the 15 of Sept. to the Cournor and inviting the minister of mines, the member of parliament for the district and our local directors as well as the interested parties which would include Messrs Murdock, Timmins, and Kendall. Do you think we could have a good sized gold brick by Sept. 15? I would hate like the devil to have a party up there and be disappointed in our mill recovery.

Near the end of 1937 Rainville advised J.P. they had spent about $100,000 on the start-up of the mine. He wanted to sell treasury stock to help pay for this expense and wanted some of the principal shareholders to contribute to this effort:

As you know Cournor Mining Co. is in debt to the banks and Quebec Mining Co. for over $100,000 owing to the fact that operations have exceeded the estimates to this extent.

I am endeavouring to make a deal with a Toronto stock exchange house to dispose of a block of Cournor stock. As I need some cheap stock to give them to sweeten the treasury stock which I am endeavouring to dispose of, I have asked the syndicate to give me an option on 15% of the stock they own at their cost price which is eleven cents per share.

I am now asking you if you would care to join in and help out by giving me an option for three months on 15% of your stock at .11 per share. There is to be no profit for me in the deal.

J.P. Norrie

J.P., who had considerable stock in the Cournor venture himself, agreed to contribute 15% of his own stock to help with the financing. He also explained to Rainville exactly why he felt that the mine would eventually come through economically. He used the old Bussiers production as an example:

We started the old Treadwell Yukon mill in Oct. 1932 with only four crosscuts and no ore developed and gradually building up the mine to a production of 42,759 tons of .1804 oz per ton in 1933. Eight months of the year production was close to 800 oz. per month with a peak of 913.55 oz.

Following this Quebec Gold took over the management and with a new program of exploration no new ore was found of any consequence and to keep the mill going much barren material was treated which reduced the average to .13 oz. per ton.

When Quebec Gold took over the proposition, the mine was in good shape from an ore angle and had the Treadwell continued I am sure the ore picture would have become better each month as you will notice continued progress during the year 1933.

Anyway I believe the 1933 performance can be duplicated whenever Frank can find one or two stopes to sweeten up the grade. The Perron ore picture is exactly parallel to the Cournor as if we did not have the 215 stope. At the present time, the production would be on a par with Cournor.

I feel certain that much higher-grade ore still remains on Cournor and that Frank will bump into it in the ordinary course of events. In fact I believe the old records of 913 oz. per month could be bettered as this represents only 75% recovery from amalgamation only.

I believe that with a production of 750 oz. per month some money can be made. However, you must realize that this is a new operation just under way and many small additions have to be made along with corrections in mining and milling which all runs into money. It is unfortunate that some of the ore which Frank developed did not turn out to be high grade as expected, however, I am sure it will come and that the proposition will gradually make some money.

I know Frank is doing a good job and if anyone in the world can make a go of it he can and I feel he will. I will go over your letter with Frank in a few days and will write you further.

J.P. Norrie

At this time it seemed clear that the mine could only reach a profitable level by sinking the shaft further. This would require a further input of cash. Rainville revised his previous financing plan to account for this expenditure:

I believe we are convinced that if we sink the shaft to the 900 foot level we could open up sufficient ore bodies which would permit us to double our production.

Frank informs me that his costs would be considerably lower if tonnage were doubled and that we would make a profit on a large tonnage.

To accomplish this, we would have to dispose of our treasury stock which amounts to 333,000 shares at .60 per share. This would provide sufficient monies to pay off our loans and carry out the program of enlargement which would include all the development work.

Of course, .60 is a ridiculously high price, but I believe we could sweeten this up with 1,000,000 shares at .15 per share. This would bring the cost down to about .265 per share for the whole 1,333,000 shares. I do not see why this stock would not be worth .50 with a mill of 350 tons.

Rainville

In September of 1938 the Beaufor south claims were found to contain small ore bodies and veins. The only way Beaufor could profitably mine these was to use the Cournor operation. This fitted into Cournor's operation perfectly and another amalgamation was obvious. J.P. wrote to Rainville:

> You have probably heard from Mr. Thomson that everyone is now in favour of an amalgamation of Cournor, Beaufor and part of Pascalis.
>
> Beauchemin now seems to be the big aggressor in this connection, due to the fact that he thinks a substantial profit is now being made by Cournor and he will have an outlet for the ore now developed in the Beaufor mine.
>
> I think until a deal is completed that Cournor should be made as good as possible with lots of publicity.
>
> In this connection I feel that I have played considerable part and if an amalgamation is consummated feel that my present holdings of Cournor stock should not be diminished.

By the summer of 1939, a little over two years after the Cournor started up again, the mine was making a considerable profit. J.P. suggested that Quebec Gold provide for the financing of the exploration of the remainder of the property now that the profitability of the Cournor was evident:

> On the Cournor Frank has a new vein of high grade in the contact near the north boundary. This seems to be of a new type and may be similar to the veins indicated in drilling on the south Beaufor.
>
> As conditions now stand, the big immediate future of Cournor seems to be in the favourable development of the south Beaufor ground. Frank should know much about this in about five months.
>
> I think much money will be necessary to explore all the remaining Cournor Beaufor contact where I feel as always that important ore will be found.

I do not believe that profits from known ore except the south Beaufor will provide sufficient money for the necessary development. Therefore I believe that while Frank has his production up now to over 50 oz per day an effort should be made to place a further $200,000 in the treasury. With the treasury in such shape Frank would be able to have a look at most of the undeveloped portions of the property.

J.P. Norrie

After all the work had been done and the mine was making money as J.P. had predicted, J.P. then had to go after Rainville for his pay and the shares promised back when the mine started up. J.P. had waited until the venture could afford these promises and now Rainville was slow in honouring them as with the National Malartic:

You wrote me outlining the indebtedness of Cournor Mining to me.

According to your letter the contents of which we agree upon there is a balance of $2,100 owing to me which you promised to pay shortly after Jan. 1, 1940. Therefore I would be glad to have a Cournor cheque to cover the balance owing to me.

I have part payment on the amount of $2,100 owing to me by Cournor Mining Co.

The number of shares allotted to me for my services in connection with Cournor was 19,000 shares instead of 30,000 as stated in your letter.

It is quite true that I was willing to accept 19,000 shares, however, against my wishes at the time it was changed to $5,700 cash, therefore I can see no reason why it should be changed at this time.

If you will look back over the Cournor operations you will remember that I was instrumental in getting Frank Buckland for you, the redesign of the mill and continual help right up to the time when I made the settlement with you. Be fair. Do you think you could have secured anyone else for twice the amount you have agreed to pay me?

Anyway I would like to take advantage of the present low stock prices and would surely appreciate receiving the balance owing as soon as possible.

The Cournor operated for a short while longer but the wartime conditions began to affect the supplies, labour quality and financing of the mine. It then closed down until the end of the war when conditions would allow for a marginal but profitable mining operation.

Pascalis Gold Mine

The Pascalis Gold Mine was owned by Ventures Ltd. Thayer Lindsley, the old friend of J.P., was president and made all important decisions. The Pascalis mine had never developed to the point where it could sustain a profitable production on its own, even though a shaft had been sunk as early as 1931. Lindsley had great confidence in J.P.'s ability and he showed this without doubt, as he had invested in all J.P.'s own companies: Inspiration, Perron and East Malartic. The fact that J.P. would direct the operations of both mines was another welcome attraction for Pascalis. So when the Perron approached them about an amalgamation, the proposition was definitely welcomed.

Geologically the Pascalis was along the contact of the granodiorite batholith, as were Perron, Cournor and Siscoe. This indicated very likely ground for gold discovery.

In 1937 an option to purchase control of Pascalis was signed.

E.V. Neelands was Thayer Lindsley's right hand man and carried out all day to day operations for Ventures:

The Pascalis interests consider it advisable that they should have their own resident engineer at Pascalis so that they may receive their own reports on Pascalis developments. I assume that you have no objection to this.

If it is acceptable to you we would like to send George Mustard to the property for the above purposes and we would be much obliged if you could make arrangements to board him at the Perron.

Neelands

J.P. advised his friend and director of Inspiration, Harry Carmichael, about what he found at Pascalis:

Developments on the 600 foot level of Pascalis are somewhat below our expectations. We have, however, over 200 feet of ore assaying $9 per ton over six foot width with some other extensions of Perron veins.

The future of Pascalis lies at depth where recent drilling has bettered the picture, and I feel that someday the Pascalis property will provide much profitable ore. I do not think enough has yet been actually developed to warrant completion of the Perron option to take over Pascalis Mine. However, it will be extended and eventually a deal favourable to Pascalis will be completed.

J.P. advised Neelands of the Perron decision not to take over Pascalis operations and Neelands was obviously not pleased:

I have your letter in which you inform me that some of the Perron directors were adverse to the idea of Pascalis developing from the Perron bottom levels.

I think it would be well for us to think this question over seriously with respect to the future because it is probable that, if no arrangement for working together can be made, Pascalis will decide to sink its own shaft with the result that the final position would be less favourable to both companies.

With respect to treatment of Pascalis ore by Perron, I shall be glad to know what you think could be done about this.

This was January, 1940, and the decision by the Perron left the Pascalis to develop their mine on their own just when the war would begin to affect all mines at just this stage of development. J.P. talked about the future of Pascalis with Carmichael:

I received a letter yesterday from Pascalis people stating that they were going to start shaft sinking at once. They asked me to supervise the work which I am prepared to do.

The work is to be financed by issuing treasury shares to Ventures and Anglo Huronian (Dorfman) at .30 per share.

Deep development work on Perron has not added much to the Pascalis picture, our best vein no. 32

will go over on Pascalis at one place, however, its tonnage on Pascalis will be small. At least four other stopes on the levels on and above 625 are cut off by the Pascalis.

I think considerable ore exists on the Pascalis property. It should be worked through the Perron shaft as I do not think much money can be made through their own shaft. There should be a profitable stock when work starts on the property. If you can get out at cost or a small profit I would sell.

J.P. always showed his commitment to any project he was directing. He did this in no uncertain terms as he put his personal money into each project:

We are enclosing certificate representing 50,000 shares Pascalis Gold Mines Ltd. capital stock in your name. Would you be good enough to endorse this certificate as shown on the face thereof, have your signature witnessed, the date filled in and return to us to be placed in safekeeping.

J.P. was still very interested in Pascalis as he had just formed a new company called Miners Incorporated just for the purpose of shaft sinking. His first project would be the Pascalis shaft if possible.

Neelands wrote J.P.:

We have your letter of Feb. 12 with respect to Pascalis shaft sinking. I note you are prepared to supply full supervision of the jobs on a basis of $600 a month. Also, that, if we prefer to contract the work, you would be prepared to do this on a basis of $70.50 per foot - stations at the same price - provided equipment as outlined were supplied.

I shall discuss this with the Pascalis directors and advise you further.

I think, in any case, we would prefer to have you put in the plant and get the shaft under way, after which possibly the contract basis would work out best. What do you think about this?

J.P. advised Neelands at all stages of sinking the shaft:

April 18, 1940. At the Pascalis property all buildings have been completed. The transformers have been installed and will be wired up in two days. The power line has been completed. Power will be available when transformer hook-up is complete.

The compressor is being installed. It will be grouted today and will be ready for operation in about two days.

The shaft has been sunk 20 feet with a portable compressor. Bearing timbers are being placed. Following this, the collar sets and headframe; also blasting set which is on the property. The BC headframe and shaft timber is here. The headframe is being framed and should be completely erected in 10 days.

The sinking hoist has been installed. Sinking should proceed rapidly when the headframe is completed. All equipment is now on the property except ventilation fan and sheaves. Both should be here any day.

I believe $25,000 public liability insurance should be placed on the property as we are operating pretty close to the town. The Perron policy was placed with the Zurich General Accident and Insurance Co. The yearly premium would be approximately $240.

July 8, 1940. An advance of 153.8 feet was made in the Pascalis shaft during the month of June. The small footage was due to freezing of the air hoist, change over from air to electric hoist and the heavy flow of water at month end. When I last saw you I thought the water had been stopped. It was [stopped] in the first two holes, however, when we drilled more deep holes, we found the inflow of water to still be great so we were forced to proceed with cementing.

The broken zone was two feet thick of open cracks which took approximately 600 bags of cement to seal. A little water was also found in a small subsidiary fracture six feet below the main fracture zone. The latter took about six bags to cement.

We have now passed through the water zone with the shaft entirely dry. I believe we will make much better footage from now on.

July 13, 1940. We have now passed well below the water seam on the Pascalis shaft without encountering further water. The shaft is now below 400 feet.

A narrow barren quartz vein occurs just above the two foot fracture zone. It is similar to the occurrence on Perron which gave us so much water trouble and filled the first Perron shaft before the water could be handled. We were able to plug the holes before the water got out of hand.

The new electric hoist is working well and will help add to sinking speed. The Wisik Hoist was delivered to Martel with instructions to install it in its original condition.

Two important ore zones extend from the Beaufor to the Pascalis ground. One of these is a stringer zone on which the Beaufor have a fairly good stope. The stope is cut off by the boundary. The other is a quartz vein which was a narrow stringer on Beaufor. It widened out to four feet of good grade quartz just at the Beaufor-Pascalis boundary.

Sept.17, 1940. The Pascalis shaft was 775 feet deep yesterday and is being advanced at a good rate. Some difficulties securing good shaftmen last month have been overcome.

We are installing three unit heaters in the various buildings with steam from a small boiler belonging to Inspiration which will be used free of charge to Pascalis. Second-hand pipe is being used. We will soon have to consider the purchase of a cage. I believe one cage would do in the meantime. An order for this probably with Wisik Iron Works should soon be placed as deliveries are slowing up.

It might be well also to excavate a small sump on the 667 foot level before lateral work is commenced. Would you consider contracting the lateral work on

Pascalis? We would be prepared to do it at the same price that Pascalis previously paid Perron.

P.S. a cousin of my family (Alex Trueman) was killed in England last week while piloting a fighter plane.

Sept. 25, 1940. I believe the cost of underground work done by Perron for Pascalis was $17.55 per foot, plus the cost of rails. The power cost in this was approximately .60 per foot. We would be prepared to do the forthcoming underground work on Pascalis for $17 per foot and pay everything except power and rail costs.

(The shooting down of a large bomber with a Lewis machine gun by the Canadian mining unit was certainly good news.) The shaft is approximately 820 feet today.

Oct. 2, 1940. We have a price of $463 FOB Pascalis on a Wisik cage for Pascalis. The Perron cage would have to be altered and would cost Pascalis $300. Which do you think we should buy? I prefer the former.

The power cost in Perron drifting and crosscutting is .60 per foot. It would be a little higher at Pascalis as the peak would have to be absorbed by this work.

Assaying would be included in the contract figures I give you on Pascalis drifting and crosscutting. We are getting prices on two-stage electric pumps for the 654 and 1,067 foot levels of Pascalis.

I believe the minimum size should be a 100 gallon pump on the 654 and a 75 gallon pump on the 1,067. If you think larger pumps should be installed please let me know. The sump on the 654 should be made large enough so that pumping can be done alternately and thus keep down the power peak. We use 125 gallon pump at Perron at a cost of $1,975 FOB Pascalis. We favour Fairbanks-Morse pumps.

We came very close to a grave accident in the Pascalis shaft a few days ago. The bucket is supported by two chains which are attached to a lug rivetted to the bucket. When a full bucket was just passing the surface one lug broke. The bucket tilted, broke off the other lug and fell down the shaft. On the way down it took out a few dividers, pulverized two sets of guides and landed upright on the other bucket at the bottom. The shaft men had the bottom bucket on its side scooping muck into it when the other one fell cutting a two foot hole in its side. One man had a foot slightly bruised being the only injury. All dividers, rock, etc. hung up on the timbers and bulkhead.

The lug of 7/16 inch steel had been bent by heating with an acetylene torch. A torch should never be used on steel where human life depends upon it as by heating in this way the steel seems to crystallize badly which caused the break in this case. We have advised the bucket manufacturer of the defect in his product and the Dept. of Mines of the near-serious accident.

Nov. 6, 1940. At Perron we are spacing our levels at 125 foot intervals below the 1,125 foot level.

Thus instead of the two deeper levels on Pascalis being at 1,267 and 1,467 feet they would be at 1,292 and 1,542 to correspond with Perron levels.

I have been in touch this morning with the Can. Wire and Cable Co. They can supply three quarter inch cable.

The extra depth should be beneficial to Pascalis in that more of the favourable area can be developed, more particularly the downward extension of the 32 vein which enters Pascalis ground at approximately 1,150 feet.

The 32 vein on the Perron 1,125 foot level is standing up about as well as the levels above.

The drift on it is now about 100 feet long of better than .3 oz per ton grade. We expect to be in it on the next two levels.

There should not be much water below the 1,067 foot level. It can probably be handled by bailing a few buckets each day. A bailer can be made from one of the sinking buckets.

J.P. Norrie

Once the shaft was sunk through Miners Inc., J.P. continued to supply the underground development. But in January of 1942, the war regulations were causing supply delays etc. and Pascalis decided to manage their own mine.

They hired Frank Buckland as recommended by J.P:

My directors have been giving quite a lot of thought to the general Pascalis situation and, after full consideration, it has been decided to employ Frank Buckland to take charge of this operation.

Mr. Buckland has his own crew who are available to take over the Pascalis work just as soon as this can be done conveniently to yourself and, unless it is inconvenient to you, we would like to make the changeover as from say the end of the current month. However, if this notice should be considered too short, and if thereby you would be caused any serious inconvenience, please do not hesitate to advise us.

We wish to thank you for the excellent way in which you have carried out the Pascalis work, and our president, Mr. Thayer Lindsley, expressly instructs me to pass on his congratulations on the way in which this job has been handled.

Effective immediately, Mr. Buckland may be regarded as being employed to look after our interests at the property.

In May, 1942, most small mines and mines at early stages of development closed down. J.P. gave his assessment of these mines to an investor from Saint John, New Brunswick, A.M. Colwell:

Probably the best gold prospect that we are interested in at this time is Pascalis. This property will shortly be closed down due to the war. The shares are now around .10 but should go back to .04 or .05 after the shutdown, perhaps less. I would watch this stock closely and pick up a few shares around those prices.

I do not believe Canadian base metal stocks will make money. We are interested in the development of Calumet, which may make a little money during the war, however, in peace time I do not think it would be profitable so I do not advise you to buy it.

Senore Gold Mine

The Senore property had been staked for many years before J.P. was able to bring enough interest to the property for any substantial development.

This property was staked in the early 1930s by J.P. and Lee Coffin. The claims were originally part of the Coffin Mining Company along with East Malartic claims. J.P. and Coffin ended up with them after Fred Bradley died. Eventually the Senore claims became one of Inspiration's first acquisitions in 1934.

As the Perron, Beaufor and Pascalis demonstrated the potential along the granodiorite contact zone, the Senore, in 1937, became a property worth investigation. J.P. gave his report to Inspiration and the mine directors in July, 1937, but it was two years before work proceeded.

It was a further two years before the recommendations that J.P. gave would be carried out:

Mr. A. D. Hoagland, mining engineer and geologist with wide experience in this area was engaged as mine superintendent on this property. An accountant with local experience has also been engaged.

A road 3,400 feet long was cut from Perron No. 5 shaft to the Senore shaft site. The road is now being rock-faced with rock from the Perron shaft at a cost of approximately .20 cents per foot. Equipment and supplies are now being hauled by tractor.

An office building size 24 by 24 feet has been built 300 feet north of the shaft site.

Sounding to bedrock was done on the corners of squares over an area of 400 feet by 600 feet to determine the most suitable shaft site. The shaft has been spotted near the western end of the ore zone in the low ground where the depth of overburden is 38 feet thick.

Some trouble with surface water may be experienced.

A small boiler, hoist and derrick have been rented for the overburden operation. The purchase of a second-hand permanent plant is being investigated in two places. It will not be installed until bedrock is reached which should be in approximately two weeks if no serious difficulty is encountered with the overburden.

Three months later the shaft was completed to a depth of 218 feet and a limited amount of drifting had been completed. J.P. saw enough to recommend that the shaft be sunk two levels deeper. The influence of the Perron and Pascalis was evident as they proved better mines at depth:

The property is equipped with a complete electric mining plant capable of carrying development work to a depth of 1,000 feet. A 45 foot headframe, hoist, compressor room, transformer house, blacksmith shop, dry room and an office have been built on the property.

The property is supplied with electric power from the transmission line of the Northern Quebec Power Co. A truck road has been built to the shaft site from the Perron Gold Mines.

A two compartment shaft has been sunk to 220 feet with a level at 200 feet. On the 200 foot level a 560 foot shear zone in altered granodiorite has been explored over an east-west length approximately 600 feet.

The present east face is still 500 feet from the granodiorite greenstone contact. I believe important ore will be found in this area.

All the ore found to date is immediately east and west of the shaft. I believe a mine can be developed in this area and recommend that the shaft be sunk two levels with 400 feet of lateral work on each of the new levels.

If the proposed levels are at 125 foot intervals, 250 feet of sinking will be necessary, in which case $30,000 would be required for the work.

Figuring that bunch in Toronto would appreciate some good news, I took the liberty of calling them yesterday.

My guess of the stuff, as far as we could see it in the shaft, is that it is not very startling but at least encouraging.

Conditions in the Senore mine closely resemble those at Perron where remarkable ore occurrences have been developed at depth. I strongly recommend that depth development be proceeded with on the Senore property.

J.P. Norrie

The war took its toll again, however, and the property lay idle until June, 1945, when a shareholders meeting was called to consider a takeover bid:

Shareholders Meeting Report June 18, 1945

Senore Gold Mines owns 20 mining claims situated in the townships of Senneville and Pascalis in Quebec. Your directors at a meeting held on May 22, 1945, accepted a proposal made by Lesjack Exploration Co. for a sale of said mining claims to a new company to be incorporated that will have an unauthorized capital of 3,000,000 shares of $1 par value each, and will defray the expenses of its own incorporation and organization. Senore Gold Mines will transfer to the new company when organized all the aforesaid mining claims for a consideration of 1,250,000 fully paid shares of the new company.

The agreement provides that Lesjack Exploration Co. will upon the completion of the said sale subscribe and pay for 200,000 shares of capital stock of the new company at .15 per share and 114,285 shares at .175 per share for a total of $50,000, and that certain options shall be given Lesjack Exploration Co. for the purchase of the total, including the said 314,285 shares of 1,700,000 shares of the new co., which options if fully exercised, will, together with said sum of $50,000 provide the new company with a total of $625,500 for exploration and development work.

The agreement provides that 1,000,000 of the shares of the new company to be issued in consideration of the transfer of the said mining claims as aforesaid shall be issued and allotted to Senore Gold Mines and shall remain in escrow as long as Lesjack Exploration Co. is entitled to exercise their options above referred to. The balance, namely 250,000 shares of the new co., are to be issued and allotted in such name or names as Lesjack Exploration Co. shall designate. The said shares shall also be held in escrow, subject to release upon terms mentioned in the said agreement.

Your directors have authorized the officers of Senore Gold Mines to execute the proposed agreement subject to the approval of the shareholders at a special meeting to be called for the purpose of considering same, and notice of which is forwarded to you herewith. If this agreement is approved by shareholders and carried out, your directors believe that the new co. will be provided with sufficient funds to finance the whole development.

Directors of Senore, J.B. Cane, R.S. Lampard

Chapter 22

Politics

by George Norrie

J.P. Norrie was a farm boy. In the early 1900s there were some prosperous farmers, but the Norrie farm was only large enough to sustain one family, and at that, just barely. An education was always the way to improve one's lot, if you could afford it, but Henry couldn't afford it and J.P. didn't have enough money to educate himself. He worked for a time at the CNR in Truro, but watching his father waste his life as a farmer made him look for a way to better himself.

Politics was a way of life in those times, and one thing was certain: Liberals supported Liberals and Tories supported Tories. This applied not only to political concerns but also to other matters; in J.P.'s case, an education. Having completed high school at Colchester County Academy, he approached a neighbour and well-known Liberal farmer, Davidson Hill, for assistance. Hill gave him the break he needed. He gave him enough money to attend Mount Allison University in Sackville, New Brunswick, where J.P. earned his Bachelor of Science in Engineering. He then continued at Nova Scotia Technical College in Halifax for two years where he graduated in Mining Engineering in 1913.

J.P.'s summer employment was with the Joggins Coal Mines near Amherst, where the men mined the coal on their belly with a shovel, all day long. J.P. was qualified to take charge of a working face.

Upon graduation from Nova Scotia Technical College J.P. became Deputy Inspector of Mines for the Province of Nova Scotia from 1913 to 1914; then manager of the Moosehead Reduction Company Ltd., Moosehead, Nova Scotia in 1914-1915.

During the First World War he moved to the U.S. as Manager, Great Falls Mine, Potomac, Maryland 1915 to 1916.

He then became Concentrating and Smelting Superintendent of the Rome Brass and Copper Company, Rome, New York, 1916-1917.

From 1917 to 1919 he was Superintendent and Metallurgist, Monarch Metal Co. Ltd., Hamilton, Ontario.

From 1919 to 1921 he was Superintendent and Metallurgist, Electro Tin Products Ltd., Brantford, Ontario.

He ventured into the northern Ontario mining country as Superintendent of Porcupine Davidson Gold Mines, Ltd., South Porcupine, Ontario 1921-1923 and was Underground Superintendent, Sylvanite Gold Mines Ltd., Kirkland Lake, Ontario 1923-1924.

Then research work on tin smelting with American Smelting and Refining Co., Perth Amboy, New Jersey and Williams Harvey Tin Co., Flatbush, Long Island, 1924-1926.

His association with gold mining continued in 1926 when he became Manager of Stadacona Rouyn Mines, Ltd., Rouyn, Quebec.

After that, Malartic Gold Mines, Ltd., Malartic, Quebec 1927-1929; Dominion Explorers, Ltd., 1929-1930; and manager of Bussiers Gold Mine of Treadwell Yukon Co., Pascalis, Quebec 1931-1934.

During this period of his life he married Hanna Grace Hartling from Port Dufferin, Nova Scotia and they had four children: Pauline, Jim, Virginia and Harry. There is not much known from those days, as the children were separated from their father after their mother's tragic death in 1923, and raised on the home place by their grandparents.

The Day Wall Street Exploded

In putting this book together the authors searched for the details of the following story for several years with not much luck. It took place in September, 1920, when J.P. was a young engineer working in the tin industry in Brooklyn, New York. Then, 81 years later, September 11, 2001 shook the world and this story fell out of a long-forgotten history lesson.

The family talked about this story many times before 2001. It must have been one of the highlights of J.P.'s life. He used to tell that, on September 16, 1920, he was to meet a gentleman on the steps of a building right where the bomb went off, but for some reason he was an hour late. The man he was to meet was killed in the blast. J.P. escaped but he must have seen the area right after the event.

Rescue crews and crowd on the streets of Manhattan after the anarchist bombing in September, 1920

Globe & Mail
Sept. 15, 2001 [Reprinted with permission]

The head of the New York Chamber of Commerce described the Wall Street explosion as "an act of war". The Attorney-General of the United States said it represented nothing less than "a gigantic plot to overthrow the capitalist system." In the aftermath, the newspaper writers called for tighter controls on immigration and greater vigilance at the nation's major points of entry.

These sound like responses that have been thundering since the World Trade Centre disaster on September 11, 2001, but they were uttered years ago in the wake of what one historian describes as "the first modern terrorist event" in the world.

It happened at one minute after noon on Thursday, September 16, 1920, at the intersection of Wall and Broad Streets in Lower Manhattan, about eight blocks southeast of the site of what would be the World Trade Centre.

A powerful dynamite bomb filled with cast-iron slugs exploded, killing 40 people and seriously injuring an estimated 250 more. Almost all of the dead and wounded were taken to neighbouring Bellevue and St. Vincent Hospitals - institutions that, years later, treated the victims of the World Trade Centre horror. The bomb was brought to the symbolic heart of American capitalism in a horse-drawn wagon by, it is commonly believed, an Italian-born anarchist.

On the north side of Wall Street stood the Sub-Treasury Building; beside it was the U.S. Assay Office; opposite were the offices of J.P. Morgan and Company, and the construction site for an annex to the New York Stock Exchange.

The bomb sent shock waves through Lower Manhattan and caused glass to shower down on passersby. Paul Avrich, the chronicler of American anarchism, wrote of the scene: "Awnings 12 storeys above the street burst into flames, people fled in terror as a great cloud of dust enveloped the area. In the offices of J.P. Morgan, people fell dead at their desks, buried in plaster, bricks and glass. Outside, scores of bodies littered the street. Blood was everywhere."

The explosion and resulting inferno, Evrich wrote, was "the worst of its kind in American history." It destroyed most of the House of Morgan and caused, for that era, an unheard-of level of damage, more than $2 million.

A ferocious worldwide hunt for the perpetrator followed, with particular attention paid to individuals of Italian and southern European extraction. The Federal Bureau of Investigation compiled what was eventually a 7,000-page dossier on the matter, complete with reports from spies and informants. But amazingly, the perpetrator was never caught. Indeed, the man generally believed to be responsible for the act died peacefully in Italy in 1963.

After the explosion of September 16, Americans realized that "there was this new kind of violence in their world. They sensed that things had changed, that there was this rupture," says Beverly Gage.

Moreover the American people "knew" back then, just as they "knew" about the World Trade Centre, that something like the Wall Street attack was possible. They knew that out there, somewhere, there were intelligent, sophisticated cadres with a cogent world view "and this desire to use violence" against the established order.

The mass media of the day anticipated that use. Politicians fretted about it. And when the "predicted" finally occurred, it was greeted with both tremendous shock and "this strange sense of inevitability."

This event has been called the Sash Weight Bomb Disaster. Evidently what, in the *Globe & Mail* article, were called cast iron slugs were in fact sash weights. Sash weights were pieces of lead installed in the framework of a window to assist in raising the window.

Quebec Politics

Politics is an affliction that never affects those who keep moving. J.P. was infected while still in his youth but the symptoms did not have a chance to grow and mature during the early part of his career. He moved from job to job every two years and there was never enough time for him to learn the local political game, or at least no need to.

When he settled down in the small communities of northwestern Quebec, however, he had to deal with local politicians as part of the development of each mining project. Mining licences and claims, government roads, power supply, timber rights, municipal concerns all involved some level of government permission or contribution.

It also didn't take very long for his personal persuasion to become well known. He was a Liberal and outspoken about it. So when the Duplessis government was in power in Quebec, he did not have much influence. They were a Conservative government with elements in the party which surfaced years later in the Parti Québècois.

J.P. wrote this little ditty and sent it to his daughter Virginia:

Mr. Duplessis dumps his poke
to try to end the oxen's yoke
by spreading gravel far and wide
on Abitibi's countryside.

One major decision involving the federal government concerned the location of a CNR spur line from Senneterre to Rouyn Noranda. It was thought that this line would pass near the dozens of mines along the greenstone belt with obvious benefit to all the mines and the new communities that were growing up around them. When the rumour appeared that this line would follow a southerly route, J.P. took his concerns to the federal Liberal government and the Transportation Minster, C.D. Howe.

November 12, 1936
Hon. C.D. Howe
Minister of Transport
Ottawa, Ontario

Dear Sir:

I have been advised by Mr. W.R. McDonald, M.P. of the County of Pontiac to lay before you our reason for requesting that the new CNR branch line from Senneterre to Noranda be located near to the Perron mine. In the original survey this branch was located within one mile of the Perron mine but for some unknown reason was re-routed to serve the Fleming and Golden Manitou properties, both raw prospects and as yet very doubtful.

The Perron mine is now on a production basis of 175 tons per day. On December 1, when the present addition is completed, the capacity will be 350 tons per day or 100,000 tons per month, and we think that even this tonnage will shortly be greatly enlarged.

Adjoining the Perron are three very fine prospects, one of which will surely reach the production stage, and the other two bear every evidence of it.

The directors of the Perron mine were very much astounded to find that this branch had been re-routed, in view of the fact that Perron developments had much to do in warranting the building of the branch, and would be glad to have your influence in maintaining the original location of the branch.

Thanking you for anything you can do.

J.P. Norrie

November 23, 1936

Dear Sir:

I have now looked into the question referred to in your letter.

Before the location plan of the Senneterre-Rouyn branch line of the Canadian National Railway was approved, the change in line from the original tentative location was pointed out to me. The reasons given for the change were two-fold: the first was a question of geology, and before a final decision was made by the CNR, that question was submitted to the Mines Department. The second was: that by the more southerly location of the line, it would bring it three miles nearer to a large stand of white pine timber in the Grand Lake Victoria area, eventually meaning large carload railway traffic.

Yours truly,

C.D. Howe
Federal Minister of Transport, Ottawa

November 30, 1936
Hon. C.D. Howe
Minister of Transport
Ottawa, Ontario

Dear Sir:

I have your letter re the location of the Senneterre - Rouyn branch of the CNR and note the reasons given you for the change from the original tentative location.

The writer has had the widest experience in the development of the Pascalis area having been responsible for the development of most properties in the area during the last six years and I am sure the geology of the district is much better served by the original location of the line. The timber limits of the Grand Lake area are so far removed from any point of the new branch that a difference of three miles seems to me a very minor factor.

I have heard rumours in this area that one of the factors in the change is political. If such happens to be the case I would like to say that I am a staunch Liberal supporter and have done much for the party not only here but in Nova Scotia where I have a wide Liberal acquaintance. I firmly believe that the branch line should be re-routed to pass close to the Perron mine and other mines of the Pascalis area and trust that you will use your influence in this cause.

J.P. Norrie

Dear Sir:

I have your letter with reference to the location of the Senneterre - Rouyn branch line of the CNR.

The survey of this line was conducted by the engineering department of CNR which, as you are probably aware, is entirely independent of this department. It is quite impossible, therefore, that political consideration should have played any part in the final location.

If you feel that the proposed line will not give adequate service to certain areas, I would suggest that you bring your representations to the attention of Mr. S.J. Hungerford, President of the CNR, Montreal.

C.D. Howe

Common sense prevailed and the line was soon built where it was originally planned.

Nova Scotia Politics

During the early part of the war J.P. became increasingly involved in the decisions that were being forced on the gold mines by the War Regulations Board. These restrictions, discussed in later chapters, threatened not only the gold mines but their supporting communities.

At the same time his roots from home were showing themselves. His time was not as committed, with much of the exploration now on hold, and he was beginning to build up the farm, his home place, which he had been dreaming about since he had left Nova Scotia many years before.

He also began a political dialogue with the local provincial government around 1940. This included Angus L. MacDonald, Premier and J.A. MacDonald, Minister of Agriculture.

January 25, 1940

Dear Mr. Norrie:

I was very much interested in your letter of January 20 and your references to the resolution which passed in the Ontario legislature a few days ago respecting the manner in which the federal government is carrying on the war.

It seems to me that if I felt the war was being conducted improperly, I should have gone to Mr. King and pointed out wherein I thought there was room for improvement, rather than to make the matter a public issue, the discussion of which will, no doubt, reach the ear of the enemy people and give them some aid and comfort. That, I say, would be the way I, personally, would attempt to deal with a case of this kind.

As to whether there has been slackness in the administration, that is a matter that would require pretty full investigation. It is true that at the beginning of the war, particularly, there were some soldiers not properly equipped with clothing, and possibly the accommodation was not all that it might be. On the other hand, I think it has to be remembered in this regard that had the government a year ago undertaken to vote sufficient money to become thoroughly prepared for clothing and equipping men, it is doubtful if such a policy would be supported by the majority of the Canadian people. In a dictatorship, of course, it would be quite easy to order the factories to get so many rifles ready, so many uniforms and the like, and whether they were needed or not would make little difference. In a democracy one has to take account of public feeling and leaders very often can only go about as far and as fast as the public will permit.

I have not read the full report of the speeches in the Ontario legislature, but I shall do so. From the newspaper accounts here one of the arguments that was made was that Colin Campbell, a minister in the Ontario government, and now a lieutenant in the army, did not have a greatcoat. In my days in the army every officer had to buy his own uniform and pay for his own meals. There was no such thing as free uniforms. I should think that Colin Campbell would be able to get a greatcoat some way or other.

Another thing that has to be remembered, as you well know from your experience in business and industry, there are bound to be, in the beginning, difficulties. Everything cannot be foreseen, even with the best organization in the world, and I have no doubt that even in England there are many things that, perhaps, could have been done in better fashion than they have been done.

I have been thinking for some time that when our house meets on the 5th March I would introduce a resolution of loyalty to the empire, and I believe that it would be very easy to incorporate in such a resolution some such idea as you suggest.

I remember you quite well, and Judge Archibald, who is a Colchester man, and I often speak of you. I do hope that you will be down this way before very long, and when you do, please be good enough to call to see me. I hope, also, that some day you can so arrange your affairs as to take up permanent residence in this province. I think that on the whole, while we have not, perhaps, the wealth or opportunities for gaining wealth that certain other provinces have, still as a place to live, it is difficult to beat this country.

With cordial personal regards and renewed thanks, I am

Yours truly,

Angus MacDonald

It is hard to imagine what agriculture meant to many people in those days. To think that Jersey cattle would be considered so important as to create such concerns as related in these letters seems astonishing today. But in those days fine breeding stock were just that important. J.P. took his concerns to Angus L. and his Minister of Agriculture, John A. McDonald. Angus L. would shortly be appointed to the federal cabinet as Minister of the Navy because of his experience with the navy around Halifax.

June 29, 1940

Hon. Angus L. MacDonald
Premier of Nova Scotia
Halifax, Nova Scotia

Dear Mr. MacDonald:

I heard over the radio last evening that the people of the Channel Islands along with most of their famous cattle had been evacuated from the Channel Islands to England so I wired you suggesting that these cattle with their owners be brought to Nova Scotia farms. As you probably know these islands are divided up into small farms each farmer owning a few head of cattle in which he takes great pride not only in the fine lines of the animals but milk production as well. The island of Jersey was the home of approximately 11,000 head of Jersey cattle, some of which were sold each year to Jersey breeders the world over and many of which came to Canada and the United States. I have about 40 head on my own farm some of which were imported. I am not so well informed on Guernsey cattle. However, I believe it would be a distinct advantage to the province of Nova Scotia to have these Jersey and Guernsey breeders with their herds transplanted there.

I would be willing to help in this venture by supplying a haven for say 50 animals until suitable farms could be found for them.

I expect ample transportation space would be available in empty vessels returning to Halifax.

With best wishes, yours truly,

J.P. Norrie

August 6, 1940
Mr. J.P. Norrie
Amos, Que.

Dear Mr. Norrie:

Since your telegram several weeks ago, I have made several inquiries and talked with a number of people about cattle on the Channel Islands. The most authoritative information has come today from Mr. A.M. Shaw, Director of Marketing Services, Department of Agriculture, Ottawa, who has written:

"Some time ago you discussed with me the question of securing some of the Jersey cattle moved from the Channel Islands. Following our conversation I wired Canada House [in London] to obtain information in this regard and have just received a telegram in reply. While I understand that a considerable number have moved into Great Britain, apparently they are being absorbed into the British herds and there are no surplus animals available at the present time."

Yours truly,

John A. McDonald

In late 1941 a provincial election was imminent in Nova Scotia and the local Colchester Liberals wanted J.P. to consider running there. He had been financing the local Liberal candidates but they wanted him to run there himself.

In late 1945 there was another provincial election and again the local Liberals wanted J.P. to run. Angus L., as well, tried to persuade him to run in Colchester County even though it was a strong Conservative area.

Dear J.P.:

I regret you felt that you had to decide not to accept nomination.

We have to pass up the best opportunity ever to win the county. Everybody concedes your election in event of your running.

You would need to be here only three weeks before election. You better reconsider and give a favourable reply. I know you have lots to do but the busy men have to do the work.

Sincerely,

John Reid, MDCM, FACS

September 21, 1945

Dear Mr. Norrie:

It was a great regret to me to see in the last paragraph of your letter that you find it impossible to run in Colchester County. I was talking to Hugh Dixon, the president of the Liberal Association, yesterday and he still had some hope you would run. However, I believe that we can carry Colchester County and Dixon tells me that the organization is getting into shape, and that already it is better than the federal organization was in the last contest.

They are having their convention on Thursday, September 27 which will give them nearly a month for the candidate to get around the county. It would be a great help if you can spend the last ten days or two weeks with us. I have in mind now to ask Mr. McMillan to have general oversight of Colchester, Cumberland and Hants Counties, and I am sure his organization ability can be put to good use in these areas. It would be better if I had had a couple of months instead of a few days to get the organizations around the country going, but the feeling among all the members was very strongly for an election this fall, and on the whole I think conditions are more favorable now than they might be next summer or next fall.

With very best personal regards and good wishes,

Angus L. MacDonald
Premier

A Dream Come True

James Norrie had very definite ideas about the farm he would have some day. Now he has it. (*Family Herald* Aug. 11, 1943)

At the head of the Bay of Fundy whose waters, boasting the world's highest tides, do their best to make an island out of Nova Scotia, is the town of Truro.

Right now Truro is somewhat overshadowed by the military camp at nearby Debert, of which most Canadians have heard. But the town is important in its own right, because of its strategic location and because the land around it is some of Nova Scotia's finest farming country. A little over a mile out of Truro is Fundy Jersey Farm, one of the showplaces of the district, owned by J.P. Norrie.

Mr. Norrie's enthusiasm for things agricultural, is now well known around Truro. Another mining man who has brought fame to his home province, Nova Scotia, through agricultural achievement, is E.H. Horne, of Enfield, which is between Truro and Halifax. Mr. Horne wandered far afield as a young man and discovered what is today the famous Noranda mine in Ontario's gold belt; retiring he came back to build a farm and establish the noted Monte Vista Hereford herd.

Fundy Farm is the dream of James Paul Norrie, come true. He was born on this farm in the gay 90s and had a good start in life. As a boy he worked hard on the farm, went on to take a college education at Mount Allison University. His boyish mind envisioned the farm much as it is today, and his greatest delight while at college was his trips home when, donning overalls, and hoe in hand, he made for the great out-of-doors which he loved. From Mount Allison he got his B.Sc., then entered Nova Scotia technical college, where he became a mining engineer.

J. P.'s dream farm, built on his old family farm, Truro, NS, 1940.

Mr. Norrie was Deputy Inspector of Mines for Nova Scotia in 1913-1914 before he left to enter the actual mining field, which has led him all over eastern Canada and many New England states.

But he never lost his liking for the farm. In 1938 he returned home and has successfully mined the earth again, producing not gold or iron, but the necessities of life in a big way. Two managers have been important factors in this, both doing a grand job. C.B. Fawcett [J.P.'s brother-in-law] of upper Sackville, New Brunswick, was there from the time the farm improvement began until he left to farm on his own, when he was succeeded by W.H. Todd of Burk's Falls, Ont., since last September. The staff consists of 12 men and every kind of farm equipment, including an eight-foot tractor binder.

The buildings include three barns, four silos, an implement shed, repair shed and six residences. They are painted white, trimmed with green and have red roofs.

The 1943 acreage is 100 acres of hay and ensilage grasses, 70 acres of grain, 10 acres of soy beans and five acres of turnips and other roots. There are about 80 acres of pasture.

The dairy herd comprises 75 purebred Jerseys. Official ROP records during 1942 showed the herd with a milk average of 97,754 pounds and a butterfat average of 509.5 pounds. The high cow for the year, nobly born Abbie Jean, with 12,097 pounds of milk and 641 pounds of butterfat, is a Nova Scotia bred cow from Yarmouth.

The large dairy barn is 134 by 32 feet, and is of metal and concrete construction. An up-to-date milk house has just been built, containing a wash-up room, bottling room and walk-in cooler.

Great precautions are taken to protect the dairy herd from disease by the most modern methods. An outsider must not walk along the feed lane, lest he scatter germs. At the entrance is what appears to be a mat on the floor. This is loaded with disinfectant to help protect the herd from any disease which may be brought in from outside. In the summer months electric fly-catchers are installed throughout the barns. These are small-scale "electric fences", which incinerate any fly which comes in contact with them. The barn's interior is white-washed twice weekly and the windows are very large to admit as much fresh air and sunlight as possible.

J. P. and Ed Horne at an agricultural field day in Nova Scotia. Horne, a fellow Nova Scotian, discovered the ore body that gave Noranda its start.

At the rear of the barn is a four-room maternity-and-general hospital for the herd. These rooms are white-washed and clean enough for anybody. The hospital has a room for the herdsman and is equipped with stove, medicine locker, table and equipment for attending sick cattle.

The herd is fed a balanced diet, including ensilage, consisting of corn, soy beans, etc. The animals are milked by machine and the milk is run over an aerator milk cooler, cutting down the bacteria count. A 400-foot well with an average temperature of 42 degrees cools the milk. Then it is carried to the upper storey on a dumb waiter, where it is stored in a walk-in refrigerator. The dairy herd is accredited. An animal, to be brought in, is placed in isolation for 60 days, and tested for Bang's disease and tuberculosis. If an animal breaks out of the pasture it is isolated for 60 days before rejoining the herd.

The Norrie residence is constructed of wood, of colonial architecture, two storeys high. The basement has a tiled floor, comprising furnace room, billiard room, children's playroom and storage rooms. The hot water system is operated by an oil-burning furnace. On the first floor are living room, dining room, breakfast room, kitchen and laundry room. On the upper is a master bedroom suite, guest room, girls' and boys' rooms. The house is situated on a terraced lawn, with a flagstone walk leading up to it. The property is landscaped with ornamental shrubs and roses.

The farm has ten Belgian horses, imported purebreds averaging a ton each. At the head is the stallion Baron d'Oultré, imported in 1936, when six years old. He showed his quality by winning the grand championship and Belgian medal at Toronto's Royal Winter Fair. In 1937, "the baron" was grand champion at Quebec.

J.P. at the farm in Truro with his Jersey herd

The majestic "baron" stood in his stall, wearing his marcelled mane with an air of true nobility, and not deigning to turn his head. The pasture lies just behind the barn and looks like an enlarged picnic ground. A little brook, capable of great expansion, babbles through it. Under a grove of trees stood a two-month-old colt, and judging from its proportions, it has had a good start in life. Numbers of calves lounged about, staring with their big soft eyes and looking for all the world like deer.

The new dairy had just been completed and I wanted to knock as I came to the door, so much like a residence does it look. Glistening white, without and within, it has every facility for preserving milk for the farm dwellers as well as for the town route. Around the front door is a well laid out flower garden. This building is of tile and concrete, 46 feet by 28 feet, with red asphalt roof. The outside walls are of white stucco and the inside walls are surfaced with white enamel paint, the floors being concrete. A large driveway and parking area makes for convenience, as well as improving the general landscape effect.

Chapter 23

The Sinews of War 1940-1945

by George Norrie

In the early 1930s Canada and the world were mired in the Great Depression. People were out of work in almost every industry and the mining industry (gold mining included) was clouded by the same economic woes. The price of gold was $20 an oz. and most mines could not operate economically, and closed down. J.P. had found this out first hand at Malartic Gold Mines when they closed down in 1929.

When Franklin Roosevelt came to the presidency of the United States, he brought with him the "New Deal", one of the principal features being that it set the price of gold at $35 per oz. and let it be known that the United States would buy all the gold it could get at this price. The gold mines in Canada instantly boomed. J.P. took full advantage of this policy during the 1930s, but as the decade came to a close and war was on the horizon, new, unexpected and unusual conditions began to make questionable the price of gold, the distribution of gold, and the need for a gold standard at all.

Economist William D. Ross of Toronto wrote in August of 1940:

Canada as a major gold producing country has a very definite stake in the future of gold. Everyone will concede that the chief obstacle to the smooth functioning of the gold standard today is the unsatisfactory distribution of the metal between the various nations. The United States now holds far in excess of half of the world's supply of monetary gold and this lack of balance is being increased daily by the chronic, one-way flow of metal into the United States.

The existence of this unhealthy condition is traceable to two distinct causes. First, the tremendous growth of economic nationalism since

1930, particularly in Europe, has brought in its wake chronic instability, both political and economic, which has upset the flow of short-term capital throughout the world.

Normally short-term capital movements help to adjust the international trade balances, but to perform this function effectively an atmosphere of reasonable stability is necessary. In the past decade [the 1930s] the usual incentive to such movements, the possibility of a speculative profit or a higher rate of interest, has been replaced by a much more powerful motive: fear. Mobile capital sought safety in flight and the greater proportion of this refugee capital fled to the United States.

The second contributing cause of the unequal distribution of gold is associated directly with the foreign trade policy of the United States: her unwillingness to accept the obligations of a creditor nation. It is absolutely essential that a creditor nation import sufficiently more than it exports to counteract the effect of a continuous net balance of interest and debt payments…

In 1937, Canada paid to the United States $385 million in excess of the amount received. In the same year United States imported from Canada only $34 million in excess of exports, leaving Canada with a net deficit of $351 million, exclusive of gold. The action of the United States in seeking to maintain a net balance of payments in her favour has produced a chronic flow of gold into the country, impoverishing other nations and leading to very undesirable effects.

…in the case of Canada the adverse balance of payments has been increased by war demands for American materials and the supply of foreign exchange is being exhausted rapidly. Loans to Canada by the U.S. would be of irresistible benefit at this time to counteract the war drain and bring payments between the two countries into balance. [Lend Lease.]

Other comments questioning the gold standard appeared in various newspapers, as the forthcoming election in the United States allowed for political posturing by prominent political senators such as Vandenberg and Brydges:

> The gold standard is doubtless the best and let us say that it has been an ultimate value, alongside silver, from the earliest history; that it has been found to be the best and most durable small compass medium through which international balances can be settled and that there is not much danger that it will be dethroned from its place of pre-eminence for any considerable length of time. Nevertheless these are extraordinary times. There are no precedents to go by and anything may happen. We believe that gold's immediate future depends upon what flag victory finally perches upon and what kind of a peace is subsequently made. Then will come the question of distribution of the gigantic accumulation in United States and we cannot see how that can be done without levelling of tariff walls, which is against the political faith of a vast majority of its people. Gold's only real value, however, is as chips in the game of world trade.
>
> Much stress has recently been laid upon public condemnation of the monetary policies of the New Deal by presidential candidates on the Republican side. In particular, Senators Vandenberg and Brydges have definitely stated that in their opinions purchase of gold from the whole world must be stopped. We believe, however, whoever may come to power, that the power will sober him and soon convince him that abandonment of the present policy of paying $35 an ounce would cause incredible chaos throughout the world, and that the reaction in the United States itself would be no less than catastrophic. Finally let us say that in our opinion the gold standard will be reinstated by the civilized countries of the world, but whether this year, next year or ten years from now, no man can tell. We are hopeful that a re-establishment will come when the English speaking

nations of the world get together in a defensive and offensive alliance. And that must come some day, if civilization is not to perish from the Earth.

Gold mines pulled Canada through the Great Depression more than any other industry. This was because production from these mines was such a large part of total Canadian industrial production, a much larger percentage of total output than in the United States. Those employed in gold mines made up, also, a large percentage of total Canadian employment. In the 1940s these mines contributed to the balance of payments with the United States and the importance of their contribution grew with every day of the war as reflected in another newspaper report:

Further evidence that Canada very urgently requires the full output of its gold mines is contained in figures from Ottawa which suggest that our unfavourable balance of trade with the United States in 1941 reached $400 million. The unfavourable balance on dividends and interest payment amounts to $320 million, bringing the total to $600 million which must be found.

Aside from net tourist traffic receipts, the only thing Canada has currently to offer the United States on its $600 million deficit for the year is the gold production of $200 million.

Obviously the government's task of financing would be considerably worsened if those papers and people who cannot see "what good the gold mining industry is in wartime" were able to work their ignorant way upon the country.

Again, it is obvious that the need for gold output is large and growing larger, which serves to dispel the fears of those who believe gold mining will be deliberately reduced in scale. Canada accumulated fair exchange credits in 1940 but these cannot last forever in the face of our heavy buying of war goods across the line.

The contribution of gold production to the war effort in South Africa was far greater than that in Canada. This news article by South Africa's finance minister explains:

How does the Union of South Africa pay for its war participation? From what principal source does it draw revenue with which to maintain a large force in North Africa?

The fact that we have, today, up in the north a very large and splendidly equipped army, is to a very large extent due to the contribution of the gold mining industry.

It was with that in view that Field Marshall Smuts some time ago, in speaking of those who are carrying on the gold mining industry at the present time said, "Your duty lies clearly here. If you were not carrying on here, it is doubtful whether we could have and maintain an army at the front. You are doing work of national importance. You are entitled to the gratitude of the country. In a special sense, yours is a real service."

I repeat those words with emphasis but I want to add something to that. The service which is being rendered by the gold mining industry at the present moment in this connection is not only in respect of the maintenance of South Africa's own war effort. Of no less importance to the prosecution of this fight in this cause of freedom is the production of gold as gold. I do not think I need to tell you that the furtherance of that cause depends very largely indeed on the purchase of goods in increasing volume, which have to be paid for in dollars; and for that purpose it is essential that the production of gold must be maintained.

Now, in part, that problem has been eased by the Lend Lease Act. But only in part. Gold and a good deal of gold is still essential if the accounts between the sterling and the dollar are to be settled. Therefore South Africa in producing gold is today producing one of the essential, one of the vital, one of the sinews of war. I would therefore say again to those who are carrying on the gold mining industry today that their work is fundamentally important.

There are uncertainties in regard to the future of gold. Of course it is true that the answer to that question lies primarily, one might say exclusively, with the United States. That question is of vital importance to us in South Africa; but we shall have precious little to say in answering it. The United States has "got the bear by the tail and it just cannot let go" therefore the unchallenged position of gold as a medium of exchange and a standard of value will be maintained.

The gold standard was obviously of critical importance to many countries – South Africa, Canada and the United States in particular. Other countries also depended on gold for its international security and value. This fact allowed opportunity for war-time propaganda which J.P. points out:

The gold stocks recently have received a serious setback, mostly due to baseless rumours which some people think are coming from paid Nazi agents. You know one of Hitler's chief weapons is to try to undermine the financial setup of the democracies. The most effective way of doing this would be to undermine gold. U.S. must buy large amounts of the products of neutral countries with gold. They can ill-afford to see it discredited.

The supply of labour was the first wartime restriction the gold mines were forced to deal with. J.P. wrote to his friend Harry Sparks at O'Brien Gold Mines, who was the president of the Western Quebec Mining Association, about this serious problem:

I have just received a letter from Mr. Larochelle with a copy of a letter to Mr. Bateman and Oliver Hall re: the shortage of men at the base metal mines and the suggestion that the gold mines be further infringed upon in supplying labour.

It is certainly a fact that the inducements offered by the war industries, the draft and other causes related to the war have depleted the working forces of the gold mines to the point where further depletion of gold mines' labour will prove to be serious.

I expect it is unnecessary to go into the oft-repeated argument that gold is still necessary to pay our war debts and relieve our country from the necessity of accepting Lend Lease from the U.S. and thus protecting its solvency. The next worse condition to Hitler domination would be foreign domination of any kind.

I believe that further encroachment on our gold mining labour would seriously interfere not only with the operation of the gold mines themselves but the communities in which the miners live. The well-being of these communities must be protected as part of our Canadian entity and I believe it is up to us to impress this point on those who would further cripple our country.

I understand that tungsten is higher on the strategic list than nickel, copper, zinc or other base metals. My opinion is that the only supply of this much needed mineral will come from our gold mines. Prospectors have been exploring the country far and wide in the search of tungsten. To date I have heard of few discoveries outside of the gold mines and I believe few will be found. Gold and tungsten are inseparable brothers so I believe government aid should be solicited through supplying more, rather than less, men for gold mining and consequent rise in tungsten production. Tungsten production can also be increased from the gold mines by increasing the price. As an instance I would like to draw your attention to Central Cadillac where quite a substantial tungsten production could be expected with a little help in labour and supplies. Further labour difficulties will probably terminate the operation.

I am a great believer in the probability that our gold mines will eventually be called upon to supply a mineral that cannot be found anywhere else and the idea should be started now before it is too late.

Since J.P. was influential in Quebec mining he wrote the following letter in Oct., 1945, to provincial and federal politicians: the Federal Minister of Mines and Resources, Hon. T. A. Crerar; the Premier of Quebec, Hon. Adelard Godbout; and Louis St. Laurent, the leader of the Quebec Liberal Party:

My purpose in writing you is to secure help in preventing any further reduction in gold mining in our province. I am enclosing a copy of a brief, outlining what gold mining restrictions up to the present have done and the effect of a further 20 per cent cut in production at Perron Gold Mines. The facts in this brief pretty well cover gold mining in general in Quebec.

The pressure to further restrict production or close our gold mines is coming from the United States where conditions are altogether different from ours. Although the United States is a large gold producer most of it, i.e. about 75%, comes as a byproduct in the production of strategic metals, chiefly copper, and few actual gold mines exist, so restriction in gold mining there means little to them.

In our case many millions of dollars have been spent in building up a great industry chiefly on low grade ores with a narrow profit margin and on which very large communities receive their sole support.

Of course we are all in favour of an all-out effort and the gold mines have been giving their all to the cause in spite of restrictions which have already almost crippled the industry, and we have been cooperating with the government in supplying men to the base metal and other operations. Our cooperation has been rewarded by the advent of force to further reduce operations immediately. A recent order forbids the local employment officer to send men to gold mines. This can eventually lead to only one thing, i.e. close down.

In many cases it is impossible to impress on the Ottawa authorities the fact that the harm done by further reduction in gold mining cannot and will not materially help the war effort. Our local Mine Managers Association has tried on several occasions to point out the grave consequences to come from further reduction.

As gold mining is one of the greatest factors in the economic life of the country and promises to become much greater, it will be necessary for us all to carry on a rigorous and continuous campaign to prevent further crippling of the industry and in this I believe your influence is necessary at this time. A recent order from the Department of Labour in Ottawa forbids the local employment office to supply workers to gold mines, advising them that if work cannot be secured in Quebec base-metal mining, they must go to Sudbury.

If you could spare the time I am sure a visit to our gold mining districts would satisfy you as to the grave consequences to the mines and communities that a further reduction in gold mining operations would bring. If you could arrange such a trip I will be glad to accompany you and help point out the problems facing us. In any case I will be glad to help in every way possible.

I understand the Dominion government paid the transportation of 5,000 men to the west this fall to help harvest a wheat crop which will not be used for many years to come. Why not restrict farming and let the gold mines go as gold is immediately important to pay our war debts.

The time has now come to show the United States that from now on we must look after the security of our own economy and in this your voice will be of untold benefit.

J.P. included his comments on the effect of restrictions on the Perron Gold Mine, in particular, and other mines under his direction to mid-1942:

Following the appeal of the Federal Minister of Finance on July 17, 1940 for greater gold production and upon resolution of the Board of Directors of

Perron Gold Mines, the gold production of the mine was increased from an average of $140,000 per month to $171,000 per month. This increase meant a corresponding increase in the number of employees at the Perron mine, and added interest in gold spurred on the mining and exploration efforts of the Cournor, Pascalis and Senore properties, all of which supported the town of Pascalis.

Scarcity of labour and supplies has now forced the closing of the latter named properties, eliminated entirely all prospecting and exploration work in the area and caused a reduction in Perron production to $120,000 per month. This means that 500 working men have left the district to work in war plants in southern Ontario and Quebec. Many of the local merchants have failed and closed up and most others are on the point of doing so.

The best type of employees of the mine have gone on war work so that those remaining are less efficient and accident conscious.

At the present rate of production 200 tons of steel of a value of $16,000 are necessary to keep the mine in operation and of this amount only $7,200 would have to be imported. Except for a few non-strategic items, the mine has on hand enough supplies to last nearly two years.

From the figures submitted it seems apparent that little has been gained to date in furthering the war effort by the reduction made, certainly not enough to compensate for the drop in gold production and harm to the community.

At the Perron mine we are producing quite an amount of much needed tungsten so badly needed in the war. With little encouragement in the way of labour we could materially increase its production.

When the war started the gold mining industry of 149 producing mines supported 600,000 people and on this basis comprised in excess of 5% of the Canadian economy. From a revenue and tax producing angle it rates much higher than agriculture and several other industries.

The pressure to cut production a further 20% comes from the United States war heads who probably do not understand the importance of gold mining in Canada's daily life, as they probably compare it with American gold production which forms a very small part of their own economy and comes chiefly as a byproduct in the treatment of base metal ores.

The gold mining industry up to last December was the bulwark of the Canadian war effort. Temporary relief in the way of Canadian dollar exchange should not prompt our own leaders to further sabotage an industry which has done much in Canadian peace and wartime economy and which with encouragement in the immediate future will be in a position to again assume its greater productive role in absorbing large numbers of returning Canadian manpower about which we should be giving current thought.

T.A. Crerar acted on the letter immediately but there were some facts not completely clear, such as the percentage of gold produced in the United States as a byproduct of larger mines. The major problem, however, was the amount of U.S. steel and iron needed by the Canadian gold mines:

The difficulty is not primarily one of manpower, although that is involved, but is mainly a question of materials. As you know, the United States has issued an order for the closing of their gold mines. Are you not in error in your statement that 75% of the gold of the United States production is a byproduct from other mines? The information I can get indicates that about 15% of it is in this class. However, that point is not so material. In increasing measure Canada's effort in this wretched war is making a heavier drain on most essential materials, and this applies with special force to iron and steel. We have to get much of this from the United States and the attitude of those in that country having to do with the allocation of these supplies is definitely that they cannot continue to supply materials for the maintenance of our gold mines when they have shut down their own because of shortages. This may be rather an arbitrary stand but I can assure you that the position of Canadian gold mines has been made quite clear to our American friends.

J.P. advised Crerar why he believed the production of gold was primarily from the large copper mines and also pointed out that the supplies of steel and iron could be at least controlled:

I have no data on which to base accurate figures, however, I believe my figures of 75% of U.S. gold coming as a byproduct from base metal mines is approximately correct. Large quantities of gold in the United States come from the high tonnages of low-grade copper ore mined. This fact is more or less confirmed by the figures that only 6,000 men are to be made available by the closing of U.S. mines. We know that a very few straight gold districts occur in the United States.

Most of our gold mines have sufficient supplies on hand to last for at least another year so that the amount of strategic steel - chiefly alloy steel - required will be very small.

The American intention to close their own gold mines did not materialize to the extent anticipated by the War Production Board in the United States. This was primarily because of the fact that the men and materials supposed to be released for strategic war production were not released, and for the reasons suggested by J.P.

Crerar advised J.P. of this in early Nov., 1942:

The American authorities have somewhat relaxed their orders to close down American gold mines and I think there is more of a disposition there now to recognize some of the problems which would be involved in any general shutdown of Canadian gold mines. I am hopeful that a reasonable, sensible course can be followed in this matter and I shall certainly do all in my power to that end.

Adelard Godbout also advised J.P. that he had discussed the matter with Louis St. Laurent and that Quebec was well represented in the federal cabinet.

MacKenzie King, as history has shown, was a master politician. Evidence of this is the fact that he was elected prime minister for so long. King owed much of this to the support he always received from the province of Quebec, like so many prime ministers since that time. The influence of Louis St. Laurent on the Canadian government's understanding of the importance of gold mining to the war effort had a very definite positive effect. The situation did not result in total shutdowns, as feared, and production of gold struggled along for 1943 and 1944. In late 1944 the end of the war was in sight and gold properties reacted with a great surge of investor interest.

Chapter 24

Wartime Gold Mining

by George Norrie

During the early part of the war, 1940-42, the Seventh Malartic's problems were typical of the problems confronted by gold mining in Canada. The Canadian government made it clear that gold production was wanted, even demanded, for balance of payments. The government was purchasing ever increasing amounts of war supplies from the United States, and the sale of gold to the U.S. treasury compensated. However, gold was not regarded as a strategic metal. This meant that machinery, parts and supplies needed for strategic metal production would take priority. Exploration was even less necessary and no supplies were to be spent on this. Labour was also under regulation; producing mines took priority.

In the early part of the war, the first two years, these restrictions were not onerous; in the last two years they became very limiting.

Seventh Malartic Mines

The Seventh Malartic was incorporated in 1940 by Glen Wilton. He was to become J.P.'s personal friend and lawyer. The company was started just in time to be confronted by these problems which had already closed down several mines in which J.P. had large interests.

J.P. was always trying to match up properties which alone could not support a mine but together would make a very attractive company. He also required partnerships in the financing of these projects. In this case he brought together the Amity Group, the Twinmac Group, the Bob Potter Group and J.P. Norrie et al to set up Seventh Malartic Mines Ltd.

The following was the share distribution of the new company determined by the acreage contributed by each group:

Share distribution of Seventh Malartic Acreage:

Amity Groups -	1,270 acres
Twinmac Group -	600 acres
J.P. Norrie et Al Group -	1,567 acres

3,437 acres: 1,000,000 shares / 1 acre: 290.95 shares

Amity group:1,270 acres; 369,506 shares
Distribution as follows:

R.S. Potter	200,000 shares
A.J. Davis	56,502 shares
M.A. Thomson	56,502 shares
J.P. Norrie	56,502 shares

Twinmac Group:600 acres; 174,572 shares
Distribution as follows:

R. Roux	43,643 shares
A.J. Davis	43,643 shares
M. Thomson	43,643 shares
J.P. Norrie	43,643 shares

J.P. Norrie et Al Group:1,567 acres; 455,922 shares
Distribution as follows:

Prospectors	145,922 shares

Divided between Messrs Davis, Thomson & Norrie
300,000 at a cost of $1,000 or 333.33 each shareholder

Final Distribution:

R.S. Potter	200,000
A.J. Davis Amity Group	56,502
Twinmac Group	43,643
total	214,693
M.A. Thomson Amity Group	56,502
Twinmac group	43,643
total	214,693
J.P. Norrie Amity Group	56,502
Twinamc group	43,643
total	214,692

The Seventh Malartic Company was formed in late 1940 and consisted of the amalgamation of the above groups of claims.

J.P. wrote Andy Davis in Oct. 1940:

The owners of the Twinmac claims are now R. Roux, M.A. Thomson, J.P. Norrie and yourself. One quarter each.

I telephoned Mr. Thomson today that I had completed a deal with Bob Potter for the adjoining ground 600 acres to the east and 670 acres to the west for $2,500 cash and 200,000 shares in a company to be formed. One third each. Mr. Thomson, yourself and myself. Please send me your cheque for $833.00. I believe we should incorporate a company at once as we now have the complete group of 3,437 acres.

With the large mines that J.P. found and developed into producers, such as the East Malartic and Malartic Goldfields, the claims were staked before there was a great amount of interest in the ground staked and J.P. staked the claims to include all the ground he wanted in one block. After it became clear that all other ground was valuable and had gold mine potential, many mining companies and prospectors staked ground and the claims were smaller and not necessarily all that each group wanted geologically. This made it necessary for the amalgamation of groups of claims so that all of a particular geological area could be included in a mining project.

The Seventh Malartic was one good example of this type of amalgamation. J.P. not only had the engineering knowledge to choose the ground correctly, he had the respect and trust of the mining community so that when he suggested an amalgamation, all participants were eager to commit their claims with the assurance they would be treated with honesty and fairness. They also knew he would ensure a committed effort to find a mine as quickly as Inspiration could carry out the development. They also knew that Nesbitt Thomson were there to supply the financing if and when the property became a mine.

Many people in northern Quebec were quick to invest in J.P.'s projects even at this wartime period. Rosario Roux was mining recorder in Amos and a friend of J.P.'s. He was obviously in a position to know all participants in that part of Quebec; they all had to record their claims. Rosario invested in this J.P. Norrie property.

Bob Potter, one of the property owners, told J.P.:

Mr. Roux phoned me this morning and was very anxious to secure 20,000 shares of Seventh Malartic Gold Mines stock for a couple of his friends, and I have written him today giving him an order to you for the above amount.

Glen Wilton was becoming more involved in J.P.'s legal work. He was a lawyer located in Toronto and did the work forming the Seventh Malartic company. His invoice for this work was $600. He also enclosed a newspaper clipping from the *Goldfields Press*:

A new prospect property sponsored by J.P. Norrie will shortly be developed. The *Goldfields Press* understands that the name of the new company will be Seventh Malartic Gold Mines Ltd.

The property extends west from the Bourlamaque boundary to the Smelters option which ties onto the Crossroads Mines.

Whether 7th Malartic refers to a lucky roll in a dice game or whether this may be the 7th success of Malartic mine-maker Norrie, is not known at this time. But suffice it to say that if Jim Norrie's mining achievements continue, Seventh Malartic mines may be another producer for northwestern Quebec in the not too distant future.

The property is located in an interesting section of the camp. Dome is drilling adjacent claims.

J.P. gave his first report on the property Feb. 3, 1941:

The property of Seventh Malartic Mines, Limited of 3,437 acres is lots 47 to 62 inc., range 5 and lots 44 to 62 range 6 Dubuisson township, Quebec.

The property is located three miles from the town of Val d'Or by motor road. Power and telephone connections are available.

Camps to house a crew of 50 men have recently been erected on the property.

The band of greenstone rocks known as the Cadillac break in which most of the productive mines are located crosses the property for a distance of three and one eighth miles. This band is bounded on the south by Temiskaming greywacke and on the north by sediments and tuffe which with the greenstone are intruded by numerous dykes of diorite and porphyry.

The property is almost entirely covered with overburden so that extensive diamond drilling will have to be done to explore the favourable areas.

In view of the fact that the productive Cadillac break crosses the property there is every reason to expect that valuable ore bodies will be found in it.

It looked as if work was terminated on the property until the war's end; however, since the Perron mine was located fairly close and had a mill which could process this gold ore, the Perron optioned the Seventh Malartic treasury shares and took control of the property. Even after the war the independent development of Seventh Malartic would require a large ore body, and the Perron could use their existing mill in present time, and also could spend some money on Seventh Malartic exploration because they were a producing mine.

Control and operation of Seventh Malartic Mines was taken over by Perron Gold Mines by the option of the balance of Seventh Malartic treasury shares Feb. 1, 1943.

Thus the Seventh Malartic exploration continued only because an existing gold mine carried out the work.

Dubuisson Gold Mines

The Dubuisson property involved two separate properties. The largest property of Dubuisson Goldfields Ltd. of approximately 1,300 acres in Dubuisson Township, Quebec, adjoined the property of Malartic Goldfields to the east.

As in the latter, the property was entirely covered with a heavy mantle of overburden so that scientific prospecting methods coupled with diamond drilling for locating contacts, intrusives and ore zones had to be used.

An accurate magnetic survey indicated the easterly extension of the productive Malartic greenstone band over the full width of the Dubuisson Goldfields property with intrusive diorite and porphyry dykes similar to the ore found on the Malartic Goldfields property.

The presence of the greenstone band was further developed by 11,960 feet of exploratory drilling chiefly along the contacts of the band. In this work one mineralized diorite dyke assaying 5.50 over eight feet was intersected.

The firm of Nesbitt Thomson subscribed $150,000 to explore the extension of the Malartic Goldfields ore zone on the Dubuisson property where a good measure of success was expected.

René Chénévert of Montreal was the legal counsel and one of the principals of Dubuisson Gold Mines, who, along with J.P., was trying to develop the Dubuisson claims in June of 1938:

J.P. to René:

Re: the 160,000 shares. When I secured an option on this stock I took it direct to Thomson and told him he could have it if he made a deal to finance the property. If the stock was available then, it should be now, and this was my understanding when I last talked with you in Montreal.

I have since talked with Thomson about this and he has informed me that if we raise 140,000 shares in addition to the 160,000 shares it will be okay.

I have already explained this to Mr. Davis and he is willing to give up 15% of his stock but definitely refuses 30%.

I am also willing to give up 15% of my 200,000 shares but, in view of the fact that I am giving my time for nothing which means years of experience in this area, and was a result of bringing the company to its present stage, I think it is very unfair that I should have to give up more.

My opinion is that a deal with Nesbitt Thomson should be made at the earliest possible moment and I suggest that you send complete statement to each vendor covering the share situation and asking them to put up 15% of their stock.

I am very anxious to have a deal with Nesbitt Thomson completed while the iron is hot. At the same time I do not feel like giving 30% of my holdings to attain it.

If any recent strings have been put on the 160,000 shares in question I suggest that this be cleaned up quickly so that a deal can be consummated at the earliest possible moment.

Nesbitt Thomson participated in the Dubuisson financing but within one year the property was closed down due to wartime conditions.

The other claims owned by Dubuisson were south of East Malartic and Malartic Goldfields and J.P. explained their potential to R.H. Scarlet of Grasset and Co., Toronto:

These Dubuisson claims lie south of the Rand Malartic property and for the most part between the properties of East Malartic and Malartic Goldfields.

The Dubuisson claims are totally covered with overburden; however, I do know from general geological information that a favourable structure occurs on these claims and that diamond drilling would probably lead to an ore body.

I believe my general knowledge of the area could be used to advantage in development of the claims.

Amphi Gold Mines

The Amphi was another group of claims adjoining the East Malartic and Malartic Goldfields mines which were coming into gold production. This property had potential as explained by J.P. in that it was within five miles of the East Malartic mine. As usual other groups of investors, such as Col. MacDowell, held large interests in the Amphi claims and J.P. along with Inspiration had the opportunity to participate as discussed with Mel Thomson, March 22, 1938:

Col. MacDowell has been in the office two or three times recently in connection with a property known as the Amphi Gold Properties Ltd., west of the McIntyre property, north of Sladen. He is very anxious to make a deal, and I was wondering what you thought of the location, etc. and if it would be wise to have Inspiration take a share in it for diamond drilling purposes.

You might think it over, and on your visit to Montreal on Friday we can discuss it then.

J.P. replied:

I believe the Amphi property of Col. MacDowell has some merit and we would be willing to drill it for 75% cash, balance in shares or interest.

Nothing materialized with Col. MacDowell and no work took place on the property until April of 1940. Howey Gold Mines had then taken control of Amphi and contracted Inspiration to drill the property. Since the Amphi was so close to the East Malartic and Malartic Goldfields, National Malartic and Sladen Malartic, the geology of these mines was available to Inspiration. When Quebec Gold hired Inspiration this information was very important when planning a diamond drill program.

J.P. explained what might be expected from the drilling:

I went over the Amphi drilling with Mr. Murphy. I would say that the drilling is being very carefully done. The area is large and will take much more drilling to eliminate it. It would be hard to squeeze in either the Malartic Goldfields or National Malartic showings between the cross section drilling on Amphi and not pick them up; however, we know that good mines may be made from much shorter lenses and the intermediate drilling now being done may find some of them. I believe the West Malartic occurrences are quite substantial and are confined to lenses of about 100 feet in length. If a few higher grade ore sections could be added to the ones you already have, I believe underground work would be justified. These may be found in the intermediate drilling now being done and the deeper drilling which I believe should be done.

The diorite ore occurrences on National Malartic seem to be better on some margins than others thus future drilling may prove that you are on the upper margin of something better lower down.

Mr. Murphy is sending me a map of all drill holes. I will then study it with the geologists at East Malartic and Malartic Goldfields and we may be able to give you some idea worthwhile. In any case I believe Mr. Murphy is doing a good job.

Roy Birks, president at Howey Gold Mines, was very pleased with some of the drill results:

No doubt you are aware of the fact the Amphi last drill hole cut 14 feet of 8.00 ore. Because of this we have decided to do some further diamond drilling, and no doubt your drill crews are on the job. These results may give us the information of which we are in search, and I sincerely trust that we will have sufficient information to warrant underground development.

The Amphi was inevitably caught up in the wartime restrictions. This property would have to await the war's end.

Marbenor Malartic Gold Mines

The war was at its peak and the effect on gold mines and gold exploration was devastating, as has been seen with Senore, Cournor, Pascalis, Dubuisson, Amphi, Partanen Malartic and Seventh Malartic.

J.P. loved the search for new mines more than any other facet of the business. He was still doing his best to develop the properties that he could. As with the Seventh Malartic, as long as the development was associated with a gold producer, some work could be done. He had time on his hands and could always bring the principals of various properties together in preparation for the end of the war, whenever that might be. Marbenor Malartic was one of these.

The Marbenor property was of much interest to J.P. geologically. The discovery of a float boulder just north of Malartic Goldfields was intriguing because float boulders had led to the discovery of East and Goldfields.

Promoters and brokers were always trying to work with J.P. because his opinions and association with properties gave the public great confidence in a project. Ed Benson was one of these promoters. He had been involved in the promotion of National Malartic but he and J.P. had not profited greatly from their work. Now he was bringing Bill Marshall into the company to assist. Hence the name Mar – ben – nor.

Benson wrote to J.P.:

This letter will serve to introduce my friend of many years, Bill Marshall.

So you may know something of his background, I can tell you that he has been a successful market operator and unlisted broker for 20 years. In other words he supplies the knowledge and experience for the market end of things while I look after the direct sales.

As you can know I have an excellent connection with many outstanding business and professional men, and I used this connection on the National to such an extent that I personally sold over 400,000 shares, in other words I put the legs under it and you did the rest by finding the ore.

Unfortunately, however, neither one of us profited to any appreciable extent, and others made all the big money, while we should have had a good piece of things.

Now, Jim, I'm all through with working my guts out for someone else when I don't share in the pie, and Bill and I have made up our minds that a combination of the three of us, yourself, Bill and myself, can take hold of a Malartic deal, such as you have suggested, and make ourselves some real money.

I am proud of my reputation for square shooting, and I'll vouch for Bill's and, I know enough about you to realize that you will "play ball" so let's take our coats off and go to town on this one and clean up enough to put things on ice for a while.

Bill believes in you, your knowledge of the Malartics and mining, just as I do, and I have a hunch that the three of us will do all right for ourselves as a combination.

Just one thing more, Jim. If it is humanly possible to throw the new property into the Bart Malartic please exert every possible effort to do so, as it will keep our present Bart Malartic shareholders happy and give us a real chance to raise additional quick money from them. Also it will save a lot of time as we can start shooting right away without waiting for an incorporation of the new company. Even if we have to phone Potter or wire him a long telegram to Alaska and do the same with our friend in the navy it is still the best thing to do.

Whatever we do, you are "in" on a three-way split with us and I'm sure we'll all do very well out of the jackpot if we play our cards right.

I know you and Bill will get along fine as you both speak the same language.

The company charter was taken out and the vendors allotted shares according to the acreage contributed by each party:

...we have taken out a charter for the company, the charter being dated Oct. 10/41 for the company having an authorized capital of 3,000,000 shares of a par value of $1 per share as contemplated by agreement.

We have instructions to issue 502,200 shares of the new company to the vendors named in the agreement with W.J. Marshall as follows:

Inspiration	103,500 shares
Dubuisson	96,600 shares
Potter	150,900 shares
McLean	151,200 shares

We have examined firm subscriptions for shares of the capital stock of the company, which will substantially exceed the provisions of the 100,000 shares at ten cents per share as set out in the agreement.

J.P. always invited Thayer Lindsley of Ventures to participate whenever a property seemed promising and Lindsley almost always accepted. He hadn't gone wrong yet although he should have taken a larger stake in East Malartic:

Thanks very much for your letter about the Marbenor property. It sounds very interesting, and I will take on a few shares.

Someday when we are a little further ahead on the Pascalis development work, I want to have a talk with you regarding the possibilities there so, when you are in town next, I hope you will give me a ring. I would enjoy having a talk with you regarding Pascalis and several other matters.

Often there was a dispute over claims. This always delayed property development and was costly for everyone concerned. Sigma Gold Mines was located several miles to the east and J.P. notified A.O. Dufresne, Deputy Minister of Mines, of the implications of a dispute:

During last fall we did much work in the vicinity of and to the north of a 50 ton gold bearing float boulder located… Mr. Ed Goranson, geologist at East Malartic studied the striations in an effort to locate the source of the boulder.

To explore my own theory, first I consolidated the ground of Mr. Potter and others in Dubuisson and Fournière townships and incorporated Marbenor…

I was informed yesterday that the staking of claims covering lots 1, 2… has been disputed by Sigma Gold Mines in that the tags cannot be found on the claim posts. Mr. Potter informs me that although the staker of the claims has recently joined the army and his address is now unknown, he can produce the person who accompanied the staker to prove that the ground was properly staked.

I have given you all this information to prove that we are doing a big job in the Dubuisson area… and if the claims are taken away from Mr. Potter, my plans will be seriously disrupted.

Nesbitt Thomson were still heavily involved in J.P.'s activities; however, they fully understood that new properties would never be likely to reach a position where they could reap their usual quick returns until the end of the war was in sight.

There were, nevertheless, promoters and investors who wished to position themselves so that they could be prepared when the war finally did end. Marshall and Benson knew this but had taken the property on in any case.

Norman Vincent, mentioned in the following letter, was to become partners with J.P. in several properties adjacent to Marbenor.

J.P. to W.J. Marshall:

…I do not believe the lots of Vincent are on strike of our Marbenor findings to date. The diorite dyke strikes in that direction. However, we think the values are in a more favourable cross-formation which strikes east.

I can make a deal on the Goldfields ground. However, before gathering up ground we should await the cross-section drilling to our north boundary.

It is very likely that the boulder came from the zone we have found in drilling. It looks to have come about that distance. We did not find any rock in the core similar to the boulder. However, this is possible even if its margin was in the zone found.

I realize the market is in terrible condition and is likely to remain so. Anyway, we will do our very best. I am very anxious to see yourself, Benson, Hamblky, Peaker and Trent click in a mine. You have handled the Marbenor in fine style and I am going to try hard and line up a real property. I hope it will be on Marbenor. If not, somewhere else.

I have another dandy on tap in another district which we have been holding for a considerable time...

By mid-summer 1942 war restrictions had caused the shut down of Marbenor. J.P. wrote to Benson:

I believe Mr. Marshall and yourself did a good job on Marbenor and under ordinary circumstances we would have been going ahead there now. It will be a good one to work on after the war.

I was very much impressed with your friends Seymour Trent and Court Peaker. Please advise them that I will try and hook them onto a good mine after the war. We are in a good position here to do it and have some nice ones with good surface showings right now.

We are going through tough days in our drilling business so please send us a cheque for the balance of our account as soon as you can.

Norbenite Gold Mines

Norbenite was another property named after J.P. and two promoters: Norrie, Benson and Knight. Nor – Ben – Nite.

Paul Croteau, who assessed the Norbenite property, was another prospector and friend of J.P. The Croteau brothers had been prospecting for J.P. for many years and J.P. wanted and valued Paul Croteau's opinion on the Norbenite in 1943:

Norbenite cores were examined by the writer, also the plans. The formation is greenstone, of the soft, schisted variety common to Malartic. Ore is in altered dykes of cyanide-diorite intrusive. Quartz is sugary and carries some free gold. There is one continuous ore body for a length of 500 feet, still open to east, with several parallel shoots. The picture here looks good with possibilities of a large tonnage. The mining problem will be similar to Malartic Goldfields. Norbenite should make a profitable mine of large tonnage.

Paul Croteau

Ed Benson was part of the Norbenite financing as he was with Marbenor. Norman Vincent also had a large share in this property:

Name Norbenite finally secured. Vincent now claims his understanding with you for release of his vendors' stock pro rata with ours. Glen, myself and Knight had definite understanding that 400,000 promotion shares were to be released first and that balance vendors' shares were to be released at the discretion of the purchaser and Securities Commission. This was my understanding with you also. It would be unreasonable to suppose Knight would create a market and buy back Vincent's or other vendors' stock. This can always be arranged later as you know. I believe Vincent is merely trying to take advantage of Glen's absence to see if he can get free stock.

Otherwise I think he is merely trying to embarrass us by his attitude. Suggest you contact him and straighten him out. Will you come to Toronto on return and when?

The financing of all properties was impossible or slow to say the least, because of the times, and it was early 1944 before drilling proceeded. The property showed great promise as Paul Croteau predicted:

...hole No. 6 on Norbenite was spotted 300 feet west of hole no. 3 in the diorite zone in which values were secured near the south boundary of the property. This hole cut over a foot of material which looked very much like the boulders for which we are looking and although it looked good the material did not assay. The drill has been moved further west on strike where I think we stand a good chance of finding the source of the boulders.

Hole No. 14 is drilling 500 feet west of hole 11 on the same zone. Free gold can be seen in the core. The hole is still drilling.

This is a hot section - too much gold around not to click somewhere.

The financing was going so poorly that Norman Vincent asked J.P. to influence the brokers to allow him to sell some vendors' stock:

Re: Norbenite. In as much as there is now no bid at all on Norbenite, it would seem that I am in a difficult position to dispose of the stock and should be at least free to offer shares in the market because Ed Benson and Draper Dobie have apparently pulled their bid...

Although the property was virtually closed down, Norman Vincent was still willing to pay .15 per share in August of 1944. The end of the war was in sight and this property was a sure bet:

Enclosed please find cheque for $6,000 re: 40,000 shares of Norbenite.

Purdy Mica Mine

With the war looking as if it was going to continue for an indefinite period, gold had been identified as a mineral which was not of strategic value, at least as far as its use for machinery and direct war purposes was concerned. However, the production of gold was still essential as a means of keeping the balance of payments under some control, and the U.S. was buying all Canada's gold, thus reducing the Canadian debt for war materials.

If Inspiration was going to develop properties successfully during the war it was unlikely they would be gold properties, but would have to be directly strategic to the war effort.

But in 1940 when Mel Thomson sought J.P.'s advice on the Kelly Mica prospect, J.P was less than enthusiastic. On June 27, 1940, he wrote Mr. Thomson:

> I have gone into the mica question pretty thoroughly and find that manufactured products are making serious inroads into the mica market. If mica was needed for war purposes we might be justified in doing some work on the Kelly Mica property, at St. Simeon, Quebec. I believe this is not the case so I recommend that nothing further be done on the property.

One of the main reasons J.P. was skeptical about the mica business was the fact that developers would have to deal with so much government approval, both Canadian and American. First the Canadian Department of Natural Resources had to pass judgement on whether the mica was of a quality high enough for war uses, then the Metals Controller had to be advised of its strategic value, then the Department of National Revenue needed to be informed of its value for income tax exemptions on war materials, and finally the federal government had to advise the U.S. Metal Control Board of the mineral's strategic value.

In April, 1942, a large deposit of muscovite mica was found by Justin Purdy in Mattawa township, Ontario. This prompted Glen Wilton and

Norman Vincent of Toronto to stake ten claims surrounding the Purdy discovery. Even though J.P. was still skeptical, at Mel Thomson's urging, Inspiration promptly took over Wilton and Vincent's staking and Donald Sirola was sent to examine the Purdy discovery and to size up the surrounding area. On Sirola's recommendation Inspiration sent four of its prospectors (Paul and Wilfred Croteau, William Hanson and J.W. Workman) to carefully prospect the entire area within a radius of ten miles and as a result 24 more claims were staked for Inspiration.

As J.P. noted later that summer:

> Realizing the importance of the Purdy mica discovery, which is declared by many to be the finest discovery ever found, we made an offer in July for purchase of the property. Our offer was not accepted, although considered along with many others, both Canadian and foreign. Late in September, very attractive offers were made by two leading American mica manufacturers. Then during the course of these negotiations we made a slightly amended offer to Purdy, which was accepted after much persuasion. The accepted offer is as follows: Inspiration to pay Purdy $20,000 in cash on completion of the agreement and 30 per cent of the gross production until he has been paid a total of $50,000 and 15 per cent of the capital stock in a company to be formed. The initial payment to Purdy has been made. The Purdy Mica Mining Company Ltd. has been incorporated and 300 shares of the capital stock delivered to Purdy. The Purdy property was officially taken over on November 5, 1942.

A photograph of one of the sample sheets from the Purdy Mica mine is still on display today at the Royal Ontario Museum in Toronto. The sheet is much larger than a normal man, and is a reminder that this particular mineral played an important part in the Allied efforts during the Second World War.

This story does contain one small vignette of personal conflict that is worth recounting. Norman Vincent, the mining promoter who became involved with J.P. and Glen Wilton in the development of various gold mine properties, around this time ran afoul of Mel Thomson because of a series of what Thomson considered dubious newspaper advertisements to attract investors to projects which J.P. was advising Vincent on.

Thomson didn't want J.P. to have anything to do with Vincent and made his feelings clear to J.P. and Wilton at a Purdy Mica directors' meeting. J.P. was also angry with Vincent as can be seen in this letter:

Nov. 25, 1942
Mr. Glen Wilton

Sorry I did not see you before leaving Montreal. Near the end [of the meeting] the last ad of Vincent was presented and from then on fire flew in all directions. The other directors are convinced that you and I are pretty close friends of Vincent's and could prevent such things.

Vincent definitely promised me that such ads would not occur again, so little can be based on his word. Anyway the last part of the meeting was quite unpleasant for me, and lasted much longer than I expected.

J.P. Norrie

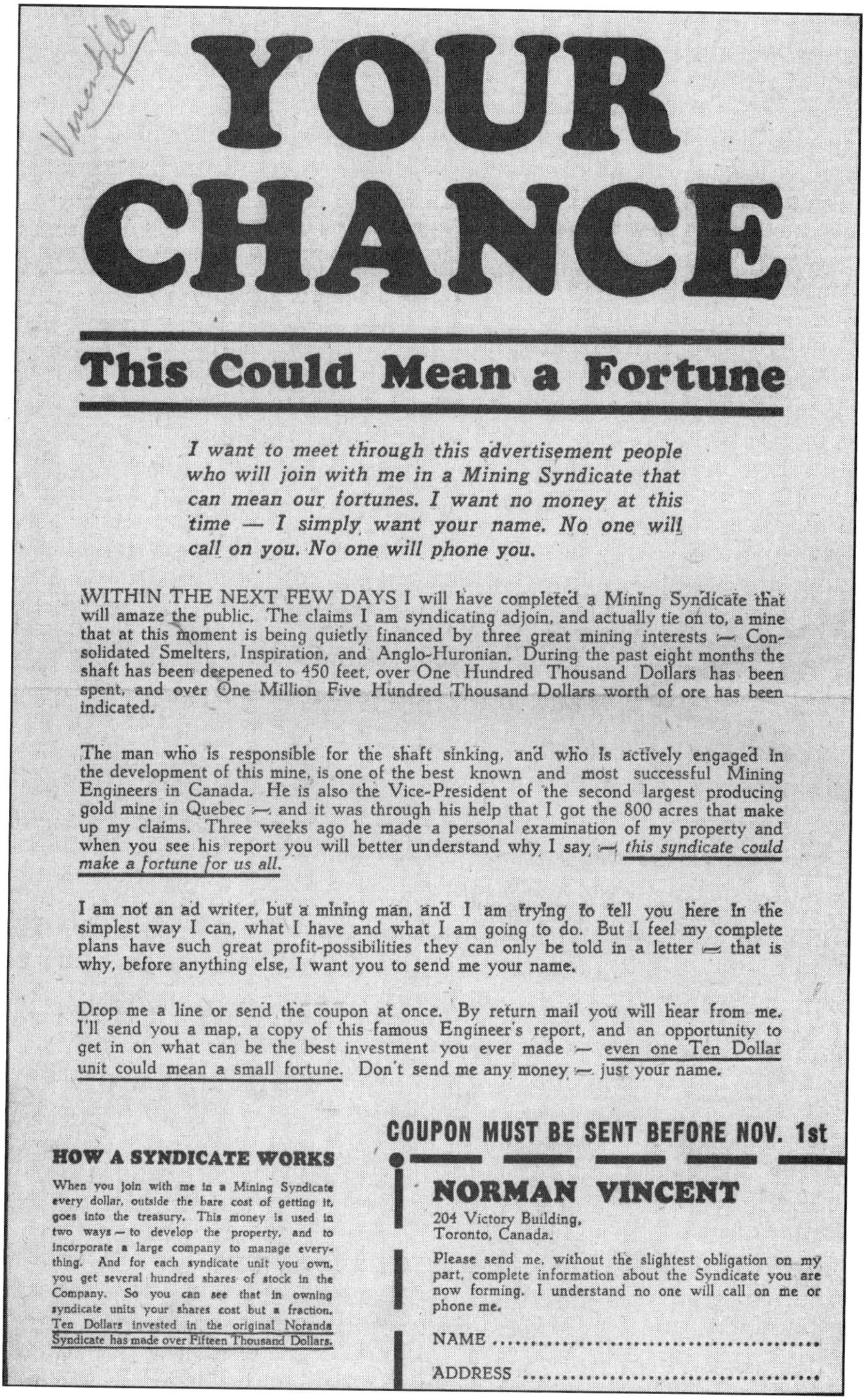

YOUR CHANCE

This Could Mean a Fortune

I want to meet through this advertisement people who will join with me in a Mining Syndicate that can mean our fortunes. I want no money at this time — I simply want your name. No one will call on you. No one will phone you.

WITHIN THE NEXT FEW DAYS I will have completed a Mining Syndicate that will amaze the public. The claims I am syndicating adjoin, and actually tie on to, a mine that at this moment is being quietly financed by three great mining interests — Consolidated Smelters, Inspiration, and Anglo-Huronian. During the past eight months the shaft has been deepened to 450 feet, over One Hundred Thousand Dollars has been spent, and over One Million Five Hundred Thousand Dollars worth of ore has been indicated.

The man who is responsible for the shaft sinking, and who is actively engaged in the development of this mine, is one of the best known and most successful Mining Engineers in Canada. He is also the Vice-President of the second largest producing gold mine in Quebec — and it was through his help that I got the 800 acres that make up my claims. Three weeks ago he made a personal examination of my property and when you see his report you will better understand why I say — *this syndicate could make a fortune for us all.*

I am not an ad writer, but a mining man, and I am trying to tell you here in the simplest way I can, what I have and what I am going to do. But I feel my complete plans have such great profit-possibilities they can only be told in a letter — that is why, before anything else, I want you to send me your name.

Drop me a line or send the coupon at once. By return mail you will hear from me. I'll send you a map, a copy of this famous Engineer's report, and an opportunity to get in on what can be the best investment you ever made — even one Ten Dollar unit could mean a small fortune. Don't send me any money — just your name.

HOW A SYNDICATE WORKS

When you join with me in a Mining Syndicate every dollar, outside the bare cost of getting it, goes into the treasury. This money is used in two ways — to develop the property, and to incorporate a large company to manage everything. And for each syndicate unit you own, you get several hundred shares of stock in the Company. So you can see that in owning syndicate units your shares cost but a fraction. Ten Dollars invested in the original Noranda Syndicate has made over Fifteen Thousand Dollars.

COUPON MUST BE SENT BEFORE NOV. 1st

NORMAN VINCENT
204 Victory Building,
Toronto, Canada.

Please send me, without the slightest obligation on my part, complete information about the Syndicate you are now forming. I understand no one will call on me or phone me.

NAME ....

ADDRESS ....

Mining promoter Norman Vincent angered J.P. with this advertisement

Bart Malartic Gold Mine

The Bart Malartic was not exactly on the greenstone band that had yielded so many profitable ore bodies. However, J.P. told Ed Benson, who was again interested, "the Bart definitely has a chance."

J.P. gave Benson his reasons why the Bart was worth a shot:

...I am enclosing a sketch showing the location of hole number one on the Bart property. This hole is being accurately tied in today and an accurate sketch will be made... and forwarded to you shortly.

The first ten feet of core in the hole is well mineralized also several other sections down to the present depth. Mineralization in this area does not always mean ore so do not be surprised if the ore does not kick.

You will note on the enclosed sketch that the south contact of the most favourable ore zone passes through to the north of the Bart property but so far no ore bodies have been found south of this contact on the Bart, although on East Malartic some isolated ore sections have been found south of this contact.

On the face of it, rock conditions on the Bart Malartic are not the best. However, aside from this, the Bart, as I have always said, has a chance for two reasons:

1) In an area where great ore concentrations are found such as East Malartic, other ore bodies are likely to be found under most unusual conditions within a reasonable distance. For instance, a like area surrounding Hollinger and Lake Shore. I maintain that anywhere in a radius of five miles, with East Malartic as the centre, is a good place to look.

2) We think that the granite outlined roughly on the sketch is responsible for the Malartic ore condition. The Bart property lies between Malartic Goldfields and the granite so the gold bearing solutions must have crossed the Bart property at some horizon or else the granite extends north under the Bart at depth.

I spotted the first hole directly in line with the Malartic Goldfields mine to the north and the indicated ore body to the south and as close to the latter as possible. It was moved west 200 feet to be well within the Bart property.

You will see from all this that although the Bart rocks are not considered the best host rocks, several other factors are intensely favourable.

...my opinion is that the ground to the north is in a very favourable ore area so I hope you can proceed on the plan I have outlined to you.

The first drill hole was located on the Malartic Goldfields and angled to hit bedrock on the Bart property.

Hole number one on Bart Malartic was spotted by me 200 feet west of the northeast corner of the property and approximately 60 feet north of the Bart north boundary on Malartic Goldfields property.

The latter was done so that the 45 deg. hole would pass through overburden on Malartic Goldfields property and enter the rock as close to the boundary as possible. We expected overburden to be about 60 feet deep. As a matter of fact, it was 120 feet on the incline of the hole so that the hole entered bedrock some distance inside the Bart property.

The hole was 250 feet deep last evening. The first ten feet of core was well mineralized vein material. Several other mineralized sections have been encountered on the way down. The core is being logged today and assays will be made as quickly as possible. These will be used in spotting further holes.

It was planned to cross-section the property from north to south in the plane of the hole to pick up favourable ore structures. If assays are secured in this hole, further holes on strike will be drilled to trace up the ore zones.

The property did not show enough mineralization and in the next two years the claims were allowed to lapse.

Clarnor Malartic Gold Mine

Interest in the Clarnor Malartic began back in 1938 along with the National, Partanen and Seventh Malartics. It was then called the West Shore Malartic and showed some promising showings to Andy Davis:

...the west shore Malartic Gold Mines property is situated directly on the strike of the Siscoe, West Siscoe and Kiena properties on the west shore of Lake Siscoe. All properties mentioned have good ore occurrences, particularly the Kiena which adjoins West Shore.

With a limited amount of work they have some fair ore showings. I believe the chances of finding more are very good.

W.C. Crawford is very anxious to have you act as a director. He says he will not take up any of your time and will give you an option on a block of stock. He also said that he will hand over the management of the property and diamond drilling to Inspiration.

I know you are busy and have little time for additional duties, however, if you care to act as a director I think we can relieve you of much extra work and at the same time we will get some free information in a district where gold mines might occur...

J.P. was spending considerable time on new properties, which didn't seem to please Mel Thomson. Thomson wanted to make sure J.P. was taking care of Nesbitt Thomson business through Inspiration. But the search for new mines was J.P.'s first love, and neither Nesbitt Thomson nor even Inspiration always wanted to participate:

Mel Thomson to J.P.:

I had a chap in to see me the other day and he mentioned the fact that there is a group going around selling West Shore Malartic Gold Mines, and that you would be in full charge of mining operations.

I do not know what interest you have in this property, and I do not want to feel that I am too critical, but it appears to me that with the work you have on hand with the mines in which you are personally interested, and in which we are interested, your whole time would be pretty much taken up. I hardly think it fair for you to take charge of outside properties when the only reason for them approaching you is the use of your name in the selling of stock...

Investors were continually asking J.P. for his investment advice. F.J. MacNamara wrote to him in early 1939.

J.P. replied:

Re: West Shore Malartic property I have been lending what assistance I can to those making an effort to finance work on the property. As you probably know, the ground is well located and anywhere in this immediate area, there is a good chance of finding mines. I do not believe West Shore Malartic have sufficient showings at this time to even suggest that a mine occurs there. However, I believe that money properly spent has a good chance as anywhere on the area of finding a mine.

You can probably get in cheap in a fair gamble. Your chief worry would then be to see that you get a run for your money.

A.J. Nesbitt also wanted J.P.'s opinion.

J.P. replied:

The property of West Shore Malartic has had considerable work done on it and has some scattered values but not sufficiently interesting to warrant further work at this time.

Extensive operations on the adjoining Kiena property have recently been discontinued due to limited ore developments. At one time these seemed to have enough importance to warrant work on the West Shore property. At the moment this does not seem to be the case.

The Westshore Malartic lay idle until the end of the war seemed at hand. In 1945 the West Shore was reorganized and called the Clare – nor Malartic. Even Mel Thomson now wanted his share of this one:

Enclosed please find cheques made out to yourself for $10,000 each from Inspiration, P.A. and myself, for the purchase of 100,000 shares each of Clarnor Malartic Mines Ltd.

I phoned you today, but was advised that you were on your way to Toronto.

J.P. himself took 500,000 shares at .10 per share. Andy Davis, as well, took 100,000:

Enclosed please find my cheque for $10,000 in full for 100,000 shares of Clarnor Malartic (Clare Norrie Mines) stock.

I understand the company is just being formed, and that it will be a little while before the stock is issued, and I will appreciate being advised before the stock is issued to me so I can tell you what denominations I would like it in and in what name...

The employees, who always had first-hand information, also wanted as much of Clarnor Malartic as they could get. J.P. wrote to Mel Thomson:

In answer to yours of the 13th, I attempted to distribute Clarnor shares to the employees of Inspiration, Perron and Goldfields. The requests were so great that out of my full allotment I was only able to supply one half of the amount.

Citralam Malartic Gold Mine

The Citralam Malartic property was located adjoining the Norbenite property to the north. Since the Norbenite geologically was so promising, the Citralam was in demand, as well. The Norbenite had been idle for several years but the war was ending and investors wanted to be ready for the inevitable gold mining revival. J.P. was consulting engineer with Norman Vincent Corp. and they owned several properties including the Citrilam . J.P. personally invested in the Citrilam which demonstrated his commitment to the property:

Attached please find confirmation covering sale to you of 50,000 shares of Citralam Malartic Mines Ltd. @ .18 a share, which amounts to $9,000.

We also beg to confirm that we have this day granted you an option on 50,000 shares of Citralam Malartic Mines @ .20. This option expires on April 13/45.

Activity on this property began in the summer of 1945 but J.P. would not live to see the results.

Vicour Gold Mines

If there was one characteristic that epitomized the life of J.P. Norrie, it was his commitment to his friends. Whether it was the prospector who found the mine, the investors who bet their money on his judgment and foresight or the employees who made the dream come true, each mine story could not be told without that fact becoming evident.

Dr. Conrad Wettlaufer

The Vicour was one of these mines and the story can be traced back to 1927 and the Sylvanite Gold Mine in northeastern Ontario. J.P. was underground superintendent at the Sylvanite and about this time met Dr. Connie Wettlaufer from Buffalo, N.Y. Whether J.P. met Connie at Sylvanite or through Harry Oakes at Lakeshore Mines is not clear. What is clear is that J.P. did everything in his power to help Connie Wettlaufer until his own death in 1945.

Connie Wettlaufer's contribution to mining is explained in this news article in the *Canadian Journal of Finance*, March 29, 1945:

It has been said and with a good deal of justification that the gold mines of Porcupine and Kirkland Lake were financed "south of the border". In the pioneer days, prior to 1914, Canada had not the wealth, within herself, to undertake operations running into millions such as were necessary to produce gold from the rocks of the Precambrian Shield. While a certain amount of British capital found its way to Porcupine, the conditions encountered differed so widely from what the English were accustomed to in South Africa that comparatively little money was attracted to the Ontario field.

The most prolific source of venture capital near at hand was in the United States and it was to the United States that the pioneers from Porcupine and Kirkland Lake went in search of the dollars which would make their gold prospects into mines. Perhaps there is no one city in the United States which, for its size, has contributed as much to the development of the northern Ontario gold camps as has Buffalo, and in 1912-13-14 this bustling city on the south shore of Lake Erie was the Mecca for many a prospector rich in claims but poor in money.

When the late Sir Harry Oakes and William H. Wright made their sensational gold discovery on the shores of Kirkland Lake, Buffalo was their natural "first port of call". After much hard work and intensive selling Oakes persuaded an influential group of Buffalo capitalists to come to Swastika to examine what is now the Lakeshore. Old timers still recall with watering mouths the entertainment which was afforded the visitors from Buffalo. However, the gold showing stood up under expert examination and the Buffalo gentlemen put up the necessary capital which proved to be one of the most satisfying, speculative investments ever made.

Some years later these same gentlemen, having had their appetites thoroughly whetted by their success in Lakeshore and the neighbouring Wright-Hargreaves, acquired a property adjoining Wright-Hargreaves and formed the Sylvanite Gold Mines. The property went into production in 1927. The Sylvanite has always had efficient and businesslike management but it never showed promise of being anything but a "good little mine".

J.P. began his gold mining career in the Porcupine mining camp and then moved to the Sylvanite where he was underground superintendent, 1923-1924.

Connie Wettlaufer became a partner in the Vicour in 1935 and asked for J.P.'s advice on what he should do:

I have been reading the newspapers and keeping in close touch with your development in Quebec. For years I have been preaching that you have more experience in Quebec and know more about the geology and formation than any other engineer I know, and I would therefore rely on what you had to say.

I have been unable to interest any mining company to take an option to develop the Vicour. Mr. Dorfman of Anglo-Huronian has been sitting back always saying things are getting better in Quebec every year.

Now I am going to ask you what you would do if you were in my place. I have been financing the Vicour for over two years, keeping two watchmen there cutting a road and wood and looking after the plant and buildings. I must say that everything is in first class condition. There are accommodations for 40 men with all equipment and you know what a fine plant we have.

Will you be good enough to give me your honest advice what you think I had better do. Can you help me? I would be glad to pay you a commission if you can help me make a deal. You have expressed yourself in your report and on several occasions that you think you could make a mine of the Vicour.

J.P. answered Wettlaufer with his advice on small mine development and what he should do at that time, Nov., 1937:

If we were dependent upon any large mining company to develop either East Malartic or Perron there would have been no mine at either place. The shot on each was too long.

I believe the Vicour should be financed by a good broker and with such the property will get a break.

Just now it is impossible to do any financing through the public. I believe we are at the beginning of a good market again when this class of financing can be done. When market conditions improve I advise you to get busy with some reliable broker when I feel that some real work can be done on the Vicour. We have a very competent organization here which can be of service when you are ready.

During the late summer of 1938 market conditions were ready for financing. Wettlaufer had the partners agreeable to participating with their vendors' shares:

I am in receipt of your letter of Sept. 1 in which you state you have agreed with McDowell on some of the financing of our claims in Bourlamaque.

I was looking over your report on the Vicour this morning and note what you say about the geology. I have had several people who are interested ask me why we did not have more of a geological report, so the thought came to me that you had your geologist with you when you made the examination and I believe it was Mr. F. Buckland. Could you give me a separate report regarding his opinions of the geology of the Vicour? I think that would help considerably. Would appreciate it very much.

The Consolidated Mining and Smelting Co. some time ago had their men go in, clean out all the trenches and make a thorough sampling of the surface. Of course, the mine then was full of water. The men spent about three weeks doing this work. Am inclosing copy of their assays.

Mr. Outhred (Consolidated) about six weeks ago took another look at the property and recommended it to Consolidated. Some time prior, however, Mr. Archibald became interested in the Barber Larder, which they are financing. They therefore were not ready to go in on the Vicour at the present time, but liked it very much.

A while ago I had a talk with Mr. Dorfman of Anglo-Huronian who hold 500,000 shares of Vicour for the work they did at .25 a share. As the Vicour stands there is left in the treasury 1,560,000 shares out of a 3,000,000 share corp'n, there being, practically speaking, only two stockholders, Anglo-Huronian and Quebec Goldbelt Mines. There are less than 12,000 shares issued to outside stockholders. Mr. Dorfman of Anglo-Huronian agreed to put back into the treasury of the Vicour their shares, pro-rata, if Quebec Goldbelt would do likewise, so the Vicour would have 1,000,000 shares to the vendors and 2,000,000 shares in the treasury, which of course would make a most desirable setup for any company to take on the financing.

Have you any suggestions to give me or can you help me in any way to finance the Vicour, with the understanding you would be personally compensated?

What I am writing you today I want you to be good enough to treat very confidentially. I am convinced that you know more about the Quebec country than any engineer I know. And if you would put forth your efforts and influence, I think we could get somewhere.

Would appreciate it very much if you would let me hear from you as soon as possible as I consider this an opportune time and Vicour is worthy of recommendation on a reasonable basis for the treasury shares.

C.E. Wettlaufer

Some work was done in 1940 and 1941 but war restrictions took their toll. Inspiration recommended the closedown of the Vicour with the intention of starting up operations whenever the war's end was in sight:

We are of the opinion that operations on the Vicour property should be discontinued when the diamond drilling program has been completed. We understand from Mr. Stee that notice had already been given to the monthly employees; if not this should be done as of Mar. 31/42.

We note that at a meeting of the Vicour directors held on Mar. 14/42, resolutions were passed altering the existing options on treasury shares so that shares which had been taken down at .35 and .40 per share would be regarded as .20 stock and additional shares will be issued for the amounts already expended. All further expenditures to the completion of the present diamond drilling program to be met by the issue of .20 stock.

We are agreeable to selling the plant, equipment and supplies when operations have been discontinued.

We note that the remaining treasury shares are to be optioned to the present participants at a price of .60 per share for a period expiring six months after the termination of hostilities, which may mean at a time of any unsettled conditions, and although this date was mentioned at the last meeting of the participants in Montreal, we would like to have the option extend to at least one year after the termination of hostilities.

J.P. suggested to Wettlaufer that a deal with Perron could possibly be worked out, as with the Seventh Malartic. Since the Perron needed more ore, they could develop the Vicour during the war:

I believe a deal on Vicour could be made with Perron. They would be the logical people to run the mine as conditions at Vicour are best understood by the Perron staff.

If you are in favour of this perhaps the option now held by Anglo, Smelters and Inspiration could be taken over by Perron. I would appreciate your views on this.

The deal with Perron never did come to be. However, the war's end became evident in late 1944 and gold properties began to attract interest again. J.P. wrote to Wettlaufer:

The Vicour deal seems to be complete now. I think enough money will be made available to do some real work and perhaps add to the present ore reserves...

I have just returned from Toronto and was discussing Vicour with various parties concerned. The ones who propose to do the financing thought there has been undue delay in getting it going.

My opinion is that every opportunity should be taken of present bullish market conditions and that all agreements, etc. should be completed and the financing gotten underway at the earliest possible moment.

Now is the time for Vicour to get some money so try and put some heat on the situation.

It took J.P. four years to bring the partners together to begin exploration work on the Vicour. Just as the first real work was completed in 1940 the war prevented any possibility of a mine for five more years. Mining properties of this speculative nature were far too risky for most large mining companies as J.P. told Connie Wettlaufer in reference to even the East Malartic. Nesbitt Thomson, as well, wouldn't touch the Vicour. This property belonged to J.P.'s friend, however, and one way or the other it was going to get its shot. Friendship was obviously a stronger motive than money, any day. The Vicour had already produced its gold, the gold of friendship, but finally the war was over and the day was at hand when the Vicour would produce the other gold. But then, this was never meant to be, due to J.P.'s sudden death on October 9, 1945.

J.P.'s Last Mine

Croinor Gold Mines
The Financial Post Jan. 27, 1945

How simply and inexpensively large sums of money can be raised for new gold mining enterprises was effectively demonstrated the past month by Croinor Pershing Mines which has outlined one of the better new Quebec gold discoveries this year.

Ninety four letters were sent out to Croinor shareholders under date of Dec. 21 offering rights on the basis of further shares at .75 cents for each two shares held. By Jan 15, less than a month later, a total of roughly $1.1 million had been sent in or approximately twice the 525,000 it was planned to raise by sale of 700,000 shares as rights. Actually 800,000 shares were allotted with 600,000 placed in the treasury.

Not only did individuals send in cheques for larger sums than they were entitled to receive in shares under the rights offering but utter strangers sent in marked cheques for up to $15,000, in fact marked cheques came in from small towns and hamlets all over the place, it is stated.

Now the Croinor has nearly $650,000 in the treasury and the management is hopeful that this sum will be all that is necessary to raise by the sale of shares to see the company through to production. Authorized capital is 3,000,000 shares with only 2.2 million shares outstanding. Vendor interests received a total of 900,000 shares, 500,000 shares were sold at varying prices and the latest offering of 800,000 shares has netted the additional $600,000.

Expenditures have been so small to date, probably not more than $25,000, but the company has been able to outline rather important results in work completed.

The company's main showing is the best so far disclosed in Pershing township to the east of the Siscoe-Lamaque area, has a length of 850 feet averaging $10 a ton across an average width of 10 feet with both ends open. An additional length of 2,000 feet is indicated to the west. Two other veins have been established, one showing a length of 300 feet averaging $7.50 per ton over five feet.

In a report to shareholders accompanying the offer of rights, Pres. J.P. Norrie reports: "It is planned to purchase and erect a mining plant immediately to have everything in readiness to bring the property into production when government restrictions are lifted."

Enough work is stated to have been completed to indicate the presence of a large tonnage of good grade ore. At the time of the offering of rights to shareholders the main ore body was indicated for 700 feet averaging $10.20 per ton across 10 feet. If underground work confirms the downward vein extension with similar tonnage and grade to only the 500 foot horizon, a total of approximately 300,000 tons of ore will be exposed and a 300-ton mill would seem a possibility.

Diamond drilling has shown numerous veins in the property and it is quite possible that other ore shoots of importance will be indicated. So far all drilling completed has proved productive.

Present board of directors of Croinor includes men who have been noted for their success in Quebec gold mining circles. The president is J.P. Norrie and in official quarters, it is stated, this is thought to be the only gold mining company of which Mr. Norrie has assumed the presidency. Vice-president is W.G. Warden while remaining directors are A.J. Davis, M.A. Thomson and G. Wilton.

The company's property is rather large - a total of 2,560 acres or four square miles - and much of this remains to be explored. Success at Croinor has resulted in a great burst of activity throughout Pershing township generally and it is quite possible that the Croinor results will finally mean a discovery of still other important gold occurrences in the area.

Perhaps even more significant to the Quebec gold mining industry, however, is the apparent indication that Quebec's gold belt continues strongly eastward even beyond the rich section of the Siscoe-Lamaque-Golden Manutou-Louvicourt section.

Chapter 25

J.P. Norrie 1891-1945

My Dear Mrs. Norrie:

I just learned this minute that my dear friend Jimmy is gone. To tell you the truth, I feel terrible. He has been such a good friend and such a nice man. I will miss him terribly. I always looked forward when going north to see his smiling face when I would meet him and now that is all gone.

My heart aches for you for I know what it will mean to you with a family and the responsibilities that will rest on your shoulders. My deepest sympathy is extended to you and if there is anything in the world I can do to help you, I would be more than pleased to comply. Best wishes from a sincere friend of Jimmy's.

Most sincerely,

C.E. Wettlaufer

By Harry Norrie

Conditions at Croinor Pershing were much the same as at East Malartic, 30 miles from the rail head at Senneterre, and when the decision was made to go underground it was necessary to build a gravel road. As general manager and president of the company J.P. would visit the property at regular intervals. On his last visit he asked about the road and was told it was within 18 miles of the mine. He could have gone back to Senneterre by boat but no, he would walk the 18 miles and see for himself. For all the miles he had walked through the bush this would be his last.

From Senneterre he drove directly to Noranda to meet with Doug Marshall, Manager of Miners Inc. Late in the evening Marshall realized that Dad was in extreme pain and made a quiet departure to allow him privacy to call Doctors Kunkle and Turner. It was four days later I met Bob Hoffman in the lobby of the hotel.

Early in the afternoon on the day that Dad died in a room in a Noranda Hotel, I came into the lobby and was approached by a gentleman inquiring about Dad's condition. I told him, "If he gets past four o'clock, I think he'll be okay." Somewhere I had read that the human body is most vulnerable around four o'clock AM and PM, and even though a heart specialist had been flown in from Montreal and had finalized his examination with "no hope" I still thought after four days and such a strong body he would recover.

Leaving the gentleman in the lobby, I went to a room where most of the family had gathered. At about six o'clock a nurse came into the room, and with a simple "He's gone" turned and left the room. Darkness fell!

The inquiring gentleman that I had met in the lobby I later found out was Bob Hoffman, an American geologist from New York City, and he along with almost the entire Canadian mining community had been made aware of Dad's condition. At that time hardly a week went by that Dad was not mentioned in the *Toronto Star*, *Telegram*, *Globe* or one or another magazine.

A few weeks after speaking with Mr. Hoffman in Noranda I ran into him again in the lobby of the Chateau Malartic Hotel. During a lengthy conversation he asked if I had ever read Epictetus. Epictetus? I hadn't even read the morning papers; but then at that time in that latitude there was no such thing as newspapers, at any hour. Noting my hesitation and sure of my ignorance Mr. Hoffman went on to say that Epictetus was an educated Roman slave and in his discourses (book 11 chapter 15, *Great Books of the Western World*) had asked "Why be determined for the sake of being determined" as if there's a choice. It was Mr. Hoffman's opinion that Dad was guilty of just that failing. But I disagree. All great men are to some degree determined, but with others determination is a built-in condition like eyesight or the sense of smell: a condition from which they have no recourse, and wish none.

A couple of years after I met Mr. Bob Hoffman, he and his brothers Arnold and Irwin came out with the book *Free Gold* (but don't get carried away, there is no such thing as free gold. Visible gold yes, but no free gold).

The central part of J.P.'s biography shows the pressures of responsibility built up over the years, and while it did not change his character it cemented his mind to the requirements of his work.

There could never have been a feeling of drudgery in his life, things changed too quickly for that. At Stadacona he was still too young, only 36, and his sense of humour was more apparent than the stress of obligation. His humour never left him but it was less frequently called upon as time went by.

His year in the Arctic was just a wonderful adventure. He was not a manager there although his ability was recognized and called on. It was unknown land; he was required to record what he saw and to protect himself and his party from the elements.

From the Arctic he went to Treadwell Yukon, where again there was nothing, no roads, no power, no telephone, nothing. He was commissioned to build a producing gold mine and he almost made it, but he was not unhappy with the turn of events that led him elsewhere.

His lucky star was just coming over the horizon when he took the job as manager of Perron. Here he was introduced to the people who would provide the funding for him to broaden his personal interests while still making it profitable for them.

It was not greed that drove J.P. Norrie, never greed. It was the necessity to get things done: to start the job and finish it. It was the greed of other men that saw in him an agent to their wealth. Money to him was just a primary tool to make things better: industry.

At East and Goldfields and the 40 plus projects he took on, he left the physical work to others and used his years of accumulated knowledge to direct them, and still his sense of humour was there, only a little deeper and sharper.

The only problem was the limit of his energy which he did not recognize and which, too soon, ran out and left him.

From *Free Gold*

By Bob Hoffman, New York geologist

"It's the same diabase dyke that crosses the Noranda," stated Jim Norrie, "the one and only. The Stadacona will find all kinds of ore." [J.P. was the Stadacona's first manager.]

We smiled condescendingly. Such a statement was a cry in the dark, wishful thinking in the face of hard fact. But Jim continued to propound such farfetched theses of structure until one day his opinion became the voice of reason. A report signed by him was gospel, for was he not "king of the muskeg", a man of faith and unswerving purpose who possessed some secret weapon which cracked open the rock and exposed ore where none was supposed to exist?

Nova Scotian and a graduate mining engineer, he drifted into Porcupine and on to Noranda. Physically rugged, with square jaw and clear blue eye, he puzzled his contemporaries by his unfailing optimism, the kind not to be expected from a man of training. He was completely unorthodox, undisciplined, but everyone admitted his insatiable drive and capacity for work, a strength which became a weakness and finally killed him. He was a crank about alcohol too, anyone on his payroll caught with a bottle was dismissed on the spot. There was no nonsense about the man. He was outwardly stolid and phlegmatic, but inwardly raged unquenchable fires of ambition. When Stadacona became a mine, he merely smiled. This was just a warm-up for him, the first inning had not even started.

His attention was directed to the Perron mine, where Alex Perron, one-time Temagami guide, was striving to work a small high-grade vein. The structure was different from anything yet encountered in Quebec - massive granodiorite. But Norrie saw something more, bands of darker diorite, which convinced him that here was something which could be profitable. He took over the development, and when he announced mill construction the reaction in most mining circles was one of anger and dismay. Scandal, the worst in Quebec, was about to break. But Jim, indifferent to criticism, serenely proceeded with his plans and the expected scandal never materialized. Instead Quebec was enriched with a new mine, which since 1936 has produced more than $12,000,000 in gold.

Next he invaded the Malartic district, which seemed a reserved province of Ventures Ltd. [Thayer Lindsley], then operating the only producing mine in the area. [J.P. put the Malartic Gold Mine on the map.] Why not diorite here too? Jim did not mull over his ideas; he went to work. And wound up with three producing mines, one of which, Malartic Goldfields, leads all others in the area.

He was beginning to feel his oats now. Backed by the powerful financial house of Nesbitt Thomson, he ventured into Louvicourt and plucked a couple more fruits from the muskeg which seemed to bloom especially for him.

"Cross the street when the light is red" is the quaint manner he described his procedure to me. "If you wait for the green, you'll never get ahead of the mob."

His untimely death in 1945, at the height of an amazing career, was the result of a heart attack. I was in Noranda at the time when poor Jim was brought into the room next to mine, there to be placed under an oxygen tent and administered to by a specialist flown in hurriedly from Montreal. A group of us collected in the lobby, anxiously awaiting the verdict of the physician.

"Very restless," was the bulletin.

"He's trying to bust out of that tent," whispered Jim DeMorrest, the diamond driller. "Just like Jim."

One of the men explained that Jim had walked 18 miles in and out of the Croinor in Pershing township, where his latest prospect was making ready for shaft sinking. He had arrived in Noranda feeling unwell. The local doctor examined him and advised immediate hospitalization. Jim scoffed at the idea, returned to the hotel, and there collapsed.

"Hear about his job in Amos?" asked one of Jim's friends. "Well, he was having a furnace installed a few days ago and wasn't satisfied the way the boys were laying the foundation, so he whipped off his coat, rolled up his sleeves, and dug the hole himself."

"Just like him," nodded another. "Couldn't sit still. What did it all get him? A few extra oxygen tanks, that's all," and he pointed to the ceiling.

At that moment final word reached us; Jim Norrie had just taken the longest trail of them all.

The Last Ride Home

by Harry Norrie

Arrangements for Dad's funeral were probably carried out by Rod Kennedy or redheaded John Allen. John worked as an accountant at Perron and was a close friend. Bob Halet and I were asked to pick out the casket. We went to the funeral parlour and chose the most expensive. Mother Marg had been so doped up by the doctors for the previous four days, she was unable to be of much assistance, but that was unnecessary anyhow.

A special Pullman car, for family and close friends, was attached to the train leaving Rouyn for Montreal by way of Malartic. This was later thought to have been a mistake since the people in Amos, his home base, planned to halt the train for a short time to hold a service.

In Montreal it fell to me to visit the baggage car to make sure the casket was aboard our train. The rule of the road was that the casket had to be accompanied.

On the train there was no evidence of an excess of alcohol but there were some wild penny ante poker games, perhaps inspired by an unseen hand. J.P. dearly loved a boisterous poker game. Sometimes Marg was a little disturbed when he was late for supper at the regular hour, he being tied up at the office with extra work.

In Montreal another Pullman car was attached to the train for friends from that area.

When I arrived on the station platform in Truro a pale-faced little devil, imagining he was riding a dark horse, trying and failing to show sympathy, sidled up next to me and whispered that he was from the funeral parlour and was selling a metal box in which the casket should be placed to protect it for all eternity from destructive elements. I said, "Do it," and he faded back into the darkness with a smirk, patting his wallet.

In the usual Christian (barbarous) way J.P. was laid out in the living room of his house on the farm where many came to pay their last respects. I was not among the viewers. I was accused of harbouring fear. Not so, I was among the most respectful by not doing just that. Respect is paid to life; death makes it a little late and unnecessary.

J.P. was taken to the little church in Onslow where the minister made the customary comments. Prior to that, I don't know that J.P. had ever entered a church. Certainly not any time I had seen him. It may have happened but not for the usual purposes.

Any Christian burial is mournful and boring and practised to me, except for a moment at this one: after the minister had finished his drone, a long line of Masonic brothers passed the grave and dropped in a small sprig of evergreen, each saying clearly, "Goodbye, my brother."

"With those other shades she first smiled a little, then answered me so glad, that she seemed to burn in the first fire of love: 'Brother, virtue of charity quiets our will and makes us wish for that which we have and quickens not our thirst for aught else." Dante (Beatrice Pontinari)

The life of J.P. Norrie was a life of tremendous mining accomplishments. The list is long and the importance to the communities of northwestern Quebec and Canada is without question. Even more important is the fact that his achievements took place during the Depression years and during the war, which demanded so much more courage, determination and faith in the work he loved.

Dear Mrs. Norrie:

Mrs. Macleod and I wish to express to you, and to other members of your family, our sincere sympathy in the very great sorrow that has come to your home, in the shockingly unexpected death of your husband.

It was not my privilege to have known Mr. Norrie intimately, but close acquaintance was not necessary for those of us who have lived in northwestern Quebec, to be aware of the place he held in the hearts of thousands of people in these mining communities, several of which communities owe their very existence to his courage and foresight.

His name was a household word with many people who didn't know him personally, but who did know and appreciate that their own living depended largely upon his enterprising spirit and sound judgement. This was particularly true during the terrible pre-war Depression years, when hundreds of men, women and little children, to whom the south hadn't been very kind, found a safer haven under his protection in one or other of his mining communities. I know whereof I speak, for I have visited these homes during my ten years ministry in this area, and moreover, our own family have shared in that haven to no small degree.

Mr. Norrie's untimely call has undoubtedly been a shattering blow, and a great loss to our northern Quebec mining area. But what must it have been to you and to his children to whom he was infinitely more than an enterprising and successful mining promoter. The years ahead will be lonely for you without him by your side. None but those who have suffered a similar loss can appreciate the depth of that loneliness, but fond and sacred memories of a great soul will come to you, and remain with you, bringing a measure of consolation and peace to your heart. Those memories you will cherish, and live with until the end of the day.

The sympathy and affection of a host of friends will no doubt be expressed to all of you at this time. And we may be grateful for the gracious support of understanding friends in our times of great sorrow. But there is a limit beyond which even our nearest and dearest cannot travel with us down that lonely road, when it would appear as if we must tread the winepress alone.

Alone, yet not alone. There is One who can go with us all the way, and Who has promised that He will never leave nor forsake those who trust Him. He shan't let you down.

"The eternal God is thy refuge and underneath are the everlasting arms."

It is to Him I would now commit you and yours in this time of trouble and bereavement, trusting that you may all find in Him, and in memories of a great-souled husband and father, a goodly measure of comfort and peace, and the courage to carry on bravely as I feel sure your husband would have you do.

Please accept this inadequate expression of our sincere sympathy and affection.

Yours sincerely,

Donald MacLeod
Val d'Or Presbyterian Church

J.P. Norrie was, like all men, a man among men, and the legacy of any man comes down to just this fact: how that man has lived his life and how he has influenced those who knew him. Although he influenced the lives of thousands in the mines themselves and in the communities that grew up from the muskeg and matured into cities, his true legacy is explained by Armine Banfield:

Dear Mrs. Norrie:

This is just a short note to tell you how sorry I am about Jim's passing and to extend to you my deepest sympathy. If there is anything I can do please do not hesitate to let me know.

I have known Jim since the early days in Pascalis and have always admired him, particularly for his tremendous energy and enthusiasm. I have never known a man who had such faith in the future of this country and the courage to push forward his plans despite all obstacles. Canada has lost one of her greatest pioneers and the loss of his sound judgement and moral support will be sorely felt by many in the mining industry.

I had a long talk with Jim last August about his future mining plans during which he said that he was going to take it easy and let some young fellows run around for him. This gave me quite a chuckle as I couldn't imagine Jim going very far away from the work he loved so well.

Jim will remain fresh in the memory of a host of friends not so much for his great achievements in mining but for the little helpful things he did for all those with whom he came in contact.

I know that the next few years will be difficult for you but time is a great healer and as the years roll by you will derive much pleasure reliving pleasant memories.

Again, please let me know if there is anything I can do.

Yours very sincerely,

Armine F. Banfield, PhD, Consulting Geologist

Although there are many examples of just how J.P. shared his knowledge, wealth, and influence in these mining stories, his dear Chinese friend, Charlie Jinn, expressed eloquently the meaning of Jim's friendship:

Dear Mrs. Norrie,

I wish to thank you so much for your nice letter of August 1, 1946. The photo of Jim has overjoyed me. It is my proudest possession. All my life it

Charlie Jinn, close friend of J. P.'s.

will remember me one of the nicest man that ever lived, a great benefactor and mostly my only true and sincere friend.

As you know I am getting ready to sail to China. As soon as there is a boat available I will go back to my country and devote the rest of my life to my family.

I want to tell you Mrs. Norrie that wherever I will be, if you ever need me, please write me and I'll come to you as fast as possible.

The photo of Jim will have the place of honour at my home and every day I'll tell my children and my grandchildren the nice man Jim was. I'll tell them also about you and your family.

Jim gave me a long time ago many thousand shares of Quemartic Mines Ltd. Whenever those shares will have a value of some kind, I will sell them and I will devote the produce of it to erect a building in my home town for children which will have engraved on the front those words "Jim Norrie Memorial".

Once again I thank you. I will always remember you and your family. When I leave Malartic, I will let you know.

Yours very very sincerely,

Charlie Jinn

Charlie Jinn never made it back to China. He was murdered on his way for the money he carried on him.

Charlie Jinn was one of many, many friends. The life of Jim Norrie belongs not only to his family and friends but to all who knew him or knew of him. This book is the sharing of his life's story and legacy.

There is a sign by the side of the road as it enters Mahone Bay, Nova Scotia with this inscription: "We love the beauty about us and welcome you to share it."

The friends and family of J.P. Norrie – Jim Norrie, Paul Norrie to some, just Jimmy to me, love the beauty of this man and welcome you to share him.

Of any man who lived a life full of hardship, and after such perseverance so much success, it must be said that he lived a life of romance and meaning.

But ironically my final thoughts go back to an old story, Harry's story of the old man from Sydney (Nova Scotia) who lost all his money on the Malartic Gold Mine along with J.P. when the mine closed down:

I drove the old man to his home in Sydney - grey like himself. He eased out of the car, all bent over as though looking for earthworms in the tall grass on a sand dune. I went to say a prayer for the old man - the Lord's prayer. But then I thought of the first two words -

Our Father - and what might have been.

Postscript

Unfinished Business

by Harry Norrie

Chibougamau

To faithful prospectors the north bush is a security blanket. It will feed you a nourishing banquet, some of which you won't like but will keep you alive, and some of which is a delicacy, much sought after in the proper season. It will clothe you but not in silk. It will house you but you will have to bring along your own axe; and it will bury you: not with long-winded oration, but better yet with the wind in the trees and the sweet song of the birds.

In the late spring of 1934 I was not too gently prodded awake on an early morning by an oversized boot and a carrying voice with a paucity of words: "Get up, you're leaving." Not long after that I was on the CNR train with Frank Buckland, to "dragonfly" here and there around Chibougamau, a place so far away even the Indians hadn't heard of it, and we'd be there for six weeks.

Prospectors working for Inspiration Mining and Development had staked out a number of claim groups. Frank was going there to geologize them and I was his dog. Early explorers had pointed at that part of the country and called it the "land that God forgot". He hadn't forgotten it, He just hadn't advertised like with Spanish moss and palm trees.

The first part of the journey took us to Oskélanéo River, a small village halfway between Amos and Quebec City. It was there at Oskélanéo that I learned to fly. Not pilot, just fly. We transferred our supplies to the airbase on the river. The aircraft we were to use was a Junkers 88, set up on pontoons, an ex-German bomber, five of which had been stolen from Germany after the First World War and hidden in various places across Canada. Later

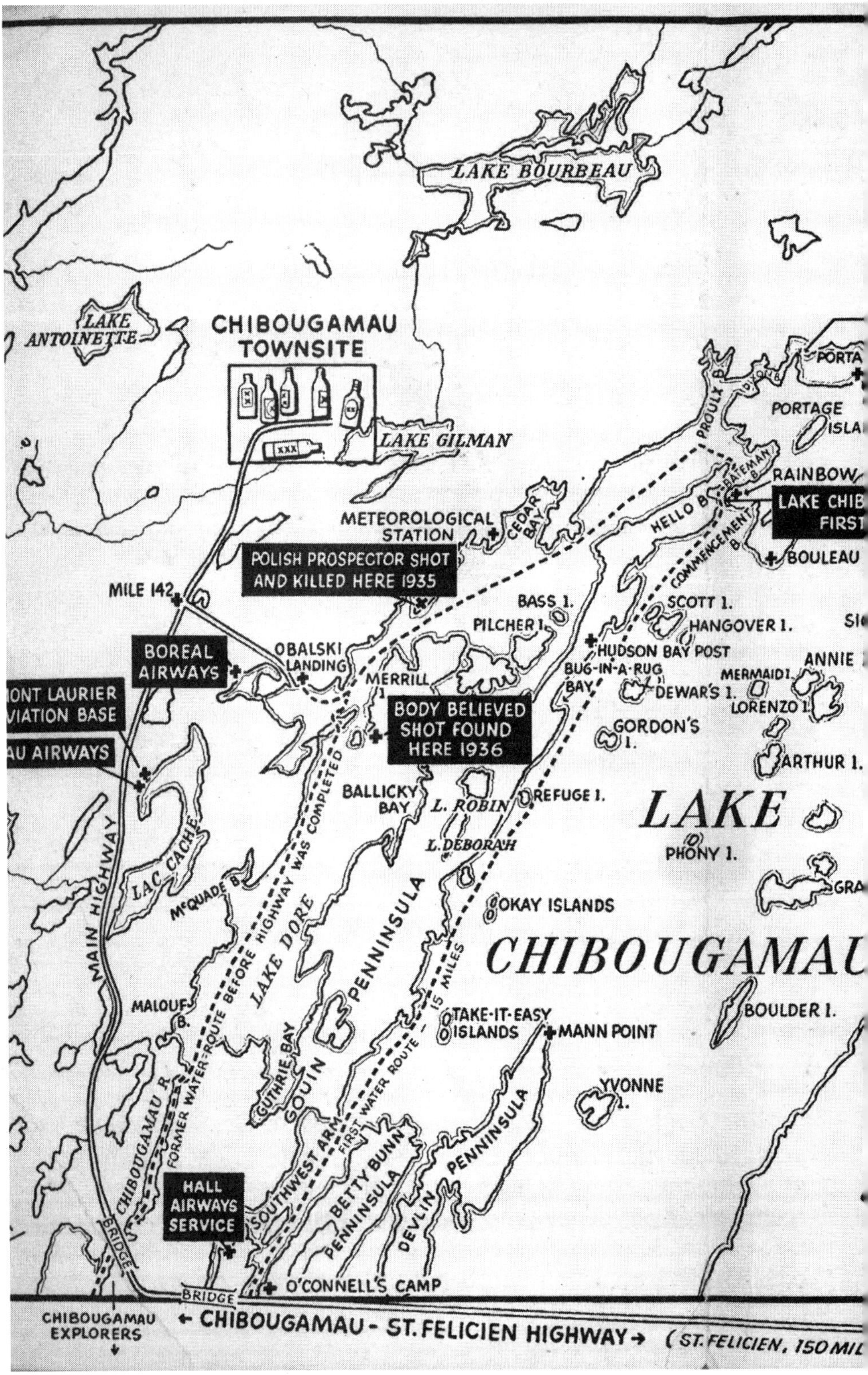
LAKE BOURBEAU
LAKE ANTOINETTE
CHIBOUGAMAU TOWNSITE
XXX
LAKE GILMAN
PORTA
PORTAGE ISLA
PROULX B.
RAINBOW
BATEMAN B.
LAKE CHIB FIRST
METEOROLOGICAL STATION
CEDAR BAY
HELLO B.
COMMENCEMENT B.
BOULEAU
POLISH PROSPECTOR SHOT AND KILLED HERE 1935
MILE 142
BASS I.
SCOTT I.
PILCHER I.
HANGOVER I.
HUDSON BAY POST
BOREAL AIRWAYS
OBALSKI LANDING
MERRILL I.
BUG-IN-A-RUG BAY
MERMAID I.
DEWAR'S I.
ANNIE
LORENZO I.
MONT LAURIER AVIATION BASE
AU AIRWAYS
BODY BELIEVED SHOT FOUND HERE 1936
GORDON'S I.
ARTHUR I.
BALLICKY BAY
L. ROBIN
REFUGE I.
LAKE
L. DEBORAH
PHONY I.
LAC CACHE
McQUADE B.
OKAY ISLANDS
GRA
MAIN HIGHWAY
FORMER WATER-ROUTE BEFORE HIGHWAY WAS COMPLETED
LAKE DORE
PENNINSULA
15 MILES
CHIBOUGAMAU
MALOUF B.
BOULDER I.
TAKE-IT-EASY ISLANDS
MANN POINT
GUTHRIE BAY
GOUIN
YVONNE I.
CHIBOUGAMAU R.
SOUTHWEST ARM
FIRST WATER ROUTE
BETTY BUNN PENNINSULA
DEVLIN PENNINSULA
HALL AIRWAYS SERVICE
BRIDGE
O'CONNELL'S CAMP
CHIBOUGAMAU EXPLORERS
← CHIBOUGAMAU - ST. FELICIEN HIGHWAY →
(ST. FELICIEN, 150 MIL

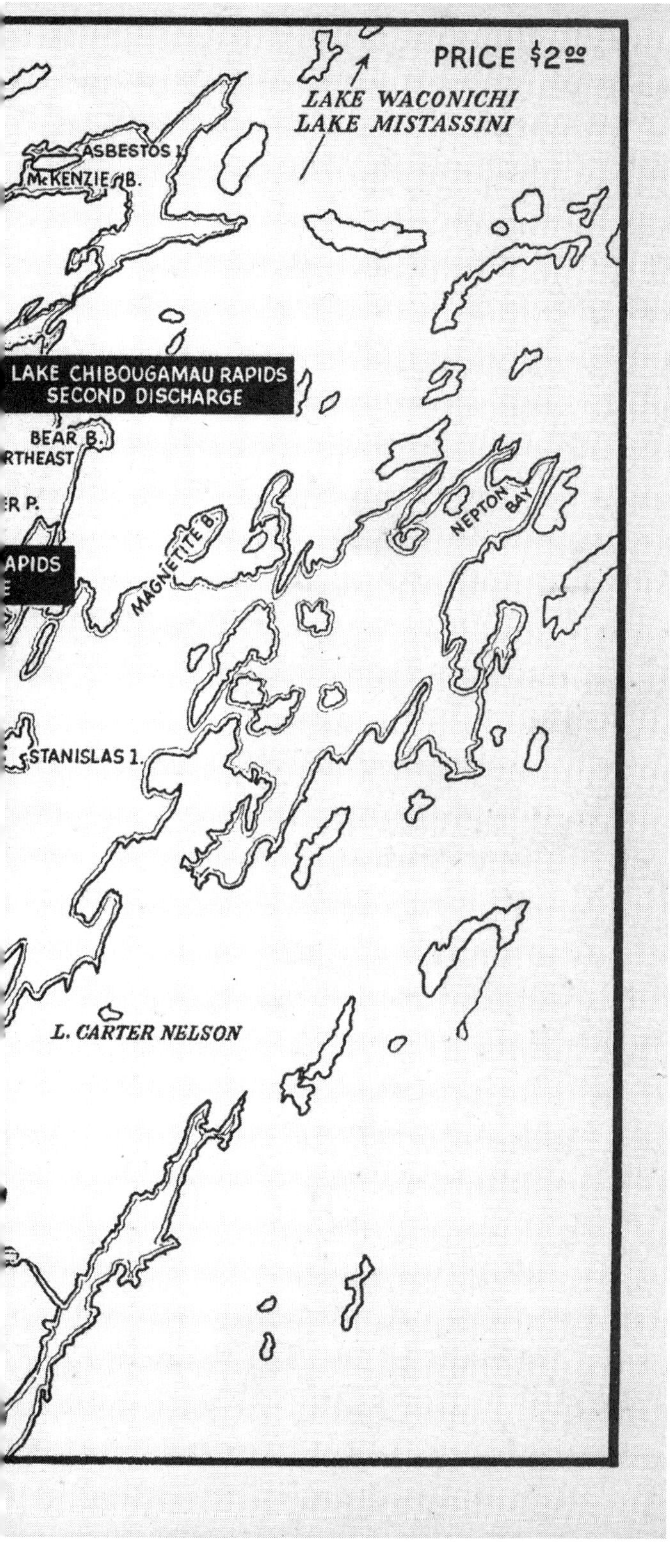

Map of Chibougamau from the 1952 book Chibougamau Adventure *by Larry Wilson. The murdered person here died in the next bed to Harry while he was in the Youville Hospital, Rouyn-Noranda after he broke his leg on the Muskeg Limited.*

I found another one at Fectear Air Service in Senneterre. They were marvellous machines. With them, I don't know how Germany lost. The body of the plane was made from corrugated metal that could carry astonishing loads given the size of the motor. That same aircraft had carted an ox to a mining camp where it was worked all summer long, then eaten.

Radio was not a part of aircraft in those days. When you were up and the weather changed, you stayed up until it cleared, or took a chance.

Our pilot's name was Red Lynbruner. Our first stop in Chibougamau was on Doré Lake where the only building might laughingly be called a store and a boarding house, run by Bill Lafontaine. After a short stay we flew on to Gwillim Lake, landing at the dock of a bush camp of a major mining company. The claims we were headed for were five miles up the lake. We unloaded the plane, but looking at the pile I couldn't figure out how we were going to get there. But Frank hauled out a long canvas package wrapped around a bunch of sticks. In a little while we figured it out, put it together and had a serviceable canoe, unstable in stormy weather but suitable.

While we were busy with the canoe we heard the sound of another plane approaching. It came in over a mountain on the far side of the lake directly at us. Just before reaching the dock it kind of side-slipped onto the water seeming not to have taxied more than a few feet. The plane was a Fox Moth with an open cockpit and a small closed-in cabin large enough for about half a person. The pilot was Eddy Bone. All I remember about his clothing was a long white silk scarf that flapped in the backwash of the propeller. I think he must have seen a movie about aviators in the First World War. Maybe he had been in both of them; but he was a hell of a pilot.

With Frank in the stern we paddled up the lake for a couple of hours until we found a likely-looking spot on top of a high bluff to make camp. Then for two weeks I dogged through the bush after Frank as he put his doctor's degree into practice, but having a previous appointment with the pilot we returned to the mining camp.

We arrived there just after the evening meal. As well as being an able bushman Frank was good at cadging meals. The cook was a little peeved but agreed to give us a little leftover stew and a few slices of bread. The cook house was quite dark and I couldn't be sure what I was eating. Near the end of the meal and on closer inspection I noticed that the black blobs I had shoved off to the side of the plate were winged: fully formed, well developed, large blue bottle flies.

Our next stop was Lake Waswanipi where we spent a couple of weeks, then on to Lake Assonitchanassabastat, a little to the south of Doré Lake, and that's all I remember about that place; its name. Many years later I returned to Chibougamau to find that some kind of cartographer had changed the name to Scott's Lake; probably his own name too.

Our last stop was on an island in Lake Opawica. It just happened that we landed at a spot where some earlier prospector had made camp and this was the day I found out just how good a bushman Frank was.

Firstly, I must say, the spot where we landed was one of the most beautiful places I have ever seen. The beach was only a few yards wide and made up of small pebbles of almost equal size. Less than a 100 feet off the beach was a curved steep cliff face, covered in places with green moss. Little trickles of water ran down making a burbling sound. The floor of the level space was a thick covering of moss with little yellow, daffodil-like flowers growing in it.

Shortly after leaving this beautiful place we stopped for a moment of rest. I leaned against a tree and applied the last few drops of fly dope from an empty bottle. I went to throw the bottle into the bush but instead, for some reason, stuck it onto a dead twig at the foot of the tree.

Frank took a compass bearing and set off through the bush, blazing the odd tree and in his usual manner, at regular intervals broke off a twig and put it in his mouth. When he had the number of twigs representing the distance he was going, he stopped and threw his arms forming a right angle. With his left arm pointing in the direction we had come from, he went along in the direction his right arm was pointing without taking a compass reading. Being his dog I didn't question what he was doing, but after we had walked the four sides of a square, I caught on; only this day it almost didn't work out. Frank stood looking around indicating he wasn't sure where he was. He seemed a little lost. But a prospector is never lost, his tent may be but he isn't. I wasn't yet a bushman, and a wee bubble of panic started to grow. As with any degree of panic the head starts to rotate. With me, after only a few rotations, I spotted that little fly-dope bottle, still in its place on the tree. Frank wouldn't have talked about it but it is likely that he was just testing himself.

Sitting around the campfire one night, I noticed a partly covered over magazine lying on the ground. I picked it up, brushed it off and saw that it had been partly eaten by some paper hungry rodent. It was a copy of the Reader's Digest. I read every last word of it, as I have with every issue of Reader's Digest since.

We were out on the lake on a clear windy day with high waves, paddling hard, when a deerfly spotted a likely target and came in and landed on my left arm close up under my rolled up sleeve. I tried blowing it away but couldn't dislodge it. Finally I stopped paddling for a moment to swat it away. Frank hollered, "Don't stop paddling." Without my effort, Frank couldn't control the canoe and it threatened to capsize. But the deerfly was still there chewing away. Blood ran down my arm. Flies are smarter than we think; taking its time, out of danger, that one enjoyed the best meal it ever had. It walked leisurely down my arm, winked and flew away with the better part of a muscle.

At each stop on our trip arrangements were made with the pilot to pick us up on a specified date. Just as our work was completed on our last stop on Lake Opawica the weather closed in, and with the aircraft unable to fly, we were stuck there. By the end of a week we were out of food, down to some useless leftover condiments, but with fish in the lake we wouldn't starve.

One night as we sat by the campfire Frank suddenly whispered, "Don't move. Don't move." A signal like that, in the deep bush, is a sure sign that a bear with four-inch incisors is ready to test them on a naked neck. Frank waddled away on his heels into the tent. "Coward!" while I, unable to move on marble legs, just sat, trying to remember if I had ever memorized an effective prayer. As the rest of my body marbleized, Frank emerged, reptile like, with a .22 calibre rifle that I had not known he always carried unassembled in his packsack. He leaned over my shoulder and fired, then took off into the bush. "Fool! Doesn't he know it takes a French .75 to bring down an angry bear that has been pin-pricked by a .22 bullet?" I waited. Frank casually sauntered back carrying a plump little rabbit by the ears. With some of the fur and most of the guts removed, it went quickly into a pot to stew with sprouts from some old onions the previous camp occupant had thrown out. We enjoyed one of our best meals, and our last for that season of prospecting. The aircraft came around early the next morning and took us back to Oskélanéo River.

Before leaving the base I asked Red Lynbruner what he would do now that he didn't have to ferry us around. His reply was: "I think I'll try flying over the south pole." A few weeks later I read in a magazine where he and another Canadian bush pilot, chosen because they were Canadian bush pilots, were to fly Lincoln Ellsworth over the south pole. (The last time I saw him was for a brief moment in the lobby of the Windsor Hotel in Montreal.)

Frank's work on this short examination trip was to check a few claim posts on each group. That would indicate that the property was properly claimed. All boundary posts would be checked if an exploration program was recommended and later instituted. The inside posts were not of great importance so long as they were near their proper location. Inside blazed lines were usually not very distinct or easy to follow. But where men were hired to stake in a certain area, considering the pay scale and the time involved, the work was usually very well done. In his geological work Frank ran traverse lines, examined and mapped outcrops for rock types. Without the proper rocks it was unlikely that metallic mineral deposits would be found. Although a number of impressive showings had been found by early prospectors it would be more than 15 years before any mines would be brought into production in the Chibougamau district. It was J.P.'s opinion that no gold mines would be found in the area immediately to the north of the CNR. The rock types were not right, and although several spectacular showings were found, none ever made a mine.

Lorenzo Blondeau was French speaking, very nice, completely bilingual, a very bashful type of guy, about five feet ten, with a good build and quite a handsome face. But when he was startled or surprised, his eyes would bug out like two large, round, white glass marbles with a spot of black in the centre. He was a bachelor living in a shack in Senneterre with an old prospector, Hughie Gilligan.

On our claim-staking trip to Chibougamau in early 1951 with Jim Hobson and Lorenzo we stayed overnight in a small hotel in lower Quebec City, opposite the Central Railway Station. After checking in we gathered in a small waiting room on the second floor where one could order booze; we started with beer.

After a while I was the first to need toilet training and found it at the end of a long hall. When I got back, Lorenzo said he was in need of the same convenience and "Where is it?" I told him to go to the end of the hall but I didn't say exactly where. I didn't think I needed to.

Lorenzo left but in a very short time came rushing back with burbling saliva and bulging eyes to tell us he had opened the wrong door. Two bare, nude figures were on the bed in very active play, and when the door suddenly opened they jumped up and went whirling around the room in tornado fashion looking for something to cover the evidence. Poor Lorenzo, shocked, just stood there gawking, feet frozen to the floor; after long moments he bolted – leaving the door open. It was days before Lorenzo quieted down, got his eyes back in place and stopped burbling.

Hughie Gilligan made a gold discovery and sold it to a mining company for a large lump sum, and spent it all in short order. That's why he was living in a shack in Senneterre.

In the summer of 1951 Bill Lipsett, a manager of Inspiration, and I decided to take a trip by car from Val d'Or to Chibougamau. I had two prospectors, Ryder Storn and Emile Perron, brother to Alex Perron, working on a block of claims that I and Jim Hobson had staked the fall before and it was a good time to visit them. There was a great amount of diamond drilling going on in the district, and Lipsett, manager of Inspiration in Val d'Or, thought it time to get established there.

When we arrived at the prospectors' camp we were told that a new gold discovery had been made just across the lake from their camp and they were very anxious to get in on it. Lipsett and I talked about it for a short time and decided to put up the money for the licences and wages of the prospectors with the further provision that if the claims were sold they would retain an overall 10% interest.

On a bright summer day, while driving an inside line, Emile stopped for lunch and a cigarette. Sitting on a windfall he noticed a rather large depression in the ground below his feet. With his axe he cleared away the moss and found the underlying soil contained a great deal of rust. He took a sample and in the evening panned it for gold. It was full of gold!

I sold the claims to a promoter in Montreal who formed a company, Chibougamau Explorers. They raised sufficient funds to carry out a diamond drilling program. That promoter went as far as he could with it, then sold it to Jim Boylen in Toronto, who brought the property to production. But with gold pegged at $30 U.S. and the high rate of inflation the mine was forced to close. In the early '70s gold was allowed to find its own level and for a short time spiked to over $800 U.S. an ounce. When it retreated to around $400, the property was sold to an American outfit who put it back into production as the Mann Mine. I think it is still in production.

If J.P. had lived it would have been one of his good ones, with little dissention from me.

I don't think J.P. ever had time to go to Chibougamau, but he sent a lot of prospectors there. If he had lived, he certainly would have had another mine at Norhart Lake in 1952.

About 25 years after our trip to Chibougamau the Duplessis government in Quebec City decided to build a road to the region.

Duplessis hated the fact that the English element, mostly from the queen city (Toronto) controlled most, if not all, of the natural resources and most else in the province of Quebec, and had controlled it all since the Plains of Abraham incident.

Duplessis' slogan was "Quebec for the Quebecers". In other words, "English out" and to some degree he made it stick. He paved the way for René Levesque.

It was Duplessis' decision to provide an access road to the many mineral discoveries that had been made in the Chibougamau area, but not to the advantage of the English. The road could have been driven in from Amos or Senneterre but that would have left the door open to the Ontario-Quebec border. Duplessis' road started at St. Félicien on Lake St. John and access led through Quebec City.

The new road started at St. Félicien and was driven straight into the wilderness for a distance of 150 miles.

I happened to be on a gold rush at mile 129 when the road was finished at Gilman Lake (a small lake near Chibougamau). After staking some claims, my partner, Jim Hobson, and I drove to the end of the road to find that the streets for the coming town had all been bulldozed out but there was not a single building: a strange sight. When I returned to the district the following year, one fairly large tent had been erected by Scotty Stephenson, to provide a store for wandering prospectors. That same year the Quebec government held an open air auction in the town to sell off the various lots, residential and commercial. The first lot at the main intersection was bought by a bank for $4,000. Later in the summer I bought one for $300, in the residential area.

When J.P. ventured out on his own by incorporating Inspiration, his first order of business was to acquire claims. Remember that Carl Erickson objected to J.P.'s wish that the Senneville township and East Malartic properties be the first two properties acquired. He wanted the company to start from a clean slate. They did include the Senneville claims, however, as well as another of J.P.'s properties at Gwillim Lake near Chibougamau. The reason Gwillim Lake claims were accepted by Inspiration was that J.P. personally gave them to the company at no charge to give the company a start. It's odd that one of the first properties in his own career should come from such a distance away from Amos. Chibougamau was one of his last interests, as well.

Alex Tait was a young engineer working for Aluminium Co. of Canada. He wanted to venture out on his own as J.P. had done many years before, and J.P. was always ready and willing to assist anyone with the spirit of adventure that had brought him so much pleasure and success. These letters tell their own story:

March 15, 1945

Dear Mr. Norrie:

Due to shutdowns in outside plants and consequent reorganization of personnel, I have been notified today that my services are no longer essential to the aluminium plant.

I know that by law I cannot be employed in an operating gold mine [gold was not a strategic metal]. However, I do not think there is any objection to my doing exploration work, more or less with the status of a consulting engineer, as I am not liable for military service. I am very anxious to get started on something like that as soon as possible as the break-up is apparently not far off. Have you anything in mind?

For quite a while I have been studying the possibilities of mining exploration of the area northeast of here, that is Chibougamau, Lake Mistassini, and further north along the height of land. I have been studying all the available provincial and federal reports and maps and am convinced that there are very good chances of finding a completely new field. Furthermore I am told by then local MP who is Minister of Roads that a highway from Lake St. John [St. Félicien] to Lake Chibougamau and Lake Mistassini will be started this summer. If this is published there will be a rush into Chibougamau as soon as there is open water.

For the above reasons I would like to stay right here in this district as it will be close to the scene of great expansion in the next few years. Very few people know that there is now a good highway from Dolbeau to the big storage dam at Passe Dangereuse on the Peribonka, and above that the river is now navigable without any portages for 100 miles. This makes a very large area northwest of Lake Mistassini quite accessible. Lake Mistassini itself is also a good highway for heavy water transportation to about 150 miles north of the Chibougamau field. I am enclosing a map on which I have outlined these features and the area of 100 miles by 200 miles that will now be quite accessible and should attract a lot of prospectors in the next year or two. I consider most of the 30 mile by 30 mile blocks that I have subdivided are favourable ground that merits intensive exploration.

Would you be in favour of incorporating a new exploration company to spend about $50,000 a year for a few years on development of this area? This company could be the vendor of any groups of claims staked on which detailed work such as diamond drilling was considered advisable and for which a mining company would then be incorporated. The parent exploration company would have to provide some of the initial funds for each of the mining companies so formed on different discoveries, above the amount voted each year for prospecting.

Would Inspiration be interested and would it be a good thing to ask some others such as Dome, Quebec Gold, Siscoe or Hollinger to take part in original financing? I would be ready at any time to give such a development my most enthusiastic efforts in any capacity. Incidentally I spent three summers in the bush up in this district 15 years ago including a trip to the headwaters of the Peribonka so I know the country and the people.

If the above is not feasible at this time would you consider a branch of Inspiration in this district to get diamond drilling contracts, shaft sinking, etc. Also to have someone on the spot to stake claims if any boom develops. This branch could be located here or in Roberval.

I would be glad to come to Montreal, Toronto, or Amos to discuss these matters if you like. In the meantime I am applying for membership in the Association of Professional Engineers of the province of Quebec in order to make everything legal. If there is anything important I should know in a hurry I would appreciate a wire or phone call collect.

Alexander H. Tait

March 17, 1945

Mr. Alexander H. Tait
Arvida, Quebec

Dear Alex:

I think that your ideas are okay and that the $50,000 could be raised for the project outlined.

Due to pressure of work I will have very little time to give to it so if you decide to go ahead you will have to work out most of the details. I think I can help you raise the $50,000 so when you are ready to start I think you should go to Toronto.

The money will have to be raised by private subscription which I think is available now and the sooner it is raised the better. I believe I can contact some people over the phone which will put up most of the money.

With best wishes,

J.P. Norrie

April 2, 1945

Mr. Alexander H. Tait
Arvida, Quebec

Dear Alex:

I note that you are prepared to go ahead with an exploration company and that you will be in Toronto on Sunday, Apr. 15.

I expect most mining people will be at Quebec on the 17, 18 and 19. That should be a good place to contact them so why do you not plan to be there. I expect to be but will be unable to visit Arvida due to pressure of work as much as I would like to.

If you go to Quebec, and I think you should, then it will be around Apr. 20 before you can do much organizing. This will take from two weeks to a month and by the time you get a crew and equipment together it will be well on in June which will be somewhat late to send in a complete outfit this year. I would not mention this point when trying to interest capital. I think you would do well to plan on say three prospectors for the summer and to spend their time in and around Chibougamau and plan on going in before breakup the next year. In fact it may develop that you would need to curtail even further due to lateness of season in which case I could provide you with work until you are ready to embark on a wider programme.

With the small capital at your disposal you will need to conserve funds in every possible way. I think you could do without an office and an office man to start with and for the present year I would plan on say four prospectors in two parties. By going slow you could give them better supervision and at the same time get your feet on the ground in the area. I would not figure on field engineers for the first season. Something substantial could be found in your first season's work which would warrant the outlay you propose for the second year. I cannot go into all this on paper. We can discuss all of these points when I see you.

I had long ago intended to organize an outfit to prospect the long fault zone running north and south in Mistassini Lake and the long belt of dolomites on the east shore. These dolomites were originally mapped as Denovian, however, being devoid of fossils they have since been classified as Precambrian. Anyway they seem to be the same formation as Hollinger is exploring in Labrador so Hollinger would probably subscribe to your venture.

To sum up I hope to see you in Quebec when we can discuss everything in detail. In any case I would not plan on more than four prospectors for this season with expenses cut to the bone for this first season.

I think I can be of help to you in raising money but can give very little time to it as I am too busy now.

J.P. Norrie

Dear Mrs. Norrie:

I was dreadfully sorry to hear the news that your husband had passed away. May I express my sincere sympathies to you and your family at such a sad time.

With Jim's passing I feel that I have lost a very good friend and one to whom I shall always owe a debt of gratitude for the interest he took in me and my affairs. I did not have to know Jim very well to admire him tremendously and I feel fortunate that I could appeal to him for friendly advice when necessary.

Since before the war we had several times discussed a project of exploration of a large part of this province in which no work had ever been done before. I felt that this was of great importance to the future development of the mining industry and was very happy that under Jim's direction I was able to get started this year. I have just returned from a four month's expedition into the region and was looking forward to telling him all about it.

I can only say that your husband's life as I knew him, and his manner towards his fellow men will always be an ideal that I would like to achieve, and I hope to do my best to carry on the work in which I felt that Jim and I were partners.

I am only one of the many people with whom Jim had similar contact but we all miss him very much indeed. His passing leaves a vacancy in the mining world that cannot be replaced, but I trust that the memory of the great things he did for the industry in the province may serve as an inspiration to those of us who follow.

Sincerely,

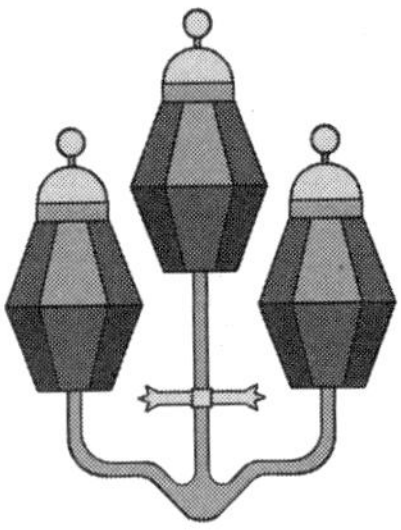

Mystery I.
Labine Pt.
ELDORADO
Glacier Bay
Echo Bay
Glacier Lake
Mackenzie I.
Workman I.
Lindsley
Eagle